GATTACA HAS FALLEN

How Population Genetics Failed the Populace

IAN A. MYLES, MD/MPH

ISBN 979-8-9892309-0-7
ISBN 979-8-9892309-1-4
ISBN 979-8-9892309-2-1
ISBN 979-8-9892309-3-8

Book cover by Ian A. Myles

First edition, 2023.

To my son, Henry.

I hope this book helps you inherit a better world.

CONTENTS

INTRODUCTION

Population genetics promised to unlock the mysteries of disease while hinting it could turn every child into an elite athlete with towering intellect. Instead, genetic fanfare stole our collective focus while we swaddled our children in baby blankets made from the chemicals found in car exhaust. Faith in genetics siphoned resources as we corrupted our air and stripped the key nutrients from our food. Meanwhile, the obsessive search for genetic explanations eroded our trust in modern medicine by stigmatizing communities and implying that most illness was preordained. For all our modern advances and successes in public health—which have collectively doubled the average life expectancy— diseases of inflammation are more common today than any generation prior. The question is why?

I was a young, impressionable college student pursuing a career in science the day President Clinton announced the completion of the Human Genome Project. On June 26th, 2000, the President highlighted the hard work researchers like Dr. Francis Collins had performed. Dr. Collins' team had successfully sequenced the entire human genome from end to end. Scientists worldwide were elated by the announcement. This breakthrough would soon establish the standard for comparing the genomes of different human populations. This approach was termed "population genetics"; the field hoped to explain diseases by identifying the genes that differed between people who were healthy and patients struggling with illness. Population genetics hoped it would open the path to finding treatments and cures that had, up to that point, eluded modern medicine. It also hoped to teach the world about what it means to be human by unraveling the very nature of our behavior.

Leading up to the announcement, genetics had already provided several revolutionary discoveries in medicine. Genetics of rare disorders had identified dozens of diseases caused by, and passed down through, single gene mutations (called monogenic disorders). Researchers identified cystic fibrosis, sickle cell disease, hemochromatosis, and more by searching the genome for differences between patients suffering with these disorders, and contrasting their DNA sequences against that of their unaffected relatives. By comparing DNA sequences of tumors and nearby normal tissue, scientists discovered numerous genes linked to cancer development. These included oncogenes, which promote tumor growth, and tumor suppressors, which inhibit it. Finally, the field that looks for genetic predictors of bad reactions to medications—termed pharmacogenomics—had also produced a string of successes capable of guiding therapy. Sequencing the genome was supposed to put these types of discoveries into hyper-drive.

Writers launched into tales of futures in which every child was born with the mathematical prowess of Albert Einstein and the cello skills of Yo-Yo Ma. Sports magazines wrote articles pondering what would become of professional sports when every child was genetically engineered to have the basketball skills of Michael Jordan. Soon thereafter, the idea that genetic manipulation could completely remake human biology began to overtake radioactivity as science fiction's chosen mechanism for creating city thrashing monsters or wall crawling superheroes. Other groups celebrating the human genome project's success were the race scientists and eugenicists, who felt that the tools needed to supposedly prove the innate inferiority of marginalized groups had finally been delivered.

For nearly 100 years, race scientists had promised that as soon as genomic sequencing was possible, the proof of white male biologic superiority would be provided. This was seen as evidence that those who did fall ill must lack the genetic stock needed to assure good health. This was the heart of what is known as genetic determinism or biologic determinism, which is defined by the American Psychological Association Dictionary as "the doctrine that human behavior and mental activity are largely (or completely) controlled by genetics". Genetic determinism fed into race science (also referred to as scientific racism) which is defined as the pseudoscientific belief that humans can be not only divided into different races but ranked by their innate physical and behavioral abilities.

Initially, their work was based on phrenology, the idea that people's skull shape could predict their intellect, abilities, and morality. Not surprisingly, the founders of phrenology decided that the perfect skull shapes were the ones that looked most like their own. Eventually, these researchers turned to twin studies, which did nothing more than explore the similarities between identical twins versus fraternal. Have you ever read some headline claiming that obesity, or criminal activity, or even whether you prefer coffee or tea is substantially genetic? If you have, odds are good that those claims originated in a twin study. These studies compare the rate of similarity between twin pairs, for example testing whether identical and fraternal twins both take sugar in their coffee. Using rudimentary math, twin studies would go on to predict the genetic contributions to just about every human disease and every human trait. The twin studies researchers claimed that intelligence was 80-90% heritable, and thus, nearly entirely genetically encoded.

Dr. Collins, whose groundbreaking work on genetics had earned him a presidential nomination to be the head of the National Institutes of Health (NIH), would push back against divisive bigotry by frequently pointing out that any two humans were 99.9% genetically identical. That is to say that if you were to read through the DNA sequence like letters in a book, 99.9% of those letters would be the exact same and in the exact same spot for everyone. When someone says you share half your DNA with a sibling, they mean that half of the remaining 0.1% is identical between siblings (so, 99.95% identical). They do not mean that you and your sibling are only 50% genetically similar. In fact, humans are more than 50% genetically identical to bananas. Genetics showed not only that humans were mostly the same, they determined that while there might be genetic signals that indicate a person's race, there are no genes that define a race or subspecies in the same way breeds are seen for dogs or other animals. Finding race to be apart from biology, biomedical academics disengaged, leaving the issue to sociology or political science.

Perhaps due, in part, to the scientific community's disengagement from eugenics debates, the belief that traits like intellect or criminal behavior were innate emerged as an excuse to justify the mistreatment of vulnerable groups. Twin studies justified programs that forcibly sterilized patients with mental health disorders, enslaved Black populations, and restricted the immigration of Jewish refugees, all under the guise of assuring that those with the best genetics would be put in positions of power and privilege.

Eventually, the initial solutions for so-called racial hygiene spread from their origins in the U.S. and England to Nazi Germany, where they became the "final solution" of the Holocaust. For a brief moment after the claims of the innate superiority of Gentiles led to Nazi policies, genetic determinism was given a bad name. Eugenics was rebranded and repackaged, but the core of the rot remained. Genetic sequencing hadn't reached the point to couple genetic claims to genetic evidence, so the proof of genetic superiority that the Fuhrer suggested remained always just a few years away. As soon as science could analyze enough people with adequate sequencing, the proof would be in hand. Meanwhile, researchers were *just asking questions* about the possibility of genetics dividing the world into the haves and have-nots.

Only three years prior to the Genome announcement, I had just completed high school and was preparing to transition to studying biology at Colorado State University, when the first major genetics-based thriller hit the big screen. GATTACA was a film starring Ethan Hawke, Uma Thurman, and Jude Law. The film's title is a play on genetic sequencing, being made up of only the letters used to represent the nucleic acids of the DNA structure (guanine, adenine, thymine, and cytosine). Jude Law's character collapses under the weight of expectations, having been told from birth that he was genetically gifted but ending up wheelchair bound after a serious accident. Meanwhile, Ethan Hawke's character refuses to be pigeon-holed into his lot in life based on genetic screens that failed to recognize his true abilities. Uma Thurman's character was depicted as grappling with the extremes of wanting more than society would offer someone with her genetic scores but also accepting her assignment as a cog in the wheel of the genetically based social hierarchy.

This was the first major thriller centered on the idea that genetics had immense powers to govern abilities, but many others relied on genetics as plot devices. *Spider-Man* starring Toby McGuire altered the comic book canon so that the character got his abilities after being bitten by a genetically engineered spider rather than the radioactive spider. The origin story of the *Incredible Hulk* was changed from being created through a gamma radiation experiment gone awry to being created via a genetics experiment gone awry. The monster, *Godzilla*, went from being created by nuclear fallout to being a product of evolution. Movies about teenagers discovering their inner hero transitioned from that of unique circumstances (even fanciful ones like radioactive spider bites) to gaining powers through family lineage.

Despite the initial fanfare of the Human Genome Project, the focus on biomedical science was quickly displaced by the terrorist attacks of September 11th and was kept out of the public eye by the invasion of Iraq, Hurricane Katrina, and financial collapse. The concern that sporting events would become obsolete quickly faded. Meanwhile, the dominant mechanism for developing supernatural powers in pop culture became the magic of Harry Potter rather than genetic engineering. Yet while genetic determinism in mainstream culture faded, race science was kept alive and well.

The genetics lessons I was provided at the University of Colorado Medical School focused on the successes of the genomic era. When I was undergoing my residency training in internal medicine at The Ohio State University Medical Center, genetics only came up in the context of transplants where the need to pair the patient to an ideal donor is key to a successful outcome. I had not forgotten the laughable claims that genetics would transform athletic competition into an obsolete vestige of ancient times, but for the most part, population genetics seemed like something that was a niche field of medicine without broad impacts. Only after I began my specialist training in allergy and immunology at the NIH did the limitations of—and harmful overenthusiasm for—population genetics become clear to me.

The rates of allergies, like just about every disease involving the immune system, had been increasing my entire life. The English physician, John Bostock, made the first description of hay fever in 1819. It took Dr. Bostock another nine years to find a total of 28 more cases of hay fever. Today, the average clinic couldn't go nine days without seeing 28 people with hay fever. Even better data comes from Switzerland. The Swiss have had centuries of mandatory military service requiring young men to undergo a physical exam. This means that the Swiss have been cataloging health and disease for at least 200 years. In 1900, the Swiss army reported that only 1% of young men had hay fever. By 1971, Aberg reported the rate had risen to 4.4% and rose further to 8.4% by 1981. By 2010 rates in the Swiss were 20% while England reported rates as high as 43%.

Asthma had seen a similar sharp increase in rates starting around 1960. Highly industrialized nations saw a large uptick in rates from 1960 to about 1990, followed by a small reduction. In contrast, developing nations had rates of asthma that hovered near zero until around 1990, when the rates began to—and continue to—increase. Rates of atopic dermatitis (more commonly referred to as eczema) increased at least 3-6-fold in the U.S., U.K., and Finland since 1970. Rates of allergic diseases were disproportionately high in Black Americans with nearly 25% experiencing eczema, despite rates in the West African nations—which would be considered the closest genetic relatives outside of the U.S.—being less than 1%. I saw this firsthand when I was deployed to Monrovia, Liberia for the U.S. response to the Ebola outbreak. The NIH developed a vaccine against the Ebola virus and needed physicians to help oversee the study. As an internist, I had grown accustomed to having about a third of my patients having at least one allergic disease. Yet, none of the hundreds of Liberians I met reported having allergies.

The rise of food allergies is probably the most obvious to the general public. No one over 40 went to a peanut-free elementary school. And yet everyone over 40 sends their children or grandchildren to, at minimum, a "peanut aware" school. Some schools flat out prohibit tree nuts (like cashews, almonds, or walnuts) or peanuts in lunches, while others have special sections of the cafeteria for those who are highly allergic. Occasionally, a rise in a disease rate might be a reflection of a newfound ability to detect the disorder. For example, the invention of the mammogram was followed by an increase in the rate of breast cancer but that was simply because the mammogram allowed people to detect cancers that had previously been missed.

For a disease like hay fever, one might be open to the possibility that doctors simply got better at recognizing the symptoms. But can you really miss a peanut allergy on a massive scale? We used to give out peanuts on airplanes. Meaning we provided peanuts in a pressurized canister that circulates the air inside while flying 30,000-plus miles in the sky with nothing more than a first aid kit and zero assurance a medical provider would be on board. Today, some hospitals refuse to bring peanuts into their facility for fear of inducing a reaction in their patients. Children eating peanuts and then ceasing to breathe within minutes is not the kind of thing that goes unnoticed by the medical community and certainly isn't overlooked by parents and caregivers. Allergic disease may have been my focus, but the dramatic rise was not unique to allergy. Similar increases were seen in autoimmune diseases like type-1 diabetes, lupus, psoriasis, Crohn's, and ulcerative colitis, just to name a few. Heart disease, kidney disease, neurologic disorders ranging from Autism to Alzheimer's, and especially cancer also saw sharp increases.

A study led by University of Southern California Professor Eileen Crimmins demonstrated that if you were an average male born in 1978, you could expect to live until age 75. However, you would also expect that for 10 of those years, you would have a serious disease, and for 3.8 of those years, you would have limited mobility (i.e., needing a cane to walk into the doctor's office to get your diabetes checked). However, if you were born in 1986—just 8 years later—you could expect to live one year and one month longer. Yet, those born in 1986 could also expect to suffer from disease for 12.3 years and lack mobility for 5.3 years. So, in exchange for living just over a year longer, the younger generation suffers over two additional years of serious illness and a year-and-a-half increase in poor mobility. Overall, the data is consistent that the average person is living longer, but getting sicker.

The massive, population wide increase in immune-mediated diseases seemed to be screaming for researchers to go forth and find the exposures that must have been spreading the disorders throughout society. And yet, the field was overrun with talk about, and research into, genetic predisposition. Allergies were referred to as highly heritable with predictions that some 80% of the disease would one day be explained by the genetic code. How did researchers arrive at this number? The twin studies deployed to mathematically claim that Jewish people were inferior a century prior were being held up as a legitimate means of concluding that people were born allergic to peanut butter. To believe that there was any meaningful contribution of genetics to allergies, let alone to believe that genetics could explain 80% of the disease, would require one to believe that the peanut allergy genes must have lay dormant in nearly every single person that ever lived and died over the course of human history until just about 1960 or so. If true, then wouldn't an environmental factor have been needed to activate those allergy genes?

The genomics era may have produced successes for rare diseases, rare drug reactions, and cancers, but for the types of common allergic diseases I was studying, the results were disappointing. The initial goal of population genetic screens was supposed to be identifying genes that might play a role in the disease of interest. So, for eczema, the idea was that evaluating genetics would reveal some shared genetic signal that must be the key to unlocking the biology of the disease. The initial genetic screen pointed to numerous potential genetic contributions. The genes that did not make obvious sense were forgotten in favor of one gene with the strongest signal called filaggrin (abbreviated FLG). This gene encodes a protein of the same name, which serves to link skin cells together to make a seal. This seemed like a slam dunk—a disease known to have a defect in the ability of the skin to make a sealed barrier was linked to a gene that encoded for a protein that helped make a sealed barrier.

Of course, most patients did not have any abnormal signal in *FLG* —
maybe 10% had anything approaching a "mutation". And for that
matter, plenty of healthy people had the same supposed abnormalities in
FLG, despite there being no evidence of disease. No one seemed
bothered by the realization that whatever *FLG* abnormalities existed
would have to have been present but silent in all the generations prior.
While broadly connecting *FLG* to barrier function was easy, connecting
the exact base pair changes to a loss of barrier was not. Mice in which the
gene was entirely deleted had no disease unless a second gene (called
Tmem79) was also deleted. Cell and mouse models that copied the
sequences that were deemed to be abnormal in patients did not produce
reliable models of disease.

When the sequences of *FLG* that were deemed to be causing eczema in
southern Norway failed to have any link with patients with eczema in
central Norway, the excuse was that there must be a unique
environmental factor separating the regions. However, no effort was ever
made to figure out what that factor might be. Every paper that came out
on eczema seemed to start with some version of the phrase "atopic
dermatitis is a complex interaction of genes and environment" and then
go on to discuss nothing but genetics. No one was claiming that a single
gene—not even *FLG* —caused most people's eczema outright, but they did
claim that genetics made one more susceptible to the disease manifesting
in a given environment. The genetics of eczema made for a lot of
publications, but none of the genetic studies ever created a new
treatment or provided a way to screen children for being high risk.

Genetic evidence became even shakier as research moved beyond
individual European nations to entire continents. Genetic sequences
deemed to be "risk factors" in one European nation would not be linked
to eczema in other European nations or countries in Asia or Africa. This
was true for eczema specifically and genetics broadly. Thus, the only way
for population geneticists to mathematically link allergies and genetics

was to segregate people into the very racial categories that had been designated by the skull measuring, Nazi-preamble, race scientists from the generation prior. The same gene-centric assumptions made to justify discriminating against people with mental health diseases were being used to claim that even if the environment induced eczema, it did so because the patient's genetics were incapable of providing protection.

The explosion of allergic diseases was not seen as a call to probe the environment for the causal toxins and then clean them from our environment. Instead, it was the patient's physiology that was seen as the root of their disease, and thus, the only remedy was to manipulate the patient's physiology with medications. Black and brown Americans were considered biologically predisposed to allergies, despite their ancestors and nearest international relatives having near zero allergic disease. The implication was that marginalized communities were genetically disadvantaged in their ability to stave off allergic disease compared to privileged groups.

Of course, no one in allergy research would have used such terminology and no one would have intended to imply such vitriol. Those implying that Black Americans must have some innate reason for having more eczema despite the mountain of epidemiologic evidence to the contrary were not secret white supremacists. They were not overtaken by discriminatory thinking but by a paradigm and dogma that filtered every scientific discovery through DNA-centric thinking. Thus, when the genes that were deemed to cause asthma in white people failed to show any link to asthma in Black people, instead of considering that genetics might be less significant than the environmental injustices, the assumption was that Black Americans must possess a different set of genes leading to their higher rates of asthma. These claims were similar to claims that Black Americans must have genetics favoring athletic talent or dancing abilities. One could have said that the population geneticists believed that genes should be treated as *separate but equal*. The claims that there would be black, white, red, yellow, and brown versions of asthma were made even though every single lung protein that these genes encode for serves the same function in every human.

Like the race scientists, the allergy-focused geneticists would put forth a stained definition of reproducibility, a term that communicates how reliable a scientific finding really is. They would argue that even though they could not find the same genes linked to asthma in different groups, the fact that they always found some gene with a mathematical association indicated that genes must be important. No one stopped to think that if you constantly get a positive result, you might need to use a negative control to make sure your assay wasn't prone to feeding you false information. Furthermore, just because your observation is predictable, that does not mean your explanation for that observation is correct. Every ancient culture in human history could predict what side of the horizon the sun would appear each morning, which direction it would travel, and where it would set. Yet, their predictions of a sky God dragging the sun across the sky in a chariot or a dragon eating it at night were more than a little erroneous.

Despite how much the geneticists may *juke the stats*, genes could only explain a fraction of why some people developed allergies and others did not. This fact was not seen as a sign that maybe the twin study researchers might have overlooked something. After all, the twin study researchers lived in an era before the scientific realization that cities should keep their sewage separate from their drinking water. It also wasn't viewed as a sign that priority should be given to investigating the environmental causes of allergic disease. Instead, it was seen as a justification for even more investment into population genetics. We were told that with just one more round of grant money, the population geneticists would run yet another genetic screen on another set of people and solve the mystery. When that failed, they claimed that more funding was needed to buy more powerful computers to analyze yet another genetic screen on another set of people. No matter the results, the payoff was always just over the horizon. With enough people and computing power, the genetic solutions that had been promised a century prior would be revealed, and allergies would be eradicated just like Smallpox.

While the failures of genetics to produce any meaningful impact on common diseases mounted, twenty years of epidemiological and environmental discoveries began to shine a light on the exposures that were causing our modern ailments. Therefore, when the large-scale *All of Us* study was designed to uncover the reasons for common disorders, you would thus think that the environment would be part of the equation. However, when Dr. Collins announced the NIH study (about 20 years after the Human Genome Project), there were no plans to survey the environment and no samples to be collected other than blood, urine, and DNA. The banner recruiting people to volunteer read, "DNA tells a story. Do you want to know yours?" But how unique could your DNA story really be? Can you imagine a library where 99.9% of the printed words in every book consisted of the exact same letters in the exact same order? Despite all the data from both the scientific and patient communities, little priority was given to investigating how the delivery of the reading or setting impacts the narrative (subsequent to the program's announcement, All of Us underwent substantial improvements to account for environmental factors; these are detailed in Chapter 21).

To be fair, genetic determinism can offer some comfort to patients. If a person's child is struggling with an ailment that was determined at birth, there is nothing the parent could have done better and no need to torture themselves with a game of 'what if?' But one of my first mentors in medical school, Dr. Eugene O'Neill once told me that "you do the best you can with the knowledge you have. If tomorrow you learn something different, don't beat up your past self. Just make sure you keep doing the best you can with the new knowledge." Parents believing their child's eczema is genetic may be more comforting than the fear that their use of antibiotics, diet, or even choice of bed sheets might have played a role. However, there is no way any parent could have known to avoid such triggers. Few of us know the types of pollutants in our air or water, let alone know which are the most dangerous. The fact that poorly known environmental exposure leads to disease is not a failure of the parents to protect their children, it is a failure of the biomedical community to empower the parents.

When the genetic determinism came from bigots or IQ enthusiasts, it would be dismissed by the scientific mainstream. When the genetic determinism came from those working on mental health disorders, a small cadre of researchers within psychiatry would disagree, even as the majority of the field celebrated. And yet, when the genetic determinism came from those researching rheumatology, allergy, immunology, neurology, heart disease, or other disorders, no controversy was raised, and no explainers were tagged along to the work to educate the lay public. I came to realize that while racism may have birthed genetic determinism, the misplaced focus on being the researcher that discovered "the gene for _____" was sustaining it far beyond what the data would support.

The analogy I came to was that of the *Glass Onion*. The phrase was coined by John Lennon of The Beatles in the song of the same name. Then, it was revisited as the title to the murder mystery movie sequel to *Knives Out*. As stated in the movie, the analogy is meant to describe "an object that seems densely layered, mysterious and inscrutable. But in fact, the center is in plain sight." The inner core of population genetics is, as it always has been, the layer of race scientists that care about nothing more than proving that white people have the genes for intellect and leadership. The layer covering them is that of the racist-adjacent pundits like Andrew Sullivan or *The Bell Curve*'s Charles Murray. Although their mask slips from time to time, this layer will try to refrain from overtly stating that Black people are inferior. Instead, this group loves to play the "just following the science" line framed around preserving specific cultures.

Just outside this layer is where we find the researchers that may avoid discussion about how intellect is distributed, but they firmly believe that intelligence is meaningfully genetic. This layer includes the clinical psychologist, Professor Katheryn Paige Harden, who wrote The Genetic Lottery and took it on a book tour chastising people who disagree with her belief that genetically screening students is the best way to improve their school performance. Others include Professor David Reich, who as discussed in later chapters, wrote an op-ed for the New York Times that he believed was an argument against racism but instead became the go-to citation for racists all over the world. This group goes out of its way to state that their findings do not support the very bigots who point to them as so-called proof of their beliefs. Yet, this layer never articulates why their research is of enough value to counter how frequently it is used to fuel discrimination against marginalized groups. They are not directly invested in proving white people are the only ones with the "IQ genes" (or a skewed share of them), but they are open to the possibility and actively entertain the idea that, any day now, genetics might prove the bigots were correct all along.

The next layer consists of those dedicated to finding genes for mental health disorders. This layer is typically lukewarm on the gene for intellect claims, at times denouncing it and at others defending it. They don't really care about looking into genes for IQ, but instead maintain that any day now, the genes for Schizophrenia, depression, or bipolar will be revealed by the latest genome-wide association study (or GWAS for short). GWAS will be detailed in later chapters, but in brief is a process of combing through the genome looking for associations between genetic differences and diseases or traits. The stigma associated with mental health disorders makes their work more sensitive than claiming heart disease is genetically encoded, but they still enjoy benefits of their work being about "brain diseases" that seem a layer removed from those working on "brain behaviors".

The final layer of the onion is the one with which I initially interacted with—the layer that thinks non-mental health diseases like eczema, psoriasis, and asthma have meaningful genetic underpinnings. I will outline in later chapters that these diseases are distinct because they each have a biomarker; a biomarker being a blood, imaging, or functional test that one can use to confirm a diagnosis. Like the layers below, this layer requires a level of purposeful ignorance about epidemiology both in terms of geography and generational change.

This book will be very harsh on the behavioral genetics, especially those claiming they have found the genes for college attendance (which they term, "educational attainment"). However, I will also argue that they are not the only source of the problem and that an entire re-evaluation of population genetics would be needed to right the ship. While the education attainment gene work has directly fueled bigotry and discrimination, those working on it mostly copy the methods of the scientists that study the genetics of non-controversial topics. Therefore, because those working to find the genes for intellect are not doing anything meaningfully different than those claiming to be looking for the genes for asthma, I struggled with whether it was fair to be as critical of their specific section of genetics as I am. After all, missed opportunities to uncover meaningful environmental contributions were widespread in all of medicine, not just education genetics.

Then, on May 14th, 2022, a man walked into a grocery store in Buffalo, New York and murdered 10 people because he had been radicalized by reading papers on the genetics of intelligence. The geneticists that spent their career claiming that Black Americans had genes causing them to have more asthma, allergies, heart disease, and nearly every other diagnosis saw no commonality between their work and those directly cited by this murderer. So, those working on educational attainment were effectively sent out to defend the entire field of population genetics from the now undeniable realization that their work could directly fuel bigotry.

Lest you feel sympathetic to these researchers, note that the behavioral geneticists did not handle the challenge well. Rather than an ounce of introspection, they replied by dismissing concerns as being reflective of an inability of racists (and anti-racists) to correctly understand their work. Worse yet, they insinuated that any effort to restrict or downplay their work would have even graver consequences for marginalized communities (without offering any specific reasons why). Even this could be dismissed as individuals who lack training in public relations being compelled to engage in online debates concerning the significance of their careers. However, what took things beyond the pale for me was how those who had been directly cited as the inspiration for a racist murderer began to state that the best way to help marginalized communities wasn't to evaluate environmental injustices but to dedicate research dollars into investigating things like the genes for Black people's asthma.

It became clear that population genetics was never going to be analogous to granting superpowers or creating an entire team of Michael Jordans. At best, the field had become a distraction that incentivized researchers to chase genetic associations for diseases that had an overwhelming association with exposure to industrialization. Bigotry and discrimination might be used to create GATTACA-like systems of discrimination, but the idea that genetics could ever be used to construct a reliable means of sorting people into future farmers or future astronauts seems far-fetched and ethically problematic. Rather than becoming comparable to GATTACA, population genetics has followed the trajectory of a different specialty that began in an attempt to explain intellect but expanded into making overblown claims of being able to predict all types of disorders. It modeled a previous field that ignored the natural variation between people across populations while never bothering to prove biologic connections between the signals they measured and the diseases they claimed to care about. A field that had its own research societies, Ivy League departments, academic meetings, and dedicated scientific journals. A field that eventually collapsed as it started to become increasingly convoluted in its practice of aggregating and caveating their findings to rationalize why their results could not replicate over time and space.

Population genetics has devolved back into the phrenology from whence it came.

Part one of this book will discuss the foundational flaws that permeated the development of population genetics and created the gene-centric paradigm that set medicine on a course to light billions of dollars on fire, then choke on the smoke it created. However, I will also outline how some researchers avoided those pitfalls to discover beneficial results in cancer, rare diseases, and pharmacology.

Part two will discuss the current flaws in study design that afflict all of population genetics. Wherever possible, I will attempt to use both examples from the more controversial topics like educational attainment alongside examples from diseases of the immune system like allergies or autoimmunity. I hope to show why belief in a gene for food allergy is just as silly as believing in a gene for iPhone ownership. I also will argue that such claims have distracted researchers from looking for environmental causes of the diseases that might be avoidable.

Part three will discuss how both the flawed methods and communication of genetic research have continued to fuel bigotry and discrimination despite the genuine efforts of these researchers to denounce such conclusions. However, I will not spend much time outlining how racists have used (and misused) science to advance their cause. This work will instead focus on how those who genuinely believed that the mysteries to common diseases and traits would be found in the genome lost their way. Readers interested in detailed discussion on how racists hide their claims behind genetics should read *The Myth of Race* by Robert Wald Sussman, *Superior* by Angela Saini, *The Science of Human Perfection* by Nathaniel Comfort, *Fatal Invention* by Dorothy Roberts, and *Misbehaving Science* by Aaron Panofsky. Much of my writing is informed by these books.

Finally, part four will discuss why shifting the paradigm away from genetic determinism is both incredibly difficult but also incredibly important. I will also outline why the gene-centric paradigm has been so hard to shift away from in medicine and behavioral science, offer predictions on how the field will play out, and add some suggestions that might steer us in a better direction.

I hope to convince you that conceptual flaws have prevented science from uncovering the true causes of the inflammatory diseases that are common today. The belief that these diseases are an internal feature of the patient has caused medicine to have tunnel vision to the point that we are watching the tide of disease rise as we stand awaiting our prescription medication saviors. Furthermore, I hope to demonstrate that the study design used by geneticists, not the data or interpretations, is what continues to attract bigots to the field of genetics.

I don't pretend that this work will induce introspection in population geneticists, let alone convince them that their field is a dead end. Rather, I write this first as a guide to the next generation of researchers in hopes of inspiring them to embark on the difficult road of environmental assessments that indeed risks failure, but offers the only real chance to impact diseases of modern living. I also write this as a caution to scientific journals, policy makers, and the general public. The obsession with genetics will get worse before it gets better and nothing short of field-wide changes in methods and standards will be able to change the course. The choice has become clear: We can either wallow in toxic substances, wondering why some people react poorly to them and insinuating that groups exposed to more toxins are inherently flawed, or we can find the determination to follow the footsteps of previous generations and stop swallowing so much excrement.

Let me address a few final prefaces about how this book will be different from what you might expect: The first is the style of citations. Typically, a non-fiction science book will use footnotes or endnotes to indicate the source of the material being commented on. For example, a statement saying, "Ohio State has won four of the six football games against Notre Dame" would have a superscript number that the reader could use to track down the citation for that comment. This book instead places the citations in narration style, so my statement would read, "According to a 2022 article by Lisa Kelly of *SB Nation*, Ohio State has won four of the six football games against Notre Dame." While this is not a formal citation, I do this because audiobooks never transcribe the footnotes or endnotes, nor do they narrate the superscript numbers. Therefore, anyone with visual impairments who relies on audiobooks, or just those who prefer audiobooks, can't track down the citations when they are traditionally formatted.

The next preamble statement is one that I hope has already been made clear: The tone of this book is not the standard academic presentation. I do not intend to burn bridges, but I'm not walking on eggshells, either. This book's main thesis is that many of the well-meaning researchers who go to work every day hoping to find the genetic underpinnings of diseases ranging from asthma and diabetes to heart disease and Schizophrenia are doing so under a fatally flawed set of assumptions. All the work built upon those flawed assumptions is so invalid that it will need to be thrown out and redone using better study design. Furthermore, I will argue their work was always destined for failure and continuing to fund their endeavors delays the discovery of real cures and preventions, erodes trust in the medical community, and fuels the type of hatred that leads to genocidal mass murder. There is no friendly way to state this to people who have built their careers on believing otherwise.

Prior to the murders in Buffalo, and before the details of how the opioid crisis came about were known (each will be detailed in part three of the book), I would have been more open to discussing the nuances of population genetics performed only in academic vernacular. However, for decades prior, population geneticists had been systematically overstating the value of genetic information for common diseases in hopes of furthering their careers. When the implied genetic determinism was taken literally by online bigots and policymakers alike, the outcomes that every history class predicted became a reality. In the time that followed the Buffalo shooting, I watched the most prominent scientific journals hand platforms to the population geneticists who were the worst offenders for promoting the types of genetic determinism that was cited by the killer. Those geneticists used that platform to espouse that a field born of eugenics and still obsessed with heredity needed to change its language, but not its actions. Over a year later, nothing had changed in their approach to research and their language had returned to what it had always been.

So, in this book I will not pull punches when discussing the practice of focusing on the imagined genetic contributions to diseases like asthma or learning challenges while completely ignoring the factories spewing pulmonary toxins into the sky and dumping neurotoxins into the water. Sugar coating it by describing such research assumptions as "unlikely to yield results" or "presenting ethical dilemmas" both ignores the urgency of the matter and disrespects those who are suffering while their tax dollars are being spent on programs that blame their DNA for the harms of their toxic environment.

I will attempt to toe the line between being provocative without being a provocateur, but know I will stray at times. Yet, know that where my words seem blunt to the point of being harsh, it will be in response to a field that once referred to certain groups as being "genetically defective and feebleminded." Today the field masks their conclusions behind the professional language of referring to those same people as "possessing a genetic profile confirming risk for reduced cognitive performance and social achievement." We live in a world where most of the headlines that described a racist murderer who killed people in the name of genetics used vague language like "a racially motivated shooting." Therefore, because the field of population genetics is doing *active* harm to medicine and society, I have lost my tolerance for *passive* phrasing. If blunt language makes you uncomfortable or if you had hoped this read to be a purely academic exercise, this book is probably not for you. But I also invite you to reflect on whether the stated priorities of improving the world through research are reflected in the belief that speaking bluntly is unprofessional, even when faced with a model of research that openly claims victims of pedophilia have genes that brought the abuse upon themselves.

Three final and quick logistical notes. First, for stories where I discuss patient encounters, every fact is true except for the patients' names, which were changed to protect privacy. Second, many of the figures in this book from the work of other researchers are screenshots of their tweets. This is done because their tweets are public record and can be shown in a book without permission. Had I attempted to show only the graphs from the original papers (i.e., if I show the data they were tweeting but without the tweet itself), I would need permission from the publisher. As I will discuss, the publishers of modern population genetics are part of the problem (especially journals under the brand Nature), and thus, I'd prefer to take a free option over one where I placate their immorality and feed their lust for treasure. Finally, while I am employed by the NIH, I am not their spokesperson and the views in this book are my own. I have spoken to several NIH scientists working in population genetics who will disagree with what I have to say. I have always found NIH to be a place of open and robust discussion for disagreement and while I do not intend to sanitize my comments to make friends, I do hope the insights provided will spark continued discussions. Therefore, this work is not representative of the Department of Health and Human Services, NIH, or the U.S. Public Health Service.

PART I

Sowing the Seeds of Failure

CHAPTER 1
A Brief Primer on Biology

BOTTOM LINE UP FRONT: *Genetics operate through RNA, proteins, and metabolites.*

Before we take on the ways in which genetics has failed to produce benefit for the public, it might be useful to outline what we currently know about genetics, in specific, and biology more broadly. When scientists talk about someone's genetic sequence, they are referring to the specific order of DNA molecules that are strung together in the genome. These molecules are referred to as nucleic acids. In DNA, there are four different types of nucleic acid, each of which are overall similar but with subtle chemical structure differences between them. These molecules are adenine, thiamine, cytosine, and guanine—abbreviated A, T, C, and G. The title of the movie GATTACA was meant to be a play on this theme, representing how one might document a stretch of DNA that had a G, then A, then two Ts, then another A, a C, and a final A. Since the DNA is in a string and is *read* from one end to the other, the obvious analogies that form are to compare it to a written language.

If the nucleic acids are the individual letters, then we might say that the word that they spell out represents a segment of a gene. But genes often have multiple segments needed to communicate correctly, so we could say that a full gene is a sentence. Just as those sentences are organized into paragraphs and chapters, genes could be organized into *regions*.

Finally, those regions are collated into individual chromosomes, much like stories can be sorted into different books.

Geneticists love claiming that DNA is the *recipe for life*: this is likely incorrect, which will be discussed in later chapters but let's go with it and use the analogy of a recipe book. For a given recipe, each letter makes up a word, which combine into sentences, which make up the paragraphs that comprise the recipes, which are sorted into various books. Every healthy human is born with 46 recipe books; you get 23 books from your mother and 23 books from your father. In most cases (but not all cases, as the discussion around trans-rights has taught us), two of these books are recipes for biologic sex. If you get two copies of the X recipe book you will (likely) be born a woman, whereas if you get one X and one Y recipe book you will (likely) be born a man. Combined, these books have about 30,000 recipes (for the 30,000 genes).

But recipes are only the starting point, not the end all of a meal. The words on the page have to be turned into raw ingredients, then combined and cooked into a final meal. For DNA, the genes are transcribed into a related structure called RNA. RNA also uses various sequences made up of four nucleic acids. For the most part, RNA is meant to be a 1-to-1 transcription of the DNA sequence, except the RNA is streamlined so that only the essential parts of the DNA sequence are included. Consider that a gene being turned into an RNA transcript is like a sentence printed in a recipe being turned into the ingredient on the counter. A full sentence saying "get two tablespoons of paprika" is transcribed into two tablespoons sitting on the counter ready to add to the pot.

However, RNA isn't the end step, either. This type of RNA is called the *messenger* because it is supposed to take the DNA message and relay it into proteins. Proteins are also made of individual molecules (called amino acids) strung together in a specific order. There are 20 different amino acids, each of which would be *encoded* into the DNA and RNA.

For the analogy, we will say that the proteins (and metabolites created by these proteins) are the final dish. So, the written recipe (DNA) is turned into the ingredients (RNA), and then those are translated into the final dish.

But even here, the dish isn't the final step. Food isn't meant to be made and set on a plate; it's made to be eaten. And we eat those meals to serve specific functions. Ideally, each meal is eaten to provide good nutrition, but if we are honest, sometimes, the goal is simply enjoyment. So, we might say that the function of the dish is either in its flavor or nutrition. Thus, putting it all together, we have an analogy where the written recipe (DNA) is turned into the ingredients (RNA), then translated into a dish (proteins), which provides the flavor and nutrients (protein function).

Imagine you are in a blind taste test between two dishes of chicken cacciatore. The first is delicious but the second is revolting. Looking at them, they appear so similar there isn't anything obviously different that would explain the difference in taste. So, a team of investigators is assembled, and you are put in charge of researching why the two dishes had such dramatically different flavors. You begin your investigation by comparing the different recipes. Most of the words are identical between them, but let's say you notice the delicious one calls for a teaspoon of paprika, while the unpalatable version says a teaspoon of "papryka". Are you done with your investigation? Is it safe to determine that the reason for the foul taste was that the second dish used "papryka" instead of paprika?

The answer is obviously no. You would want to know if the misprint caused a change in the ingredient used. If the chef read "papryka" and still added paprika in the correct amount, then there is no reason to believe that the letter change (called a point mutation in DNA) had any impact. A point mutation that has no impact on the final protein is called a *silent mutation*. Every human has an enormous number of silent mutations that help distinguish one person from the other. Then there are versions of the same gene with differences in the letters, these are called "alleles", "variants", or "SNPs". Paprika versus "papryka" would present two alleles, but the differences would be silent and thus the allelic variation would be moot. They don't have any real meaning biologically but would be the basis for identifying whether a blood sample at a crime scene belonged to the victim or the suspected perpetrator.

So, ruling out paprika as the cause, you keep looking and discover that the tasty dish had instructions saying, "Make sure to soak the chicken in brine for two hours before cooking," while the foul-tasting dish said, "Make sure to soak the chicken in urine for two hours before cooking." Just as before, this represents only a single letter change—another point mutation. But is this one as silent? You would still need to check and make sure that the change was carried out. Did the chef really soak the chicken in urine prior to cooking? If so, that would mean a change in the ingredient, which caused a change in the dish, which then generated an expected harm to the flavor. Perhaps a less dramatic change could be identified. For example, a misprint may change a line stating, "Add 1 teaspoon of paprika" to "Add 10 teaspoons of paprika". Adding 10-times the correct paprika would be less harmful than soaking the chicken in urine; however, you would expect some degree of harm to the flavor and nutritional value nonetheless.

The letter change analogy, however, would be far more subtle than what would be comparable to most of the well-known genetic diseases. For some diseases, the sentence and paragraph simply stop without finishing. This would be akin to a recipe stating, "First, preheat the oven to 400 degrees Fahrenheit. Next, place the." That early 'period' completely stops the RNA transcriber from moving forward. So, in effect, the entire recipe is lost and can't be made. These early stops are called *truncations* and depending on how early they appear in the recipe, the resulting protein may be half-finished or non-existent. These kinds of mutations can be the most damaging since an entire protein—and its intended function—may be lost. However, one failsafe in this process is the fact that everyone gets one copy of a recipe from their Mom and one from their Dad. So, even if you completely lost Mom's version, you could at least use Dad's to prepare the same protein. Disorders where you need both parents to pass along a dysfunctional version of the same gene for disease to appear are called *recessive*.

Like all analogies for genetics, this one too has some flaws in that there really are *Italian recipes* but no such thing as *Italian genes*. Furthermore, the analogy is stretched for *dominant genetic disorders* in which you only need one bad copy to develop symptoms. One could analogize this by saying that you have to make both Mom's and Dad's recipe for a given dish every time, then mix the two together. Thus, if Dad's version calls for soaking the chicken in urine, it will ruin the dish regardless of how great Mom's recipe is. This isn't completely accurate, but no analogy is.

So, the written recipe can greatly influence the dish's final flavor but so could many other factors, such as whether the ingredients are rotten, how long the dish was cooked, the temperature used, whether the oven worked properly, how long the dish sat out, whether the plate it was put on was clean, and so on. So, if you were discussing ways in which flavors can be altered, the written recipe would certainly be one of the factors, but not the only one.

It is also important to note that most of the time, when a researcher says that a gene is associated with a specific organ function, they are referring to when a major error destroys that function. For example, the change between "brine" and "urine" would completely ruin the dish (should it be carried forward). Worse still, if both versions of the recipe were broken beyond comprehension, then the dish couldn't be made at all. These kinds of changes cause massive harm, but they do not mean that a similar change could make the dish taste better.

Here is a challenge; take the sentence (gene), "Soak the chicken in brine for two hours prior to cooking" and change a single letter in a way that would make the dish taste even better. Can you do it? Again, we can ruin the flavor by changing brine to urine. We would also potentially do harm by changing "soak" to "soap" and maybe even "cooking" to "cooling." But can you give me a change that would benefit the recipe? Numerous diseases impact the health of the cells in the brain in ways that cause enough harm to manifest a developmental delay or loss of cognitive abilities. But even though losing the proper function of a gene can cause a loss of intellectual capacity, it doesn't mean that a different change can give anyone heightened intellect. Similarly, having a recipe that was truncated to the point you couldn't read past the opening instructions might cause harm, but having five extra copies of the recipe would not make it taste any better.

We will see in the chapters that follow how well-meaning geneticists, bigots, and bigot apologists each love to use *genetic* and *biologic* as interchangeable terms. They do this to over inflate the importance of the recipe in how well the meal tastes and/or nourishes. They also erroneously think that because a dish can be ruined by changing "brine" to "urine", it also implies that you can make it taste better with just one letter change. Despite its limitations, we will carry this analogy forward to demonstrate how scientists successfully used genetic information to improve the lives of people struggling with various diseases to contrast these victories with the many failures of population genetics.

However, research into why people became sick did not start with genetics. Plenty of people studied how to be the best cooks before written recipes became the norm. Thus, the next chapter will briefly review how the notion that genetics would be the main factor in determining health and disease came about.

CHAPTER 2
The Competing Camps of Early Population Science

> **BOTTOM LINE UP FRONT:** *Population science developed competing research tracks focused on either environmental or inborn causes of disease.*

No era is free from debate; however, ancient times were frequently marked by the notion that one's morality was the gatekeeper against illness. If you got sick, or certainly if illness befell the entire village, the first guess as to why this was happening was that God, the Gods, or some spiritual entity was punishing the wicked for something they had done (or even punishing the innocent for something someone in the town had done). In some ways, this belief exists even today in the form of TV evangelists who imply that God would send a hurricane upon a town because someone in their same state held a gay-pride parade. If you lived purely, such moral behavior would prevent whatever was swirling around you from being able to get in and bring about disease.

Robert Wald Sussman's book, *The Myth of Race*, expertly outlines how this belief would soon become merged with discrimination and bigotry. When European explorers visited new lands filled with people they would soon deem to be less-than, it was their behavior that initially made them inferior. It was the willingness to allow women to be bare breasted in public, the lack of English, French, or Spanish as a language, or the lack of Christian beliefs. The *barbaric* lifestyle led many imperialist-minded conquerors to conclude that indigenous populations were sub-human. The Catholic Church, on the other hand, concluded that if some individuals could be converted to Christianity, they must be considered human and deserving of respect. However, a significant caveat remained: if the population resisted converting to Christianity, it would be seen as an indication of their subhuman status, and they could be subjected to mass killings without moral concern.

Initially, morality and cultural norms were viewed as traits anyone could adopt. Ancient Greeks, for example, believed that anyone who took on Greek culture (clothing, food, language, politics, etc.) was superior to those of any other culture. One need not have been born Greek to be considered part of the elite culture; one simply needed to integrate into Greek society. Similarly, being born Greek did not assure you were considered elite if you were to abandon the Greek way of life. While tribal conflicts have existed throughout human history, the idea of biologic superiority was not invented until the Spanish Inquisition around 1400 AD. Cultural and ethnic discrimination existed well before, but the idea that societal greatness was encoded into the bloodlines did not come about until the "limpieza de sangre", which translates to blood cleansing. Prior to the Inquisition, Spain held a policy that was similar to that of ancient Greece; so long as one converted to Christianity, they could fully participate in Spanish culture and benefits (such as land ownership). People of Jewish background were thus allowed to fully integrate into Spanish society so long as they converted; this group was known as Conversos. Eventually, some in society became upset (whether through jealousy, greed, or just plain hatred) at this policy that allowed conversos to open businesses and buy land.

Those opposing this idea felt that native born Spaniards should be given preference over anyone that converted into society. Eventually, those opposing rights for Conversos won out and Spanish law began to discriminate based on Jewish heritage. Anyone with *Jewish blood* could convert to Christianity but would still face restrictions on benefits that were not applied to those who were ethnically Spaniards. The idea was thus enacted into Spanish law that certain people could inherit rights from their parents, while others could not. Apparently, sensing that bias against Conversos would escalate, some contributed to the voyage of Christopher Columbus in hopes that he would find safe places to move to in the new world.

To be fair, to some extent, the idea that complex traits as well as privileges were passed down from parent to child were in place for any culture that believed in royalty. The very idea of hereditary monarchies suggests that a ruler possesses innate leadership gifts (likely given by God) that are passed down to their children. These gifts are concentrated amongst those of *royal blood*, and thus, it was important for princes of one lineage to marry princesses of another, lest they risk *diluting* these gifts with the blood of commoners. Certainly, people can defend the idea of hereditary monarchy without genuinely believing the crown was biologically inherited. However, at least by the stated rules of royal succession, leadership was always a birthright.

The notion of inherited traits was also prevalent in agriculture and animal science. Darwin published his *On the Origin of Species* in 1859, which set the narrative for adaptive change among animals for the centuries that followed. Humans had already been selectively breeding plants and livestock to enrich desired traits like crop yield and egg output. When eventually applied to humans, people attempted to use physical traits to predict behavior and ability. Just as a svelte, muscular horse would be predicted to be better at racing than a frail one, humans with large frames and ample muscle could be predicted to be stronger than skinny ones. Eventually, these researchers went beyond the obvious to begin guessing at less viable traits. The first predictor was hair color and facial features.

This era of ethnic bigotry predates widespread exposure of Europeans to non-European cultures but certainly laid the groundwork for what would become the forms of discrimination today. The initial idea was that features associated with looking more Northern European (Nordic) were predictors of superiority compared to features more common in Southern Europeans (Germanic or Mediterranean). Darwinist approaches which viewed adaptations in bird beak shape as dictating if the fittest survived the process of natural selection were easily transferred onto human traits and behaviors. Just as giraffes might develop longer necks to reach the highest leaves, humans with the greatest natural intellect were thought to maximize their chances of having offspring. A competing school at the time led by French biologist Jean-Baptiste Lamarck, stated that organisms could acquire traits that could be inherited independent of innate biology. Darwinism won out and, in effect, created the inverse of the initial ancient Greek beliefs, in that no matter how you lived your life, the physical traits you were born with predicted your innate abilities and societal worth.

Slowly, this practice became the field of physiognomy, which is the idea that one's overall physical appearance predicted their character. Eventually, this approach was overtaken by phrenology, which focused on the shape of one's skull (and the features upon it) as the predictor of all sorts of characteristics. The slope of one's forehead, for example, might predict the propensity for criminal behavior and the location of the ears relative to the eyes could tell one's intellect. Eventually, skin color was added to the list of physical traits used to supposedly predict behavior and worth. Since skin color and many other physical features are clearly passed down from parent to child (i.e., most children physically look like at least one if not both of their parents), these physical characteristics became indicative of the value of entire populations. Suddenly, the Southern Italians were *less noble* than the Norwegians based on the shared (and heritable) features of the population. Meanwhile, non-European populations were ranked based on how dissimilar they were to the supposedly *ideal* Nordic look.

The belief that populations suffer due to their collective behavior or innate biology and not the environments they live in saw its first cracks develop when researchers began to study what we now know to be infectious diseases. Colonial visitors in Africa assumed that the local population was contracting malaria due to their innate weaknesses (moral or biological). However, when the colonizers themselves started getting sick, a new hypothesis emerged that maybe it was the environment that might dictate the risk.

In 1840, Hungarian physician Dr. Ignaz Semmelweis, observed that the part of the hospital where midwives delivered the babies had a much lower rate of infection and death than the ward staffed primarily by doctors. Semmelweis hypothesized that the act of physicians performing both autopsies and deliveries without any break in between—a practice that midwives did not perform—might be the cause. He instituted a program for doctors to dip their hands in a solution of dilute calcium hypochlorite after they performed the autopsies. Semmelweis observed that this solution removed the putrid smell and thus hypothesized that it might also help reverse what was called *cadaverous poisoning*. The mortality rate after this program dropped from 18% to under 2% in just under four months. You would think this would be celebrated, but instead, Semmelweis' colleagues dismissed him. One prominent obstetrician said, "Doctors are gentlemen and a gentleman's hands are clean." Since he could not get his colleagues to listen, Dr. Semmelweis suffered a nervous breakdown and was committed to an insane asylum. He eventually died from an infection of a wound he suffered at the hands of a beating by the asylum guards.

It wasn't until over a decade after Semmelweis' discoveries when the work of Dr. Louis Pasteur identified microbes as the cause for spoiling beverages like milk or wine did germ theory enter into the broad scientific consciousness. The introduction of germ theory inspired scientists to evaluate potential microbial causes of disease, including eventually identifying *Streptococcus pyogenes* as the cause of *cadaverous poisoning*. After the 1929 discovery of penicillin by Alexander Fleming, drugs for killing microbes and the association of bacteria with an unclean state became the norm.

Suffice to say that after the scientific concept of germ theory went viral, waves of researchers began to search for environmental microbes that might be driving diseases previously believed to be caused by evil spirits, sinful living, or bad smells. Research in the burgeoning field of infectious disease required a keen eye towards epidemiology—the study of *what befalls the people*. The example most often used as the birth story for this field is that of John Snow, the English physician credited with realizing in the 1850s that cholera outbreaks in his town were more common in those that drank water from certain wells than other wells. He famously mapped the locations of the sick and concluded that their drinking water was the source of their infections as shown below. What is said to have happened next (but appears to be debated by historians) is that he removed the pumps from the wells he considered contaminated so that no one could drink from them. Shortly thereafter, the infections began to wane and both his legend, and the field of epidemiology, were born.

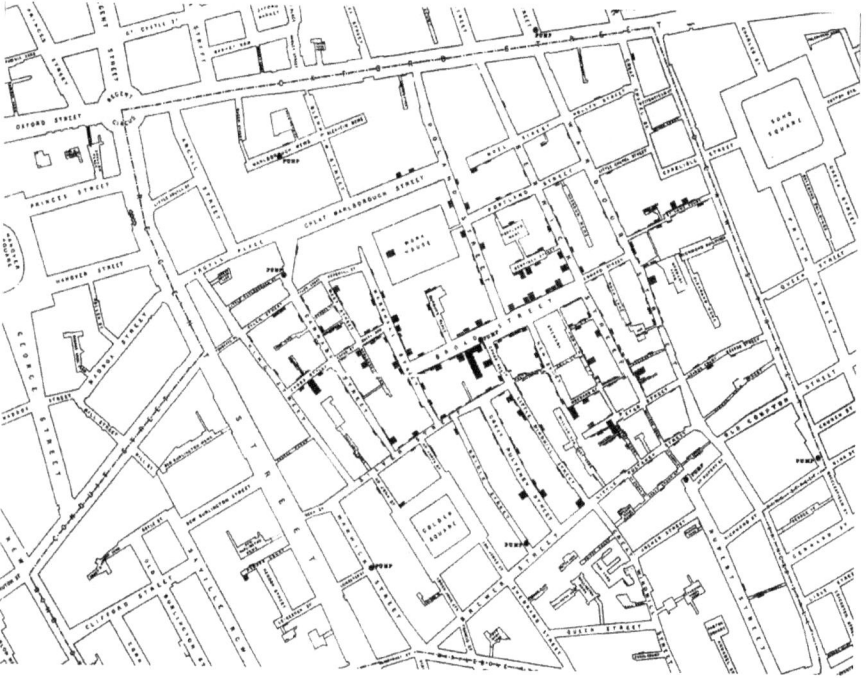

Much later, others noticed that a battalion of Army troops might train in Kansas without any major issues arising. Yet, when those troops deployed to Cuba, many would contract the newly discovered disease of Yellow Fever. Some of those troops were sick enough to require evacuation back to the U.S. The patients could return to the U.S. but even if they remained sick, they never seemed to spread the disease to family members, other patients, or the healthcare staff. The conclusion was clear: Whatever was causing Yellow Fever was in Cuba and stayed in Cuba. Thus, there was no point in looking for the causes of Yellow Fever in Kansas or in the troops themselves—at least not until there was a thorough evaluation of the environment of Cuba that appeared to determine who got sick and who did not.

This realization inspired Dr. Walter Reed to travel to the region and initially identify fecal matter and flies as a contributor to the disease. When better sanitation did not fully cure the problem, Walter Reed's panel of researchers identified that Yellow Fever was carried by a mosquito that lived in Cuba, but not in the United States. Today, there are entire departments of research dedicated to *vector borne diseases* like Yellow Fever and malaria. Furthermore, the flagship military hospital in the U.S. has been named in Walter Reed's honor.

However, not every disease had a clear signal in epidemiology. Some diseases—most prominently the diseases of mental and behavioral health—lacked any clear geographic pattern. It wasn't as if the only people that developed Schizophrenia were those who traveled to Florida. Also, with infectious diseases, there tended to be an exposure that clued people into how they got sick. For example, if you sat on a bus next to a guy with a hacking cough, and then later developed a hacking cough yourself, the connection seems clear (even if it is not always that simple). However, with diseases of mental health especially, there is rarely something obvious that predates the onset of symptoms. Today, we have a better understanding of the reality that a trigger for a disease may come years before the symptoms. Not all the mechanisms are well understood, but today's researchers understand that exposures in the womb may not manifest until adulthood. Overall, those searching for a clear infectious trigger to mental health disorders failed to find anything.

Into this mix stepped genetics.

In the absence of any clear environmental cause, the camp espousing the idea that mental health must be an innate defect of the patient infamously took hold in psychology. I say infamously because of the unique response of the researchers of that era to their diseases rather than any unique insights. The notion that diseases reflected the innate biology of the patient, and that those properties could be directly passed down from parent to child, was not a conclusion that began with behavioral health. However, unlike other fields, some mental health researchers determined that the best way to *treat* their disorders of interest was to propagate what we now term *eugenics*. Eugenics was the idea that society could be perfected through policies that would maximize ideal breeding—akin to policies used to breed larger livestock. Eugenics did not start out focusing on mental health, but the mental health researchers eventually latched on. Founders of eugenics like Francis Galton initially sold the practice to improve human performance – both mental and physical – through selective breeding.

So called *positive eugenics* involved incentivizing matches between people of *good stock*, such as a program that would assure college professors could marry the prettiest women (not surprisingly, such a policy was pitched by a college professor whom one can only assume felt his luck with the ladies was so poor he needed government fiat to help him out). So called *negative eugenics* involved the overt oppression of *inferior* groups, which often meant sterilizing those with diseases, outlawing mating outside of one's race, and limiting immigration from any populations that were not northern European Gentiles.

This era of science is somewhat paradoxical in that it was marked by two parallel, but vastly opposing, tracks of research and public health interventions. The track that focused on epidemiology and environment ushered in three of the most consequential discoveries in human history: water sanitation, antibiotics, and vaccines. Each of these have greatly extended life expectancies and reduced suffering around the world. Meanwhile, the other track—which adhered to the idea that diseases and traits were innate, inheritable properties of individuals—set forth on a campaign to marginalize anyone with these diseases or traits.

In academic circles, history has not been kind to phrenology or eugenics. However, it is important to note that this history was not that long ago. Phrenology had two respected scientific journals and multiple university departments around the world. Somewhere in the annals of Ivy League universities, there are bound editions of doctoral theses for those who received a PhD in phrenology. Departments of Eugenics were also prolific publishers in numerous prominent academic journals (and I would argue they still are in the form of *Nature Genetics* and numerous Elsevier journals). Phrenology had fallen out of favor by 1840 or so, but eugenics carried on openly until about 1940 and only then changed because of the public relations implications of being cited as the inspiration for the Nazi Holocaust. A public connection between genetic determinism and the Nazis changed the conversation around genetics research, but it did not change the methods.

CHAPTER 3
The Original Sin of Population Genetics

BOTTOM LINE UP FRONT: *Flawed twin studies set an unrealistic expectation for how important genetics are for human traits and diseases.*

As previously stated, the idea of hereditary traits initially came from animal and plant studies. Sadly, but not surprisingly, those who were willing to treat populations they deemed inferior as if they were livestock soon began to assume that those populations functioned like livestock, as well. The first real attempt to mathematically assess the degree to which a given trait was due to heritable factors was twin studies. Twin studies operated under the knowledge that, as it relates to the potential DNA differences between people, identical twins share 100% of their DNA sequence while fraternal twins share only around 99.95%. However, twin studies also assumed that identical twins and fraternal twins would differ *only* by the amount of shared DNA. Their math—termed the *equal environment assumption* or EEA—assumed that both types of twins would more or less share the same environment between them; thus, any difference in environmental experiences between two twins would be negligible, leaving only the difference in genetic similarity to explain why identical twins may be more similar than fraternal twins.

The twin studies equation first put forth was heritability (h^2) equals 2 times the concordance rate in monozygotic twins (R_{MZ}) minus the rate in dizygotic twins (R_{DZ}); $h^2 = 2(R_{MZ} - R_{DZ})$. If the rate that both identical (monozygotic) twins had Schizophrenia was 80%, while the rate that both fraternal (dizygotic) twins had Schizophrenia was 40%, the math would work out as 2x(80-40), which is 2x40 or 80. So, Schizophrenia, in this scenario, would be 80% heritable and considered 80% genetic. There was an additional calculation for *shared environment*, which was abbreviated as c^2 and calculated as $(R_{DZ} - 1)/2h^2$. In this scenario, this would come out to 12.5%. So, the overall claim here would be that Schizophrenia was 80% genetic with only 12.5% of the environment shared between the twin groups.

This formula was used to derive all sorts of claims of heritability, and by proxy, make claims about genetic assumptions. Clear signals were seen in diseases with Mendelian patterns of inheritance like cystic fibrosis and sickle cell anemia. Mental health disorders like Schizophrenia and bipolar, along with asthma and eczema, were also deemed *highly heritable*. However, more laughable examples include things like 'the love of sailing', which was calculated to be 85% heritable (and thus presumed to be 85% genetic). Even Christianity was considered 65% heritable, implying that being Christian is *mostly* genetically encoded. Given the reality that Christianity began among people of previously Jewish faiths and then spread to nearly every culture on Earth—despite those cultures having their own entrenched religions at the time—it seems dubious to think that there are *Christian genes* that spontaneously appeared 2000 or so years ago. The entire idea that Christians can and should proselytize to others in hopes of converting them runs counter to the notion that people's faiths are inborn. That being said, it's a fun thought experiment to pretend that Christian genes have existed throughout evolution but were only activated in certain humans around the birth of Christ. This could imply that, somewhere in the wild, there exists Christian mice alongside Buddhist rams and maybe even some Taoist toads.

The first flaw in the EEA logic is that the EEA isn't even biologically correct. Identical twins always share the same yolk sac, while fraternal twins rarely do. Thus, any exposure which crosses the mother's placenta has a chance to evenly distribute in the fluid around identical twins but may preferentially concentrate in one yolk sac from fraternal twins. Additionally, identical twins most often share a single placenta, while fraternal twins rarely do. This means that any environmental exposure that can influence or cross the placenta in the mother will do so far more evenly in identical twins than fraternal twins. In fact, research focused only on identical twins demonstrated that identical twins who shared a placenta were more similar to identical twins who had two placentas.

The second major flaw in the EEA comes after birth. Identical twins are so much more likely to be dressed alike and confused for one another, that they form a shared identity. Identical twins are therefore more likely to share the same hobbies, eat the same foods, and run in the same social circles than fraternal twins. So, if your identical twin wants to go sailing, you are more likely to share that interest than if they were your fraternal twin. If your identical twin wants to eat Italian food, you are more likely to want to also eat Italian food than if they were your fraternal twin. And if your identical twin wants to commit crimes, you are more likely to be their accomplice than if they were your fraternal twin.

In his book, *The Trouble with Twin Studies: A Reassessment of Twin Research in the Social and Behavioral Sciences*, Dr. Jay Joseph presents an illustrative hypothetical on the flaw in EEA. Imagine one set of parents gave birth to four babies at the same time, two identical twins and two fraternal twins. If all four grew up in the same house, at the same time, going to the same schools, the assumption is that they would all be impacted by their shared environment to the same degree. But the data indicate the exact opposite. The identical twin boys would be far more likely to dress the same, participate in the same activities, and be treated as a distinct pair that was apart from the four-kid sibling group as a whole.

Furthermore, identical twins receive environmental feedback in ways that are more profound than fraternal twins. For example, if one twin has a nervous smile that people often mistake for a smirk, thus generating negative reactions from others, then the identical twins will share environmental feedback in ways that the fraternal twins will not.

A modern example is the idea that people can have what is termed *resting bitch face*. This is the notion that someone's neutral expression might be misread as them expressing frustration or being upset. Some research has actually validated the idea (without validating the questionable terminology) that people's blank expressions might convey emotions they are not actually feeling at the time. So, if the identical twins both have faces that miscommunicate emotions, then people's responses will be more similar than they would be for fraternal twins. This means that conversations will have more similar misunderstandings in identical twins than fraternal. Since environmental feedback is more significant in identical twins, the environmental impact will also be more significant. The observations that different types of twins experience their environment in different ways renders the foundation of twin studies—the EEA—invalid.

Here is a quick exercise as an example. Take a look at the pairing of first names, with focus on their first initials, from 20 male twin pairs below:

	Identical (MZ) Twins		Fraternal (DZ) Twins	
	Twin Name 1	Twin Name 2	Twin Name 1	Twin Name 2
Pair 1	John	James	Sam	Robert
Pair 2	Steven	Todd	Jamal	Janet
Pair 3	Tom	Tim	Anthony	Jack
Pair 4	Bryan	Brandon	Greg	Robert
Pair 5	Greg	George	David	Mark
Pair 6	Harvey	Henry	Ian	Isaac
Pair 7	Danial	Nathan	Jordan	Eric
Pair 8	Chris	Charles	Noah	James
Pair 9	Jonah	Levi	Mo	Ali
Pair 10	Adam	Allen	Tim	Todd

If you ran a heritability test for first initials on the group above, the rate of concordance in the identical (MZ) twins is 70% (7 out of 10) while the rate in fraternal (DZ) twins is only 30% (3 out of 10). By the esteemed math of the early twin studies researchers, the heritability of your first initial would be 80% (equal to two times 70-30, or 2x40). Would you conclude that your first initial is 80% genetic from this data? Do you think it would be valid for a news outlet to run the headline, "Study finds that your first initial is determined by your DNA"? Of course not. You would expect that parents would be more likely to try to pair names for identical twins than fraternal. You would even expect the parents' names to influence their decision; for example, if you heard two siblings were named James and John, would you think their parents were more likely to be named Joe and Jennifer or Robert and Susan?

However, there are numerous anecdotes of identical twins separated at birth that both work for their local post office, drive 1987 Toyota Corollas, and both married blonde women named Margaret. This work was the basis for numerous *reared apart* studies of twins that had been adopted by different families. The assumption was, the similarities that manifested despite being raised separately must be, to some degree, genetic. Joseph's book also outlines the flaws in these studies. First, as outlined by Joseph, many of the reared apart studies fabricated their data or refused to publish their control group for reared apart fraternal twins. If concordance was supposed to be so important, then data from reared apart fraternal twins would be an essential part of the rudimentary calculations. Yet, the most-cited reared apart identical twin studies claimed they could not include the control group data due to *space limitations*.

But even in the studies that did not fabricate data, twins are most often reared together for some time prior to being separated; that length of time can vary greatly and cannot simply be ignored because their later environments are different. Even between identical twins, it would matter greatly if they spent their first five years together compared to if they left the delivery room through different doors. Furthermore, twin studies that enroll groups that were reared together, by their very nature, are going to be biased if some people know their twin exists. If you know each other, spend enough time with each other, stay in communication, and are close enough to sign up for the same research study, then odds are you have more environment shared than two twins that have no clue they even have a twin.

Even identical twins that—with seemingly different environments—separated on day one of birth are still very likely to share numerous features as identical twins reared in the same house; these include nationality, religion, ethnicity, birth cohort, and gender cohort. Let us hypothesize that a researcher finds two sets of identical twin girls, each separated at day one of life in 1983, one set raised together while the other pair was reared apart. If the study enrolls them in 1997 and finds that all four girls end up naming N*Sync as their favorite band, *Titanic* as their favorite movie, and *Harry Potter* as their favorite book, then what should we conclude? Should we calculate that each of these pop culture preferences are genetically inherited because that trait was shared irrespective of the environment the twins were raised in? Can you think of a non-genetic reason that 14-year-old girls in 1997 might each have those preferences? While this example seems obvious, similar concerns have been raised over reared apart twin studies performed during the 1960s, which assumed shared use of marijuana indicated a genetic proclivity to drug use rather than being a reflection of the national cultural shift of the era. You would need to know what the odds are that any two randomly selected people from the same age, sex, race, region, and possibly religion would have to be before you could label them as *eerily* similar.

In part due to the realization that the EEA was flawed, but mostly due to the aforementioned connection between the EEA and the Nazis, the EEA was abandoned after World War II. However, the math that underlies its summations has gone unchanged. What replaced the EEA was the *relevant environment theory*. This notion espoused that even though the overall environment was more similar for identical twins than for fraternal twins, the *relevant environmental exposures* were the same. If you are wondering how these researchers could have possibly known all the relevant exposures for diseases like Schizophrenia, join the club. The idea that we can *assume* the important environmental exposures are shared between any two people isn't even valid for diseases where the meaningful environmental factors are well established. Let's take the example of Yellow Fever. It may very well be that all troops on deployment were bitten by the exact same number of infected mosquitoes, and thus, only their innate biology differentiated who got sick and who did not. However, assuming this to be true without even bothering to check flies in the face of logic and science.

These 'relevant exposure' claims were further bastardized when researchers began claiming that the environmental differences between twin groups were also genetic. The EEA had fallen apart on the observation that people treat identical twins more similarly than they do fraternal twins. But the pro-twin studies researchers noted that the reason people treated identical twins as a unit was because they looked so much more similar to each other than fraternal twins. Since looks are indeed highly genetic, these researchers supposed that the differences in how twins are treated (i.e., how they experience the environment created by any interaction with another human being) actually counted as a genetic trait.

To believe this, you need to believe that a pretty girl getting out of a speeding ticket counts as a genetic trait since her looks are genetic (while also ignoring the environment needed for getting out of a ticket requires a police officer that is so attracted to her that he/she would let them off with a warning). If every aspect of the environment that is influenced by human interaction also counts as genetics, then everything once again becomes a genetic trait. This allowed the twin studies researchers to claim that they were abandoning the EEA, without any alteration in their math.

A very important realization is that heritability was meant as a forecast, not a measurement. The calculation intended to predict what researchers would find in the future, when direct measurement of genes on a population scale became possible. The fact that most traits did not demonstrate Mendelian inheritance suggested to researchers that these disorders would not be due to single genes. Therefore, the expectation was that it would take a population level analysis to identify all the genes responsible for each given trait. Schizophrenia having a heritability of 80% implied that when the day came that researchers could compare the genetics of entire populations of patients with Schizophrenia to healthy controls, they would be able to account for about 80% of the disease rates using genetic sequences. Essentially, researchers would compare the variances in base pairs throughout the entire genome. By doing this, they could identify that 80% of the causes behind Schizophrenia can be attributed to disparities in these specific base pair sequences when comparing patients and controls.

Much of the early work on heredity came from Europe, which isn't unexpected, especially considering the major universities were located in Europe. However, this still holds true today, especially for the United Kingdom. Much of the hereditary work, along with the GWAS into behavioral traits that followed, are performed by British researchers, or researchers accessing the British database called *The UK Biobank*. This database asks enrollees to fill out detailed surveys on their life and health. Then, those surveys are matched to various biochemical markers (e.g., standard blood tests) along with genetic sequencing. This allows those with access to search for genes associated with a wide range of factors, including educational attainment (e.g., graduating college), commuting habits (e.g., taking the bus to work), and even personal preferences such as their favorite type of cheese. These findings, more often than not, are published in the *Nature* family of journals—a set of academic publications run by the parent company Springer Nature based out of the U.K. or a second U.K. publisher called Elsevier.

It's hard to ignore the realization that the one research-focused nation that still believes in hereditary monarchy is also the one that puts out the most papers attempting to support genetic determinism. I know I'm a Yank for saying this, but the U.K. seems to believe that Charles Windsor is the most qualified person to "lead" the nation *solely* because he was the first male born to one particular woman, who, herself, was the most qualified ruler through a combination of being birthed by her mother and not having brothers. I suppose if you believe that the ability to lead one of the world's largest economies can be inherited in one's genetics, it is easy to believe other traits of the King are also inherited, ranging from his asthma to his proclivity for infidelity.

The flaws in the twin study calculation could have been overcome had the researchers been willing to modify the approach to match the data. Genetics studies made it very clear that certain traits with high heritability were indeed highly genetic, like cystic fibrosis. However, other traits, like the love of sailing, were clearly not. While those that ran the initial twin studies had mostly died by the onset of the true genetics era, the researchers that staked their careers on using twin studies to unravel the mysteries of various diseases were still alive and well. For them, updating the heritability calculation to account for the environment would have slowly eroded the role of genetics, and thus eroded the priority such studies would be given when funding decisions were made.

As a counter example, let us consider the evolution of a different equation that was intended as a forecast of future measurements—the equation for predicting rainfall. The first equation meant to predict rainfall was written around 1910, the same time as the heritability equation was derived. The equation for how much rain was possible was complicated math, but still within basic algebra.

Over the following years, meteorologists used this equation to guess how much rain would fall in a given area in the subsequent 36 hours. However, like all equations, they did not always get it right. Sometimes, their calculations were way off but—this is the important part—they admitted these shortfalls and updated their equations to account for the new information. As time progressed, the equation grew more complex, growing by 1980 to include higher order calculus functions.

Since the first genetic sequencing technology didn't arrive until 1980, this might be the time to start judging how valuable the heritability equation really was. From 1980, the meteorologists continued to recognize when they were falling short of perfection, and to this day, they continue to update their approach in include far more advanced calculus. One such approach looks like this:

$$\text{PWV} = -\frac{1}{g}\int_{p(Z_0=0)}^{p(Z)} q_\text{h}(p)dz \cong -\frac{1}{g}\sum_{i=1}^{N} \bar{q}_{\text{h},i}\Delta p.$$

No matter how complex these equations are, all are still forecasts meant to predict how much rain will be directly measured in the future. Because the meteorologists admit as much, they are willing to modify their forecast anytime the actual results fail to match their predictions. So, to review, the equation forecasting how much rainfall you will be able to measure over the next 36 hours *only in your immediate area* was written about the same time as the equation forecasting how much genetic sequence differences would be found for *every human disease and trait on Earth*. Each of these equations had years of successes and failures, which presented opportunities to both endorse some sections of the equation while modifying others.

So, we have now clocked four decades for which both equations have been able to directly measure their forecasts and adjust their math accordingly. And yet, the equation admitting it is only helpful in predicting rainfall in a limited area has expanded from basic algebra to calculus so complicated that only a computer could reliably perform the calculations. This weather calculation has proven immediately valuable in numerous ways over the years—from decisions to bring an umbrella to work, to making weekend plans, to informing the need for cities to prepare for flood evacuations. Meanwhile, the equation that claims to assess how much genetic sequence differences will influence everything from heart disease to political affiliation to preference for green or black tea: a) hasn't been updated in any meaningful way, b) continues to represent basic arithmetic that could be calculated in your head, and c) persists unedited despite more numerous examples of abject failure than success. To say the least, something isn't adding up.

The example of calculating the heritability of first names might seem unfair because no one would claim your name is genetically encoded. However, the reason this admittedly silly analogy has value is that the twin study researchers never modified their equations to objectively identify the criteria they use to dismiss a heritability score for first names but respect one for attending college. If one argues "the data are the data", then these data imply first initials are genetic. As we will discuss in later chapters, if one were to run a modern genetic analysis looking at people named Shamus, the results would also link their names to genetics. The data may be reliable numbers, but the interpretation is what matters.

The mentor of Professor Harden, Professor Eric Turkheimer, both defended and dismissed the issues around the EEA and heritability calculations. In his review of Jay Joseph's book, Turkheimer wrote, "The EEA is false, but it doesn't really matter" before noting "Heritability isn't zero; it isn't one" (meaning that genetics can't be said to play no role in a disease, but are not accountable for 100% of the disease either).

Turkheimer described this by stating, "Hundreds of thousands of words have been spilled about the EEA already, arguments and counterarguments and counter-counter-arguments no one should have to slog through." At the risk of adding to what Turkheimer said, I will note that the EEA matters because a disorder that is 80% genetic would be viewed and mitigated in a completely different fashion from one that is 10% genetic. No one is designing genetic treatments to deal with COPD caused by smoking, but they might for those with the monogenic variant of COPD known as alpha-1-antitrypsin deficiency. Turkheimer fails to understand that the point of deriving heritability scores was never to genuinely assess complex traits; it was to justify biologic determinism and to guide public policy. One needs to be sheltered in comfort to shrug off statistics that have been used to justify forced sterilization and ethnic cleansing as if they do not really matter. Even if every disease were viewed as a mixture of genes and environment, diseases that were viewed as primarily genetic will be met with a completely different set of responses compared to diseases viewed as mostly environmental.

It is important to note that when population geneticists point to one of their critics, Turkheimer's name is one of the most frequently cited. Yet, his penultimate comment about Joseph's book further attempted to dismiss environmental focus:

> The last chapter of [Joseph's] book is a parable about house fires. Joseph imagines a town in which houses are made out of different kinds of wood, building materials standing in for genes. The town is threatened by marauding bands of arsonists, representing environmental threats. Civic leaders waste time computing flammability coefficients representing percentage of fire variance accounted for building materials, while giving the arsonists free rein. The book closes with a plea to mind the arsonists who are the real threat, the leverage point where something can actually be done. It's a nice story, but do you notice something? It completely undermines Joseph's own argument. I mean, if you lived in a community under constant threat from arsonists, mightn't you have some interest in a house made of relatively fireproof materials?

While I presume the author thought this was a clever quip, what would be an environmental analogy for living "in a community under constant threat from arsonists" and how exactly would we fireproof our genetics? I get that Turkheimer is saying we should have "some interest" in fireproof materials and not that our entire focus should go there, but if the question is about actionable follow up and not just research, what would that look like?

The twin study era started with claims that many human behaviors and most human diseases would one day be revealed as 80% genetic. These claims inspired the geneticists that followed to dedicate their careers to researching human flammability under the belief that losing a home to marauders would be almost entirely explained by whether it was built with oak or pine. As will be discussed in the next chapters, the results demonstrated that differences in building material was of dramatically less importance than twin studies had promised. Yet, the response, even from genuine critics of the field, has become that any contribution greater than zero is still worthy of investment.

If it would take thousands of years for evolution to change our homes from wood to steel, I think we can safely say that defending ourselves from the marauders would be the most useful approach. If having combustible material in our homes would be irrelevant if not for the marauders (who never troubled our grandparents), then spending money on researching the molecular nuances of oak versus pine should not be a priority. Whatever genetic susceptibility may exist to yellow fever didn't matter for the troops that stayed in Kansas. Whatever genetic susceptibility may exist for cholera became moot in America with the advent of water sanitation.

So, we should not dismiss the communities who have been marauded by EEA-derived heritability scores or the patients who have suffered while the medical community desperately researches the flammability of their genome. Overall, twin studies' most enduring legacy is convincing the well-meaning scientists who inherited their assumptions that the solution to modern diseases would be found in the genome.

CHAPTER 4

Where and How Some Geneticists Got it Right

BOTTOM LINE UP FRONT: *When a genetics-level investigation is carried through all biologic processes, it can be extremely helpful.*

Rare disorders

A disease resulting from a single mutation in a single gene is referred to as a monogenic disorder (meaning 'one gene'). The most well-known successes in the field of genetics have been the ability to identify these types of diseases. For example, sickle cell anemia results from a single letter change in the gene for hemoglobin. Hemoglobin is a protein inside of red blood cells, whose function it is to carry oxygen to your cells and carry away their carbon dioxide waste. As a group, these diseases are referred to as *Mendelian diseases* in reference to Gregor Mendel. Mendel was a monk who performed some of the earliest experiments into how genes are passed along in pea plants. I will instead refer to the monogenic disorders as *family-style* genetics. Not only does that better serve with the recipe analogy, but it also speaks to how these disorders were sorted out by looking at families instead of large populations.

One recent major family-style genetic disease was discovered by Dr. Amy Hsu, Dr. Alexandra Freeman, and Dr. Steven Holland of the National

Institutes of Health. A small number of patients were presenting with infections that were more severe than expected and involved microorganisms that were uncommon. For example, some patients had infections in the lungs including an organism called *Staphylococcus aureus*. *Staph aureus* is a common bacterium that can cause recurrent skin infections, especially in patients with eczema. However, the patients referred for evaluation would get *Staph* infections in their lungs, liver, or even their bones. The patients also got recurrent boils on their skin that were reminiscent of the Biblical story of Job who (in addition to losing his wealth and family) was stricken with boils by God to prove to the devil that he would maintain his faith despite his suffering. As a result of these boils, the disease was initially named *Job's syndrome*.

The patients were also susceptible to infections with less-common pathogens like *Candida albicans*, suffered from abnormalities in bone development, had eczema, and very high immunoglobulin E (IgE) levels. IgE is a version of antibodies made by all humans. The role of IgE is still debated. IgE seems to protect against some parasitic infections, most of which are no longer a genuine risk to people living in industrialized societies. Patients with eczema or allergies often have high levels of IgE in their blood, perhaps as high as 1,000 units per milliliter. The patient group sent for evaluation would have IgE levels well over 120,000. With the new biochemical finding, the disease was thus renamed *hyper IgE syndrome*.

Overtime, the patients with similar symptoms were collected and their DNA was sequenced. DNA samples were also collected from their healthy relatives (referred to as *unaffected relatives*). Dr. Hsu's first task was to sort through the differences to ignore all the changes that would be irrelevant (like paprika versus "papryka") to identify changes that she thought would be impactful (brine versus urine or 1 tablespoon versus 10 tablespoons). She started by looking for genes that would be severely damaged, such as those where both copies had early *stop* signals in the patients but not in the healthy controls. Having gathered possible candidates, Dr. Hsu next needed to compare the DNA of patients with the larger population found in DNA databases. In this step, Dr. Hsu was using the DNA databases for their original purpose.

The original purpose was to guide discoveries on rare disorders. DNA changes responsible for a rare disease would not be expected to be common in healthy people. So, as Dr. Hsu found gene variations that might make sense as a cause of the altered chemistry in the patients, she would compare those DNA changes to the general public to see that these variations were uncommon.

Eventually, Dr. Hsu discovered that the patients all had abnormalities in one specific gene called *STAT3*. These abnormalities were not found in the general public, nor were they identified in the patients' unaffected relatives. As mentioned in the last chapter, there are two general categories of single-gene diseases: recessive and dominant. Broadly speaking, dominant disorders only require one diseased copy of a gene for the disease to manifest. If a father has one bad copy, he will have the disease. If he passes it on to his child, then they will have the disease as well. Sometimes, a dominant mutation occurs *de novo*, meaning *anew*. In these situations, neither parent has a diseased version of the gene in question, but the mutation arose in the embryo. The term for when a mutation that was not found in the parents arises in the embryo is *spontaneous*. As we will discuss in the chapters dealing with the environment, use of the term *spontaneous* is a misnomer. The Oxford Dictionary defines spontaneous as "occurring as a result of a sudden inner impulse or inclination and without premeditation or external stimulus." That last part, "without external stimulus," implies that spontaneous mutations are unpredictable while overlooking the ability of several toxins, medications, and infections to act as an "external stimulus" to a single-gene disease.

In contrast to dominant disorders, recessive diseases need both parents to carry a diseased copy of the gene. Cystic fibrosis is a clear example where both Mom and Dad need to have one bad copy of the *cystic fibrosis transmembrane receptor* (CFTR) gene. When people have one good copy and one damaged copy of a gene for a recessive disease, they are said to be *carriers* of that disease. For recessive diseases, having one good copy of a gene is just as protective as having two copies, apart from the potential to pass the disease onto a biologic child made with another carrier. There are other variations on the recessive-dominant binary, such as *haploinsufficiency*, in which one needs both copies of a gene to be fully functional to make enough of the protein to assure health. Having one defective copy leaves a person deficient for whatever protein the gene encodes for and puts them at risk for disease.

One final variation on the theme would be a disease like sickle cell trait. Sickle cell disease is a recessive disease requiring two mutated versions of the hemoglobin gene to manifest. Patients have abnormally shaped red blood cells (that look sickle shaped under the microscope) and resultant difficulties carrying oxygen throughout the body. As a result, they are at risk for low oxygen delivery to the tissues, which results in severe pain. Furthermore, since their red blood cells have an abnormal shape that makes it difficult for the cell to transit around the body normally, the patients are also at risk for their cells getting *stuck* in the small blood vessels, thus creating blood clots that further starve their tissue of oxygen. Having only one copy of a sickle cell allele isn't totally benign, however, like other recessive diseases. In the right environment or with the right stressors, people with sickle cell trait can run into serious health risks. However, it should be noted that sickle cell trait offers a form of natural protection against malaria infection, and therefore is a net benefit in places where malaria is common.

There is further nuance in the dominant versus recessive paradigm that is often taught in introductory genetics classes. In his books, *Understanding Evolution* and *Understanding Genetics*, author Kostas Kampourakis notes that although most introductory genetics classes teach dominant and recessive alleles under the Punnett square from Mendelian genetics, the truth is far more complex. Mendel used a model system in which he used plants with the greenest peas versus the ones with the most yellow peas. This was effective for his work in a lab setting, but in reality, the pea plants that Mendel cultivated can produce peas with a gradient of every shade in between. The same is true for eye color. We tend to learn that people have a gene for brown eyes that is dominant over a gene for blue eyes, even though the variation in the shades runs the full spectrum in between. So, the concept of dominant and recessive alleles is real, but simplified given that most traits are far more complex.

Single-gene disorders can potentially show variability in the symptoms they create, even for people with the exact same mutations. The finding that not everyone with a given mutation will have the associated disease is referred to as *penetrance*. The finding that not every person with a monogenic disorder will have the exact same types or severity of symptoms is termed *variance*.

For *STAT3*, the population genetics databases did not appear to indicate that anyone alive had two *bad* copies, which suggested that humans needed at least some STAT3 protein to survive. But cells with at least one copy of the mutated *STAT3* would demonstrate abnormal behavior indicating that hyper IgE syndrome is a dominant disorder. There is variance in the way *STAT3* mutations present, but it appears that those with damaging mutations will have some symptoms. Work is still ongoing trying to uncover all the ways in which *STAT3* mutations may cause harm to patients, including work by my own lab in collaboration with Dr. Hsu and Dr. Freeman.

Importantly, cells and mice with *STAT3* mutations can be studied in the lab to uncover why immune cells may fail to fight off *Staph* or why their bones were not forming properly. These studies identified targets for drug treatments; however, the small numbers of total patients with the disease present a challenge for designing a placebo-controlled study. Current work has focused on bone marrow transplantation, which aims to replace the bone marrow cells that lack a functional *STAT3* with cells that are fully intact. Even while a cure has not been identified, these kinds of discoveries can offer patients rapid diagnosis to get them onto the best possible treatments. Patients with symptoms of immune deficiency and eczema can be screened for *STAT3*, as well as other genetic mutations. If a positive result is identified, they can then be placed on the most up to date care plans to improve their chances of a long, healthy life.

It is worth noting that even though scientists and clinicians alike refer to single-gene disorders as *rare disorders*, in total, they are surprisingly common. Each specific monogenic disease may only impact a small number of people. (Hyper IgE, for example, likely affects fewer than 1000 worldwide.) However, there are at least 4,013 different monogenic disorders, as reported in a 2023 paper in *Human Genome Variation* by Liu and colleagues. Per the European Commission on Public Health, approximately one in 17 people worldwide has a diagnosis of a *rare* disease.

Most of the well-known single-gene diseases involve mutations that reduce the function of a protein. Hemoglobin made from a gene that has a sickle cell related mutation functions less well compared to a typical hemoglobin. These kinds of mutations are called *loss of function*. Sometimes, however, mutations prevent a gene from turning off. These would be called *gain of function*. Related to this is a category of genetic findings that would represent a *kind of* success called *copy number variations*. Copy number variations are when a gene is either completely missing (called a deletion) or might have *extra* copies (called a duplication). Deletions are perhaps easier to understand since they would clearly cause a loss of function for the protein the gene encodes. If a mutation occurred that seriously hampered the function of protein, then completely lacking that protein would cause harm as well. The *kind of* aspect of the success, however, is when there are duplications.

Having an extra copy of a gene—again, let's use hemoglobin—does not mean one's body will make twice as much hemoglobin. The main hormone that regulates hemoglobin production is called erythropoietin (or EPO for short). Since your body is constantly making new red blood cells and losing red blood cells, you have a constant EPO production level that tells the bone marrow to make more hemoglobin to put into the never-ending assembly line of new red blood cells. If your body senses that it needs more red blood cells (perhaps due to severe bleeding or being in the low-oxygen environment of high altitude), the amount of EPO your body makes will increase, which will drive up the hemoglobin and red blood cell production. But as soon as the emergency is passed (the bleeding has stopped, or you return to sea level), the EPO levels will reduce to their baseline. When athletes engage in *blood doping* by taking EPO, they are overriding the normal brakes on the system to force their body to make more red blood cells and gain a slight advantage in the amount of oxygen the blood can carry. So, here is a question: If your body had an extra copy of the EPO gene, would you naturally *blood dope*?

The answer is no because even though your ability to make EPO would double, the brakes on your system would remain. So, if a person with one copy of the gene for EPO made—let's just say—10 proteins per hour, someone with two copies would likely still only make 10 per hour, just with five per gene copy. Duplications may double the maximal theoretical protein production, but even then, there would be factors like whether you had double the amino acids around to generate those proteins. Someone with extra copies of EPO would have an advantage at altitude or in recovering from a severe bleed since they could, in theory, double their output and respond more quickly when demand increased. Population genetics have pointed to associations between duplications in various genes and numerous diseases as *proof* of genetic contributions of these diseases. But they almost all miss the key next step of showing that differences in either the baseline or maximal protein levels exist (let alone matter).

So, why did I say that copy number variations are *kind of* a success? Well, there is at least one exception known as hyper *alpha-tryptasemia* (or *HaT* for short).

Dr. Jonathan Lyons, also of the National Institutes of Health, discovered *HaT*. Dr. Lyons found that patients with *HaT* have duplication of a gene called tryptase. The intended function of tryptase is currently debated but there have been theories that it may protect against venom from snakes or other poisonous creatures. Whatever the purpose of tryptase was, today, the system is most often connected to severe allergic reactions. These reactions, termed *anaphylaxis*, are allergies so severe they can kill. According to the Allergy & Asthma Network, one out of every 12 U.S. children have a severe allergy and almost 700 children per year die from severe reactions to foods, bee stings, and drug reactions. People with duplicated tryptase do not have a higher rate of anaphylaxis, but they do have more severe reactions when anaphylaxis occurs. Said another way, whatever factors might give a person a severe peanut allergy, if that person also has *HaT*, they will be at greater risk of dying from the peanut allergy than if they did not have the duplicated tryptase.

Most importantly to Dr. Lyons' discovery is that he proved that people with increased copies of the tryptase gene actually have an increased amount of tryptase protein in their blood. The reason why the normal *brakes* fail to counter the production is unknown. However, by connecting the gene to the protein, Dr. Lyons provided his team with an opportunity to discover the meaning behind the duplication. Dr. Lyons has searched for increased copies of alpha-tryptase in people from all backgrounds and ethnicities; however, the HaT patient population is overwhelmingly considered of Northern European background. Every human has alpha-tryptase genes, but for unknown reasons, populations living in Europe around the time of *the black death* plague developed increased copies of alpha-tryptase. Therefore, while his work is ongoing, it appears that extra copies of alpha tryptase may be harmful in severe allergies but may have been protective against certain pathogens like those faced in Europe during the black death.

Cancer

The second major category of genetic success is in the field of cancer. I say success with a heavy heart given the rates of cancer have grown exponentially in America. Most people reading this book will personally know someone that has been diagnosed with cancer. The CDC reported that in just 2022, over 1.9 million new diagnoses of cancer were made and at least 609,360 people died. However, those numbers do not capture the months of nervous anxiety between being told there is a concern for cancer and when the final diagnosis is provided. Those numbers do not capture the lives cut short, families left grieving, or dreams deferred by the disease. Again, per the CDC, if 225 healthy Americans read this book in January, at least one of them will be expected to be diagnosed with cancer by December. The impact of cancer demands that science not only find new insights but find ways to transform those discoveries into improvements in people's lives.

One example of work impacting the lives of people dealing with cancer comes from lung cancer research by Dr. Robert Winn, the chair of the Virginia Commonwealth University Massey Cancer Center. When I was a medical student, I rotated in Dr. Winn's lab. My major accomplishment for that month was ruining a Western blot by exposing it to light before processing it correctly. Despite the ineptness of his medical student, Dr. Winn made discoveries into the molecular pathways by which many lung cancers form. The specific pathway is called *Wnt* (pronounced *went*, as in "I went to watch Notre Dame get pummeled by Ohio State in football"). To uncover the details of this pathway, Dr. Winn compared lung cancer tissue with healthy lung tissue from the same person. Keeping his analysis within the same person eliminated a lot of the confounders that would crop up in analyzing population differences.

Dr. Winn could see that *Wnt* signaling in the lungs was more active in newborns than adults. This made sense since it had been established that *Wnt* is a key pathway for normal organ development. Since development was complete, the healthy lung tissue of adults had quieted its *Wnt* signaling, turning it down but not totally off. However, in the cancer tissue, *Wnt* and the related signaling molecules had reactivated. This activation spiraled out of control until the cancer tissue was growing and proliferating as if it were trying to make an entirely new lung. When a cell begins to grow too quickly, the ability to *proofread* the DNA declines and the risk for additional mutations mounts. Eventually, the molecular guide rails disappear, and the cancerous tissue not only grows out of control but develops the ability to migrate (called *metastasis*).

Importantly, like the types of monogenic disorders researched by Dr. Hsu, Dr. Winn manipulated this pathway in cancer cells to verify the importance of *Wnt*. He blocked the pathway in cancer cells to see that he could reduce their neoplastic behaviors. He turned up the pathway in healthy cells to watch the transformation to a tumor unfold. Dr. Winn did not just find an association with *Wnt* and call it a day. Let me be clear, this process is far more complex than what I've just described. However, the crucial point to remember is that contrasting mutations in a tumor with the DNA of normal tissue from the same person is significantly more likely to yield meaningful discoveries, which can be confidently validated through biological evidence.

While the discovery of *Wnt* as a mediator of lung and other cancers has given clues to the mechanisms by which these cancers develop, the central role of *Wnt* in normal growth has made it a difficult target for treatment. Numerous drugs have been used in clinical trials to block the over-activation of the *Wnt* pathway in cancers. However, many of these trials have been terminated because of the toxicities generated by blocking the *Wnt* pathway in healthy tissue, as well. But research is ongoing to look for a safer way to modulate this pathway, and even without a treatment, these discoveries have provided a means to assess environmental factors. If we now know that over-activating *Wnt* is a key pathway to inducing lung cancer, then any environmental exposure that increases *Wnt* activation might be one of concern. If you had to guess, what do you think is one of the more potent inducers of *Wnt* in the lungs of humans? I hope you guessed cigarette smoke since that would be correct. Yet what about air pollution?

As discussed in later chapters, air pollution is a broad term that could include a lot of different molecules. On a molecular level, the air pollution in one city could be very different from the pollution in another city. If you wanted to evaluate how a specific pollutant might influence lung cancer, how would you go about doing so? One option would be to expose a cell to the pollutant and see if it becomes a cancer. Another might be to expose a mouse and see if it develops cancer. But these approaches might take a long time to provide results and would present a logistical nightmare if you wanted to screen thousands of potential pollutants. Yet, with an understanding of the importance of *Wnt*, researchers can screen different types of pollutants to see which induce this pathway. This would not mean that a chemical that fails to induce *Wnt* would be *safe*, but it would mean that those that induce *Wnt* would jump to the top of the suspect list for pollutants that might contribute to lung cancer.

Perhaps the clearest example of an end-to-end success in cancer genetics is the drug Gleevec (also called *imatinib mesylate* in its generic form). Researchers comparing chronic myelogenous leukemia (or CML for short) cells to healthy blood cells from the same individuals identified a specific mutation. This mutation swapped out entire sections of the genome so that the bottom portion of chromosome 9 was attached to the top half of chromosome 22. The reciprocal change put the top half of chromosome 22 onto the bottom portion of chromosome 9. When this swap (formally called a translocation) occurred, it turned two gene segments that are not supposed to be near each other into neighbors. This mis-paired chromosome was termed *the Philadelphia chromosome* because it was discovered at the University of Pennsylvania School of Medicine. The segments (called *BCR-ABL*) made a new gene that told cells to divide but lacked the ability to *turn off*. Thus, the blood cells with this mutation proliferated and became leukemia.

It took nearly two decades, but researchers at Oregon Health and Science University, led by Dr. Brian Druker, identified a drug that could turn off *BCR-ABL*. Since only the abnormal cells had *BCR-ABL*, the drug did not have major side effects on healthy cells. A full eight years after their clinical trials began, at least 98% of patients were still in remission. Again, the success was born from contrasting tumors with healthy tissue, identifying the genetic differences between them, verifying those genetic differences mattered using biological assays, and then using those assays to identify a directed therapy with previously unthinkable levels of success. One caveat on the inaccessibility of Gleevec was detailed in the book, *The Tyranny of the Gene: Personalized Medicine and Its Threat to Public Health*, by Professor James Tabery. As will be discussed in later chapters, Gleevec may have started as the model for genetic-based medicine, but it quickly became the model of pricing of these therapies in ways that made it difficult for eligible patients to afford.

For the medical impact at least, there is one nuance in cancer genetics that forces any researcher in the field to dig deeper than just a gene association. They must differentiate between a tumor suppressor or tumor enhancer. When Dr. Winn was looking for gene mutations linked to lung cancer, finding some in *Wnt* could indicate either that: *Wnt* normally blocked the development of cancer and the mutation removed a tumor inhibitor; or that *Wnt* encouraged cancer development and the mutation ramped up the pro-cancer behavior. Genes that limit cancer are called *tumor suppressors*, genes that encourage cancer are called oncogenes. The gene *TP53*, which encoded the protein p53, is probably the most studied tumor suppressor gene. Genes like *Wnt* or *BRC-ABL* are good examples of oncogenes.

Say you found any gene that was linked to colon cancer, if it were a suppressor, you would want a drug that replaced its function; if it were an oncogene you would want a drug like Gleevec that blocks its function. A mere association alone would not provide you with any of the insights needed to come up with the medications that could actually make a difference in the patient's survival. The omnipresent requirement for cancer researchers to show whether a genetic quirk associated with their cancer of interest is a suppressor or oncogene creates an unwritten requirement to connect gene-to-phenotype in ways that are sadly lacking in other fields.

Since researchers have tallied the various ways that tumor-related genes like *TP53* or *Wnt* can be mutated, oncologists can screen tissue for these mutations as a means of early detection. In fact, cancer researchers like Noah Earland of Washington University of St. Louis are working to detect mutated DNA from tumors circulating in the blood. Like all tissues, as these tumors grow, some of the cells die and their contents are put into the blood for filtering and recycling. Earland's approach can detect the mutated DNA in the blood and signal to doctors that the patient may have a growing tumor inside them without the need to probe or biopsy. As will become evident in future chapters, it is important to note that the database of mutations is for all humans. There is no such thing as a p53 mutation that signals danger in Koreans but not in Chinese people.

Pharmacogenomics

The third area of success in the genetics field is that of pharmacogenomics. One example was published in the *Lancet* in 2023 led by Dr. Munir Pirmohamed. Dr. Pirmohamed collected reports demonstrating that specific genetic differences were linked to severe side effects when taking medication. For example, the HIV drug abacavir can lead to serious allergic reactions in patients with a genotype of HLA-B*57:01. Another example discovered by Dr. Pirmohamed relates to the blood thinner, warfarin. Certain genetic markers could be used to predict the dosage of the blood thinning drug warfarin, which can protect patients against taking too much and risking a severe bleed or taking too little and suffering a clot.

By tallying these and many other drug-gene reactions, Dr. Pirmohamed's team was able to generate a screening tool for the U.K.'s National Health Service (NHS). Patients underwent genetic screening prior to getting any of the medications with known genetic risks and were offered different medications if the results suggested the drug might put them at risk for severe reaction. Overall, the program was able to reduce the rate of severe reactions by over 30%.

Because of his success, I reached out to ask Dr. Pirmohamed about what he felt made his program successful and whether his work might be a redemption for GWAS approaches. He noted that his work did not use

SOWING THE SEEDS OF FAILURE

GWAS at all; in fact, it only collated from drug interactions with distinct sequences linked to an exact drug. He was able to make a screen that was like a GWAS based on those results, but such a screen was built off the types of gene-to-biology projects like those described for *STAT3* and *Wnt*.

It is possible for the rate of a genetic finding to be different between populations, but that is different from believing that the same sequence would have different effects in different populations. For example, 6% of white Americans and 2-3% of Black Americans have HLA-B*57:01. That would mean that if you gave every person in America abacavir, you would expect a higher rate of bad reactions in the white population. However, Kolou and colleagues presented in *BMC Immunology* in 2021 that the rate of this same allele in West and Central Africans is only 0.1%, and other researchers have found the rate in Asians to be approximately 0.2%. So, if a physician intended to put a patient from either West Africa or Asia on abacavir, the most important aspect of pharmacogenomics is that the provider should test for HLA-B*57:01 regardless. Yes, the risk might be lower in Asians than white Americans because the rate of HLA-B*57:01 is lower, but the risk isn't zero. So, if you have two patients with this allele—one that is of Asian ethnicity and the other Caucasian—they should be treated the same and told to avoid abacavir. If the gene matters enough to have biologic consequences, where the patient's great great grandfather was born should not influence the medical treatment plan.

Therefore, Dr. Pirmohamed explained that his database treated humans as a genetically variable population, but a single population, nonetheless. This is in stark contrast to the types of studies we will discuss in population genetics that segregate their results by ethnicity or nationality. If Dr. Pirmohamed found a gene variation that impacted warfarin dosing that was common in Europeans but rare in Asians, he added it to the database for everyone; he did not pretend that the gene variant was a "European gene". One example of the opposite approach was published in the *New England Journal of Medicine* in 2021. Authors led by Dr. Ronnie Sebo of the Mayo Clinic pointed out that the American version of that trial (called COAG) did a better job predicting the safe dose for white Americans than it did Black Americans. Rather than realizing this was a reflection of how the databases included more of the warfarin-related alleles common to Europeans, Sebo and others argued that taking a *color blind* stance toward genetics would miss the genetic differences between Black and white people in ways that would prevent improvement in the warfarin algorithms. They argued:

> *Indeed, the algorithm used in COAG was known to be almost twice as accurate in predicting warfarin doses in European-ancestry populations as in African-ancestry populations. These data indicate that race-specific algorithms would provide more accurate warfarin-dosing prediction than a one-size-fits-all approach. But the full range of genetic factors affecting warfarin-dosing requirements in persons of African ancestry is unknown.*

Dr. Pirmohamed, who actually worked on these types of trials, would not agree. While Pirmohamed does agree that far more effort needs to be spent on enrolling non-Europeans into genetic databases, he does not envision a *race-specific algorithm.* The reason to evaluate as many different people as possible is that the total number of variations in genes that dictate warfarin dosing will vary across the population as a whole. It is extremely unlikely that any one gene change will be exclusive to only one group, so cataloging all the possible gene variations that matter for warfarin dosing will inform better care for everyone.

Let us make a thought experiment and pretend that there are only three gene variations that determined the dose of blood thinner patients might need. The three genes can be found in every human population on Earth, but the first variant is only *common* in Europeans, the second only common in Asians, and the third only common in Africans. If you run a study only in Europe, then you would only find the first variant. Any European that had the variations common to Asian or Africa wouldn't be correctly treated because researchers would not recognize to adjust their dose for the variation that they had. If you collect information from Europe and Asia, you'd still mistreat any European that had the versions that were common in Africa. So, the reason to sequence Africans is not because Black people have *Black blood*, it is to find all the ways in which human biology can impact our response to pharmacology. If you could build a database that contains every single gene variation impacting warfarin dosing imaginable, then each patient could be treated as an individual. Africans could be given proper doses based on whatever variants they had, whether those variants were more common in Koreans or Kenyans. It would not matter if the specific gene variations they had were more common in one population or the other; you could look at them with the respect for human diversity necessary to have the success that Dr. Pirmohamed has achieved.

Angela Saini's book, *Superior*, outlined the best example of poor medical management resulting from treating patients based on their skin tone even when intentions are not bigoted. U.K. guidelines had initially stated that Black patients with high blood pressure should be started on a class of drugs called calcium channel blockers, while white patients should be started instead on ACE inhibitors. A smattering of studies on calcium channel blockers had claimed that when they limited their analysis to only Black patients, the results were much larger than they were for white patients. However, subsequent evaluation revealed that the practice of using someone's race to pick their medication had done both Black and white patients harm. Work by Jay Kaufam at McGill University revealed that there may be reasons patients might do better on one type of drug versus the other, but race was a terrible proxy for whatever those reasons might be. Thus, while the stated intent was to provide the best possible treatments, in practice, the Black patients that would have done better on ACE inhibitors were denied that treatment while the white patients that would have benefited from calcium channel blockers were incorrectly placed on ACE inhibitors.

Since this book is focused on the role of genetics in biomedical treatments, I have centered on these stories of success. However, it bears comment that even though they have not led to medical treatments, various successes in evolutionary biology have come from genetics, as well. A good review of these findings can be found in the book, A Brief History of Everyone Who Ever Lived, by Adam Rutherford. The work outlines how genetics can be used to follow movements and mixtures of populations over time and space. For example, while genetics can never identify specific groups (i.e., you can't use any single gene to tell one Native American tribe from another), genetics can show when settlers from Europe arrived in America and began mixing with Natives. Genetics can demonstrate when populations in Europe mixed with Neanderthals and to what extent.

Genetics can even demonstrate social changes, such as identifying that when humans developed the agricultural ability to raise dairy cows, a mutation in the lactase gene began to proliferate in those cultures. That lactase mutation allows the enzyme to stay active into adulthood, thus allowing cow's milk to serve as a nutrient source for anyone with that mutation. The types of evolutionary genetics practiced by Dr. Rutherford also have served to disprove the biologic theory of race and push back against racist ideologies. As will be discussed in more detail in Chapter 7, evolutionary geneticists like Dr. Richard Lewontin demonstrated that there is more variability within human races than between them—meaning you are more likely to find a genetic difference between two Black men than between a Black man and a white man. These results have been helpful in the politics around genetics and eugenics; however, sadly, not every geneticist working in biomedicine seems to have learned the lessons taught by these evolutionary geneticists.

So, what do the biomedical success stories all have in common? Each of these success categories have clear comparison groups to contrast against genetic changes. Rare diseases have unaffected people from the same family, cancer studies can compare to healthy tissue from the same person, and pharmacogenomics can contrast patients who have reactions to a drug to people who took the same drug without any harmful side effects. These success stories also have clear grounding in biology that goes beyond just the genes themselves. Cell and animal models are used to evaluate RNA transcripts, proteins, and metabolites that may be affected by the change in gene sequence. Furthermore, these approaches treat humans as a single species rather than a coalition of ethnic ancestries. Mutations that significantly hinder STAT3 would be expected to lead to hyper IgE in every ethnic group. Chemicals that activate Wnt would be expected to increase lung cancer risk in every ethnic group. Gene variations that predict drug reactions in one group, would predict drug reactions in others. These world-altering success stories inspired a generation of researchers to join the field of genetics looking for equivalent success in common diseases. However, as the remaining chapters will outline, when those hopes were left unfulfilled, the geneticists contorted their ethics and their data to rationalize their career choice.

CHAPTER 5
The Selfish Geneticist

> **BOTTOM LINE UP FRONT:** *Gene-centric framing is illogical and unhelpful.*

If there is an event horizon in our modern-day genetic obsession, it would be the publication of the book, *The Selfish Gene*, and its follow up, *The Extended Phenotype*, by Richard Dawkins. I suspect many of the people working in genetics read one of these books and somehow left with a sense that the author offered valuable insights into biology. Dawkins is an evolutionary biologist, which means he works on trying to figure out how and why certain adaptations arose in species. In *The Selfish Gene*, Dawkins attempts to present a unifying theory of all evolution based upon the idea that genes basically act as if they are selfish in their quest to replicate into the next generation and expand in number. In effect, Dawkins is repackaging the *survival of the fittest* presented by Darwin but set at the level of a molecule rather than an organism.

He builds his claim around the idea that DNA is the *replicator* upon which all life is built. Since, in his analysis, genes dictate phenotype, the genes that produce the best phenotype *win*. Dawkins acknowledges that the environment can dictate the value of any adaptation. For example, a tiger's orange and black stripes would be useless in the snow while a polar bear's white fur would be useless in the jungle. Importantly, he also admits that human adaptation has been freed from the need for replication for survival. For instance, in the bacterial world, if the microbes encounter extreme cold, survival might depend on the bacteria replicating in a way that generates a spontaneous mutation to better adapt to the cold. Whereas for humans, we can just put on a jacket. However, rather than stating that these behaviors can be passed on and adapted within and between groups, he then decides to package behavior into *memes*. Today, we think of the word 'meme' as some internet image that everyone is familiar with and can use as a springboard for jokes. But Dawkins considers all culture items memes, like music, literature, or language. He then tries to connect genes and memes by stating that the genes are what encode for the cultural behaviors, invoking the analogy of a brain being programmed like a computer might be.

The book is probably most known for—and most tolerated by academics because—it attempts to explain an evolutionary advantage to altruism. Under pure evolutionary biology terms, there should be no reason anyone would become a firefighter, soldier, or any other specialty where people might risk their own lives for unrelated strangers. Dawkins argues that societies are a balance of doves and hawks with hawks being aggressive and selfish and doves being passive and altruistic. Dawkins believes that more people adapt to be dove-like but are willing to react if provoked. The hawks, he argues, are lesser in number but are aggressive enough to essentially keep the doves on their toes.

The podcast, *If Books Could Kill*, hosted by Michael Hobbes and Peter Shamshiri, attempts to dissect what they call *airport books*. These books tend to be superficially engaging with pithy, oversimplified takes on complex problems. As of this writing, the podcast has not taken on any of Dawkins' books, but his work embodies one of the themes of the popular science versions of airport books the hosts identified in pop-econ books: Explicitly reactionary ideas can gain wider appeal by sprinkling in specific content to give the allure of egalitarian ideals. The podcast episodes for the books *Freakonomics* and *The Population Bomb* are good examples. Dawkins' work does the same thing but for eugenics.

Dawkins' claim that some people are born doves and some hawks is functionally indistinguishable from the favorite eugenics claim that some people are born lambs and some lions. He claims that the world's hierarchical structure is a natural consequence of the molecular genetics governing it, which could go under the banner *biology is destiny* familiar to eugenics crowds. To be clear, Dawkins never embraces nor hints at support for eugenics policies. He never advocates for exterminating *inferior* groups or creating breeding programs for people. That is an awfully low bar for praise, but I'll give it to him because, to the contrary, Dawkins argues that a society with too many hawks would be as unlivable as one with too many doves. However, combining the eugenics endorsements with Dawkins' arguments that altruism and monogamy may have evolutionary value puts a pseudo-liberal fig leaf on a book dedicated to genetic determinism. His book can be summarized as "things are the way they are because of genes. The things you like and the things you don't like."

Perhaps in response to criticisms of *The Selfish Gene*, his follow up, *The Extended Phenotype*, begins by walking back from genetic determinism. He outlines that genes are not the lone determinant of outcomes, but that they impact probability. His example is one of a student that might have *bad algebra genes* that could be overcome by a good tutor. This *probabilistic and not deterministic* excuse has been used by pretty much every population geneticist to hide the implications of their beliefs. First of all, nearly everything in life is *probabilistic and not deterministic*. Since some people have survived a gunshot wound to the head, then shooting someone in the head would be defined as only *probabilistic and not deterministic* of their death. Try telling the judge at your trial, "Your honor, my actions of shooting the victim did not determine his death. I merely impacted the probability that he would die that day." Technically, shooting someone in the foot would also increase the probability of death but not determine it, right? The question you want to know (and the reason shooting someone in the head would get a far greater sentence than the foot) is how much of an impact on probability does an action or item have? If *bad algebra genes* were even a thing (they are not), then it would make a big difference if those genes governed 80% of your math skills versus 0.00001%, right?

Secondly, probabilities become deterministic when aggregated at scale. Smoking cigarettes does not determine whether you get lung cancer; it only increases the probability of lung cancer. There are plenty of people that smoke for decades and never develop lung cancer (often because they die of a heart attack before the cancer has a chance to set in). On an individual level, you cannot say with confidence that a smoker will develop lung cancer. But on a *population scale*, you can say with absolute certainty that smokers will develop more lung cancer than non-smokers. If there are any difficulties in understanding the concept of aggregating probabilities, imagine playing Russian roulette 50 times in a row.

By *conceding* that genes have an impact that is some value less than 100%, Dawkins and his followers are pretending that they are not promoting eugenics without actually detailing why their overall point isn't eugenical. Sure, it might be that the specific student isn't assured a bad algebra score due to *bad genes*, but what about a population level view? What if you rounded up everyone with *bad algebra genes* and either sterilized them or killed them. What would happen to the average algebra abilities in the survivors in that society? Eugenicists love to debate on the grounds of utilitarianism—that being a philosophy of generating "the greatest good for the greatest number of people." Eugenicists have long since advocated that wiping out everyone with Schizophrenia would certainly be harmful to those patients, but might be worth it if you freed the world from ever having to suffer with the disorder in exchange. If you really believed that a substantive portion of one's math skills were genetically determined at birth, then expelling or eliminating everyone with *bad algebra genes* would increase the average algebra abilities. If, instead, you think that such genes have a marginal to negligible effect, then why would you dedicate two books to demanding people define the world through genetic terms?

Bluntly put, for this and many other reasons, Dawkins' arguments have corrupted the minds of generations of geneticists that have followed without altering a single one of their base pairs.

First, it cannot be stated enough how much he builds his entire theory, especially in *The Extended Phenotype*, around the idea that DNA is the *replicator* that guides the fate of getting into the next generation (both molecules and the entire organism). Dawkins goes so far as to state that "all life evolves by the differential survival of replicating entities." If this were true, he might have a basis for the rest of his theory. But, it is not.

DNA is not a *replicator*; it is *replicated* by RNA and proteins. First, DNA is transcribed into RNA then translated into proteins. Some of those proteins and RNA then work in concert to make copies of the DNA. DNA cannot make new DNA on its own. Dawkins uses the analogy of the Xerox copy machine to describe DNA as a *replicator* in that the DNA document replicates to make a copy of itself. The obvious flaw in this analogy is that if the document is DNA, then the Xerox machine is the *replicator* and the DNA is what is *replicated*. This realization helped form a newer way of thinking called *The RNA World*.

I do not intend to present The RNA World hypothesis as commonplace, or without its controversies. Harold Bernhardt wrote a nice summary of the arguments for and against the theory in 2012 for the journal *Biology Direct*. The crux of the argument is that RNA came first, then DNA, which goes against Dawkins' obsession with claiming that DNA is the basic building block of life. The reason the DNA-first idea does not make much sense is that DNA cannot self-replicate. For a strand of DNA to have arisen that could encode for all the RNA and protein machinery needed to copy DNA, it would have had to arise spontaneously in total. Essentially, if DNA were evolving and a strand of DNA arose that encoded for all the needed *copy supplies* except just one item, then that DNA strand would not be copied forward and that would fizzle out. In contrast, RNA can catalyze its own replication. However, RNA is inherently less stable than DNA and is far more sensitive to destruction. So, it would behoove RNA systems to evolve a way of storing *learned information* into DNA. That DNA could then assure the protein functions the RNA was encoding were properly copied to the next generation.

Now, to be fair, this would be an impossible thing to test; we cannot go back to pre-cellular times and check if RNA is present before DNA. However, researchers can trace the evolution of RNA-based systems through organisms and reproduce self-replicating RNA in the lab. DNA, however, can do no such thing. DNA cannot replicate, despite Dawkins' musings and decision to base his entire book (and career) on the idea that DNA is the *replicator* upon which all else is built. Therefore, DNA is not *the building block of life* but rather life's *documentation system*. This is not to say that the documentation system is unimportant. To borrow Dawkins' flawed analogy of a computer, your hard drive is certainly valuable, and if it gets corrupted, the computer will not work correctly. But hard drives did not come prior to other aspects of computers. Nor would anyone say written recipes came before meals. People cooked first, then wrote down the instructions to assure they could reliably recreate things moving forward.

But even then, the ability to adapt isn't limited to replication. Dawkins walks right up to a potential revelation and then just allows it to peter out. He correctly identifies that human adaptation does not require replication anymore. That is, people do not have to produce offspring that are better adapted to new climates because we can just adapt our environment to address the new climate. People moving from Northern

Canada to Costa Rica can wear shorts and a tank top to adjust; they don't need to hope for a baby that has a mutation for tolerating the increased heat and humidity. Dawkins was hinting at an 'ideal' system that possesses a high fidelity between generations, while also encoding the ability to adapt and survive in any situation, without relying on gestating new offspring in the hope of better adaptation. Given that any two humans will share 99.9% of their DNA sequence despite completely different environments, doesn't it sound like such a system has evolved?

The other limitation of using observations about *how things are* as if they are clear indications of *how things ought to be* is the risk for the survivor bias fallacy. Survivor bias was best demonstrated by the example of planes returning from battle in World War II. The military ran an analysis of where the most common bullet holes were found on planes returning from their mission (a theoretical version of what that might look like has been created and circulated around the internet is shown below). The military decided that the heavily hit areas should be reinforced but Abraham Wald of Columbia University proved that would be a bad idea. Wald pointed out that the map of bullets was only of those that survived to make it home. Instead of reinforcing the areas that a plane could be shot but still survive to make it home, the military should reinforce all the other areas since being shot there must be linked to the planes crashing and the crews being lost.

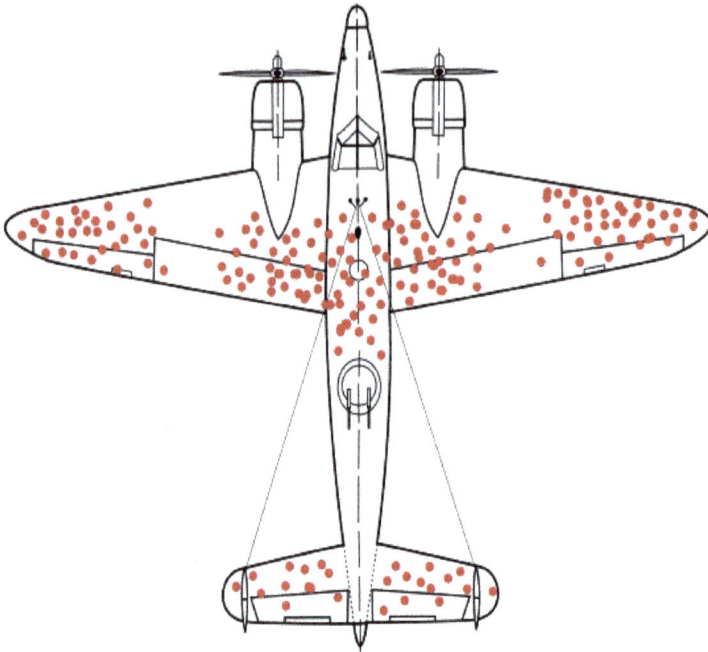

As the analogy applies to genetics, if the goal is survival, does the variation of where the bullet hit matter among those that made it home alive? The variation we see in the genome today, particularly on large population scales, would, by definition, be changes that would not preclude life or reproduction. Variation that would truly harm life or health would not have made it this far in human evolution. But appreciating Wald's logic means realizing that we have a lot more vulnerable areas in our genome than the planes did in WWII. Any two humans are 99.9% sequence identical, suggesting that any harm to those areas must not have allowed for survival. If we redid the same plane example but partitioned off 99.9% of the area, it would look like this:

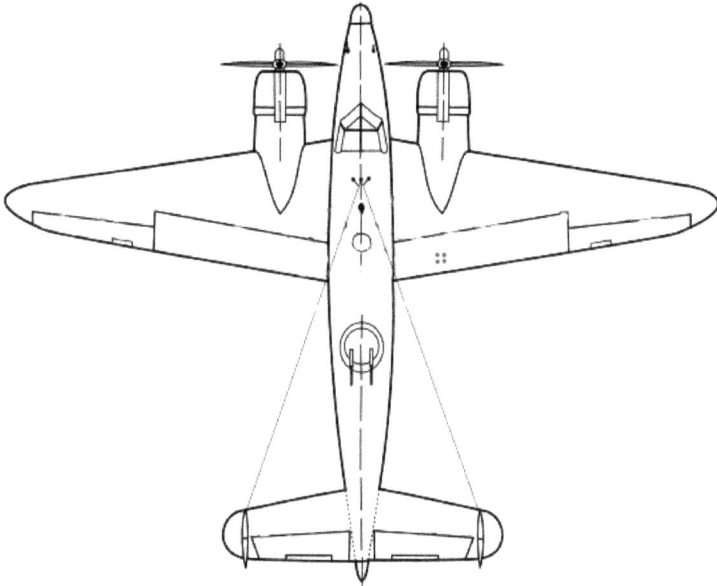

In some situations, could the variation in the location where the bullets struck within that small area make a difference? Maybe. But if you focused only on the areas that differ among survivors, you would risk making a large-scale survivor bias fallacy.

As outlined in *Superior*, even if you did this analysis on a population scale, it would not look that much more impressive. If every person in the world except for Peruvians were to disappear tomorrow, 85% of human genetic diversity would be preserved. Now, this is a biologically meaningless distinction because there is no biologic definition for *Peruvian*, only a socially and politically defined one. But, regardless, this means that the greatest population-level differences between Peruvians and others would be, at most, 15%. A 15% difference on the plane diagram looks like the image below.

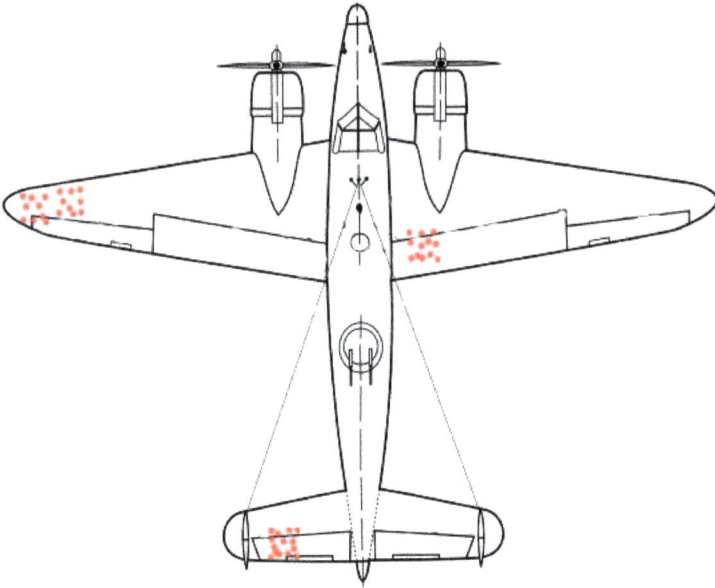

Sure, getting a bullet in the tip of the left wing might be different in some meaningful way from getting a bullet in the main flank on the right. However, if the entire theory as presented in *The Selfish Gene* is a goal of making it home alive to have children into the next generation, then focusing on the genetic differences between groups that successfully have children would be ripe for a survivor bias fallacy.

Again, this would not indicate that DNA is unimportant but would preclude the type of thinking that puts anthropomorphized DNA as the central drive of all evolution—both biological and cultural. You can see the breakdown in logic when he tries to explain altruism. If humans are 99.9% identical, and those 99.9% of genes are selfishly trying to get to the next generation, wouldn't everyone be supportive of one another? To believe that conflict arises in humans because of a selfish gene would be to claim that somewhere in the 0.1% difference, there must lie a gene for being a *hawk* that has reached the equilibrium with the 99.9% dove genes. So, if there are genes for altruism, one has to ask if there is a gene for genocide.

The Nazi guards carrying out their orders would have been killing Jews that shared an overwhelming amount of genetic material. So, if we are to believe that their behavior was encoded into the 0.1% differences between Jewish and Gentile Germans, that would mean that somewhere in the German genome, a gene for mass murder was activated right around the 1930s. However, this quickly becomes a circular logic that can always be dictated after the fact. If a prison guard were to help some Jews escape their concentration camp, then Dawkins would argue that the guard's dove genes were operating to use altruism to move copies of those shared genes forward. However, if the guard were to carry out the orders and murder his captives, then the guard's genes are acting like hawks trying to get his unique Aryan genes forward at the expense of the far more numerous and shared Jewish ones.

Of course, there is no such thing as Aryan genes and there are no genes that are truly unique to white Gentile Germans. Yes, there are gene variants that may be more common to Gentile Germans than their Jewish countrymen, but there are none that are exclusive. So, which genes would Dawkins argue were trying to replicate themselves by murdering a bunch of people that also shared the exact same genes? Although Dawkins never implies the genes are themselves sentient, he speaks about them as if they are, implying they have some sort of logic. For example, it makes sense for a firefighter's genes to encourage him or her to save an unrelated child because the shared genetic material would make it to the next generation. But what is the genetic logic of genocide? If the *Nazis' genes* had been successful in wiping out all Jewish people, the only thing they would have accomplished would have been a slight increase in frequency of a small number of genes, few of which (if any) would carry any biologic relevance.

So, here is where Dawkins would flip to discussing 'memes.' He might argue that the Nazi that executed the Jews was trying to preserve the memes of Nazi culture. Yet it is hard to discern the degree to which

Dawkins thinks memes are also just genetic markers. In *The Extended Phenotype*, he argues that genes govern behavior, and thus, any behavior that improves survival will advance the genes that encode that behavior. So, if the Nazi soldier were trying to advance Nazi cultural memes, and cultural memes are genetically encoded, then Dawkins would have to argue that Germans in the 1940s had a genocide-inducing gene encoding their genocidal memes (that amazingly, has not been active before nor since).

This confusion isn't trivial, and Dawkins uses it to attempt to argue that marginalized cultures—like minoritized groups and poor whites—are downtrodden due to the cultures that they inherit from their parents. He argues that the downtrodden are stuck in this cycle because they keep inheriting memes that encourage being downtrodden. How much of this downtrodden nature is encoded into genetics Dawkins never really commits to, leaving it to the reader to decide whether these marginalized groups are only culturally inferior or are both culturally and biologically inferior. Of course, the logic here could be just as easily reversed.

One could argue that those with oppressive hawk genes and memes will pass down these oppressive behaviors or biology to their children. Thus, perhaps the downtrodden remain downtrodden not due to their flaws, but because they are oppressed by the generational inheritance of oppressive memes of each hawk-generation. Hawks may not intend to be oppressive and they may not even appreciate that they have inherited an oppressive nature any more than they realize they inherited two kidneys and a pancreas. The idea that oppressive elements of culture are passed down from one generation to the next and that each generation unconsciously takes on those behaviors is effectively the center of Critical Race Theory—a legal framework that Dawkins has been highly critical of. I'm in no position to debate Critical Race Theory but I will say that if you adopt Dawkins' line of thought, you can't believe that doves have dove genes and dove memes without also thinking the same of hawks.

His arguments about genes and cultures which produce superior adaptations demonstrate even more hypocrisy when he discusses any culture that is not his own. Dawkins has been referred to as a *neo-Atheist*, which was a term coined by Gary Wolf that indicates that non-religious people should openly criticize religion rather than ignore it. Dawkins is one of the most vocal of the *neo-Atheist* in his attack on anyone who believes in, or even supports, religious beliefs. However, if survival and proliferation of genes and memes is the goal, and proliferation of the genes and memes are a sign of superior adaptation, then the most highly evolved cultural genes would be those for religion. In a global poll, 85% of the people in the world identify by at least one religion. To be fair, this would not mean that they all attend services or truly believe in their respective religious text; it could mean that people just identify as a Christian to fit in with their community. But the 15% that are *unaffiliated* might also be people who have religious beliefs but lack a definitive belief system to fit into. So, if genes encode culture, and the spread of those genes and memes are a sign of superior abilities, then Dawkins should conclude that the most highly evolved people in the world are Christians, followed closely by Muslims.

If a gene increased religiosity, it would give its human *vehicle* (as Dawkins calls living organisms) a better chance at survival to the next generation. A study in 2018 by Wallace and colleagues published in *Social Psychological and Personality Science* indicated that religious service attendance could add over six years to someone's life expectancy, particularly when they participated in the types of volunteer work that many religious organizations perform. This study has its flaws, but the overall data suggest that religious beliefs are found in people who live longer and multiply more. So, shouldn't Dawkins defend the religious *memes* of people as an excellent tool for survival? Instead, he wrote an entire book called *The God Delusion* and has publicly chastised parents for raising their children with religious teachings, spouting "how DARE you" online to those who would (as he put it) "force your dopey and unsubstantiated superstitions on innocent children too young to resist." So, which is it? Does the passing of memes that improve survival the way religious affiliation does indicate that such memes are valuable, or does passing on religion mean you are taking advantage of innocent children?

Even worse, Dawkins' claim that genes encode for cultural memes, even if only partially so, requires a level of detail that is completely lacking from his work. If someone wants to claim that a gene encodes a behavior, they will need to show how. As was discussed in detail in Chapter 1, genetic code could only impact the world after first being transcribed into RNA and then translated into protein. So, if someone thinks there is a gene encoding for *British culture*, what is the exact sequence that would do so, and how? How does changing a base pair in a gene found somewhere in the brain lead one to believe in a monarchy? Furthermore, not only would such claims require a clear description of the exact gene(s), they would also require a clear definition of the cultural memes.

If British cultural memes are in a zero-sum fight for survival, it would be imperative to define when exactly the British culture started and what exactly it entails today. Were the first British people those who lived in caves? In that case, then the existence of the monarchy would be an incursion against traditional British culture; it would be a sign that the monarchy meme invaded Britain and took over what were the true foundations of England. If, however, we define England as having started with the monarchy, then all the British territories should be given full independence lest we allow the meme of colonialism to soil the pure culture of England. While they are giving back the Virgin Islands, one could argue that every Englishman should get his arse out of Northern Ireland, right? And perhaps England should return all the stolen treasure from other nations like Egypt that they have stashed in their museums. After all, traditional British culture never included that area or those belongings. Okay, maybe you think British culture is truly defined after the Irish occupation but still within the British Empire days, then does Downing Street working well with the Americans represent them capitulating to the superior American memes that gained independence?

Or what about day-to-day things? Is the double decker bus a part of the British culture meme collection? Are blue jeans now part of British fashion memes or are they an invasion of American fashion memes struggling for supremacy against traditional British knickerbockers? Dawkins never bothers to define exactly how a gene could encode for culture, then never bothers to define what exactly a culture is. He does use examples like a beaver making a dam; so, the dam in this case is the meme that is defined by what he presumes to be the genes that govern this behavior.

The idea that there may be genes for animal behavior does not seem too far-fetched, even if no one has even put a gene into a mouse and made it perform the courtship dances of songbirds. However, how well does building a dam translate into human behavior? Do you think the ancient Egyptians had genes encoding for their meme of building pyramids? If there was a selfish gene that convinced Egyptians to build pyramids, what happened to it? Surely, there must be some current Egyptians who are descendants from pyramid construction workers from antiquity; are they repressing their innate desire to stack two-ton stones on top of each other? Using the example of a behavior that animals have done for as long as humans have observed skirts the issue that human behavior, especially on a population scale, is extremely dependent on time and place and ignores the possibility if we could look across all of beaver history, we might find that they didn't always build dams.

Furthermore, even the behavior of animals can be passed down through a form of *cultural behavior* completely independent of genes. Work published in *PLoS Biology* led by Dr. Alice Bridges demonstrates that bees can learn to solve puzzles by watching other bees. Their team designed a puzzle box that, if solved, would reward the bees with some sugary treat. The bees could either rotate the puzzle box clockwise by pushing the red tab or rotate it counterclockwise by pushing the blue tab—both directions gave an equal reward, so there was no difference in which way they opted for. Initially, bees had to be trained to solve the puzzle by first making sure they knew that the bright yellow circle on the box held the treat. Eventually, Bridges and her team had trained a set of bees to solve the boxes and noted whether those bees opted for the red or blue tab. Next, she wanted to discover whether the trained bees could train other bees by demonstrating the solution. To do this, she first put the puzzle box into a colony of bees that had never seen it before and watched what happened. These bees would eventually learn to solve the puzzle, but it would take a long time. If, however, she put the puzzle box in with a bee that already knew the answer, the entire colony would quickly model that behavior. Suddenly, every bee could quickly solve a puzzle it had never seen before simply by watching the bees that knew the answer already. Furthermore, if the *demonstrator bee* was a blue-tab pusher, then the colony they went to would be a blue tab pusher, too; if the *demonstrator bee* was a red-tab pusher, then the colony would also prefer to push the red tab.

Obviously, the bees could never have evolved a gene that truly encodes for "push the red tab on this puzzle box invented in 2022," so whatever learning they were doing was social. For these bees, traits and behaviors were being passed down without genetic encoding. But Dawkins' argument that behaviors would have a meaningful genetic underpinning would argue that these bees should not have been able to solve this novel puzzle in the first place. Sure, he might argue that bees have general puzzle solving skills, but the red versus blue pushing specifics suggest even that would be highly influenced by social norms. Thus, Dawkins' entire theory assuming there will be genes governing every type of behavior may not even be passable for something as easily defined as "a pile of wood blocking the flow of water in a river", and thus, falls apart completely when applied to far more abstract and poorly enumerated concepts like *classical music*, *traditional values*, or even love.

In reality, culture is as tangible as the river those beavers are damming. You can observe it, you can see it, hear it, and feel it, but it is always changing. iPhones and Twitter are now more a part of *British culture* than those Dr. Who style phone booths, but that is only because technology memes have advanced to serve our needs for communication. By painting cultural norms as a zero-sum fight for survival, Dawkins is doing what old reactionaries always do: He is defining the world in terms that *normal* is whatever he is familiar with and *abnormal* is anything that would ask him to adapt.

Just like with religion, if the goal of DNA is to generate the largest number of people spreading their culture as far as possible, then Dawkins should view those from India and China as superior. With over two billion of the world's six to seven billion people between them, clearly the Chinese and Indians are exceptional at making new *vehicles* of genes. In terms of cultural spread, only the English language has done better than Chinese and Indian cultures. Here, in the U.S., we have no king, respect no king, and you are more likely to find a *traditional English breakfast* in a trash can than at a restaurant. Meanwhile, nearly every mall has both a Chinese and Indian restaurant in the food court. So, shouldn't Dawkins be elated to see Indian restaurants popping up all over England? Wouldn't that be evidence that his theory was correct, and the clearly superior genes and memes of the Indians were overtaking the relatively weak genes and memes of his fellow Brits? My point isn't that Dawkins isn't allowed to comment on social strife; in fact, to his credit, he opposed Brexit, which was the eventually successful call for the U.K. to leave the European Union. So, Dawkins isn't purely beholden to his exact norms. However, he seems to selectively apply his paradigm that dominant culture practices are genetically encoded and that such memes become dominant through superiority. Things he likes are *natural order* and things he does not like are violations of that order.

His paradigm that genes are in a constant struggle for survival and duplications could just as easily be reframed in any number of *just-so* ways. Someone could write a book called *The Selfish Mother Nature* and pretend that nature has a sentience that guides its tolerance of the organisms living within its area. The claim could be that Mother Nature is selecting for balance as a means of protecting her overall creation. If the *hawks* become too invasive, Mother Nature will strike them down with famine or infection to cull the herd. If the *doves* become too complacent or overpopulated, Mother Nature will send in hawks to pick off their weakest and restore balance. In this framing, it is not the genes that fight for survival but the environment sorting and selecting for the genes that best serve Mother Nature's goals. This kind of Nature guided thinking is part of many indigenous beliefs all over the globe. I'm not actually contending that the *Selfish Mother Nature* theory is the correct one, but I outline it to indicate that if the goal is just to make a paradigm using a woven tale, it can be done. If the goal were instead to select the best paradigm, one would need experimentation to see which one best explains the data.

And here is where Dawkins' entire theory being entirely built around DNA's replication despite its inability to self-replicate becomes a problem explaining the data. If we were to embrace his viewpoint, which emphasizes replication as the core aspect of the unit of life, then the most minute measure of replications possible is a solitary cell. RNA can self-catalyze replication but needs other *ingredients* to pull this off; protein cannot self-replicate nor can lipids. Only a cell has all the needed machinery for replication. Even viruses, which lack most replication equipment, need to hijack a cell to access what it needs for replication.

If we look to the cell as the *replicator* in Dawkins' premise, then we would see a pattern of behavior that is far different than one of a molecular struggle for supremacy. We would see one of cooperation and consolidation. Cells can replicate, but to survive, they need to adjust to all kinds of environmental challenges. Over time, the cells learned that their odds of survival could only be enhanced in two ways: 1) to copy themselves so frequently that any challenge might be beaten by overwhelming numbers of offspring, or 2) to team up so that the challenge could be overcome by the cells themselves. The *just outgrow the problem* approach is what all microbes try to do. These are single cell organisms that have no other option when faced with, say, penicillin than to divide as quickly as possible in hopes that one of the daughter cells spontaneously develops a mutation that allows it to be able to survive the antibiotic.

Option two would be familiar to any multicellular organism. Let's call it *The Collegial Cell* theory. The first cell that would become animal cells lacked a good way to generate energy, so it teamed up with an ancient bacterial cell by internalizing it entirely. The bacteria that went into the animal cell is what we now call a mitochondria. Mitochondria have their own DNA sequences separate from those of the main genome that is the focus of population genetics. Mitochondrial genes are extremely important and lack of their functions either results in death or severe disease.

After the cells added a mitochondria, they began to team up with each other and even seemed to devise a system for division of labor. A single-cell organism busy collecting sugar for energy would have had to stop what it was doing to fend off a toxin. But a multicellular organism can assign a few cells to deal with the challenge while other cells continue their work. For example, the cells in the brain only do well when fed glucose, so what would happen if the diet shifted and made obtaining glucose harder? What if the only energy available to an organism was fat?

What would the brain cells do? In a single cell organism, those cells would have to start using suboptimal energy sources and suffer for it. But in a multicell organism, the cells in the liver can change their metabolism to break the fat down into ketones; those ketones can keep the brain operating at nearly the same level as on glucose. By teaming up, cells increase their chance of survival. The liver cells can rely on the brain cells to protect them from threats, to build adaptations to the environment, and to identify and track food sources. In exchange, the brain cells can rely on the liver cells to modulate toxin exposure and assure a more predictable metabolism in a world of shifting and intermittent food supply.

Human cells further teamed up with microbial counterparts. As will be discussed in later chapters, every human has microorganisms living on and inside of us. These—collectively called the microbiome—help perform all sorts of helpful tasks that support the health of human cells. As just one example, ancient mammals used to make their own vitamin K, but we outsourced this job to our gut bacteria. Vitamin K is needed for normal blood coagulation, so today, humans must either rely on the diet or these bacteria to help prevent us from bleeding to death after a minor injury. In exchange, these bacteria need humans to serve as their home and food source.

But here is the most important aspect of this *Collegial Cell Theory*. It, too, just like *The Selfish Gene*, is a 'just-so' story that can't ever really be tested. We can't go back to early cellular life and ask the cells about their logic for working together more than we would be able to go back in time to find whether RNA came before DNA. These are just constructs meant to help make enough sense of the world to navigate and investigate it. It could be that my cell theory is more accurate; it could be that the gene theory is more accurate (okay, *The Selfish Gene* idea is definitely not more accurate... but I'll throw Dawkins a bone on this). The point is that the world is not required to fit neatly into your narrative about how it functions. History is littered with the hubris of those who thought they had it all *figured out*.

At its core, the scientific endeavor is about generating actionable intelligence. The scientific *process* is about asking questions, designing experiments to attempt to answer those questions, then looking at the results and asking even more questions. But that is not the scientific *purpose*. The purpose is to answer only one question: Knowing what we know right now, how can I best address this problem I'm facing? Science is supposed to help you predict the future enough to guide decisions, not describe the past in flowery language. In medicine, science is meant to assess a person who comes in with a certain set of symptoms and determine which tests might be needed, which treatments should be started, or what reassurances might be offered. For technology, if you are in England and want to phone a person in Japan, you will pick up a device with a certain transmitter that will bounce a signal off a satellite in space. Then, that signal will ping a receiver in Japan to set up the communication. Maybe next year, that system will be replaced with something new. Maybe 100 years from now, we will look at cell phones as if they are archaic. But for right now, if you want to communicate across the globe, the actions you have available to you have been worked out by communication scientists in generations past and present.

So, the reason that the *Selfish Gene* theory has been such a drain on science isn't that it is less believable or less descriptive than a God-based, Mother-Nature-focused, or Collegial Cell idea. The reason the *Selfish Gene* has been so harmful is that it created a paradigm that built everything around the idea that genetic sequences would explain nearly all adaptations, both cultural and biological. Oddly enough for such a devout atheist as Dawkins, *The Selfish Gene* serves as a sermon, which preaches veneration of the genetic code for its ability to sustain life and bless us with ethics, festivals, music, and family. It attempts to explain the origins of life while promoting salvation to those who dedicate their lives to learning and incorporating its wisdoms. It offers faith that, one day, suffering will be vanquished if only we open our hearts and minds to its teachings. While it does not promise that evil will be eliminated, it does claim that the dove-like forces of good will be in balance with the hawk-like forces of evil despite their existential struggle.

While this *Sermon on the Helix* does provide support for occasionally helping *the least of these*, it sadly comes with some of the negative characteristics seen in other religions: It encourages unnecessary suffering as people ignore prevention under the assumption that the presence of disease reflects predetermined fates. It also supports authoritarian oppression, inequality, and ethnonationalist intergroup conflict as being part of *the natural order*. Plus, whenever it encounters a situation where reality is inconsistent with the central premise of a DNA-guided world, the *Selfish Gene* theory gets to claim that DNA *works in mysterious ways* that could be better understood if only we would continue to have faith and give our money to the cause.

The Selfish Gene succeeded in setting the paradigm for biomedicine to center around genetic thinking, even to the point where environmental mitigation strategies were also somehow genetically encoded. The book focused on behavior, but the paradigm it created corrupted all aspects of biomedicine. It also set the standards where there was no need to be specific about anything you were linking to genes, nor the genes themselves. Geneticists that followed could find correlations between a region of the genome and amorphous concepts like *social values* and claim causation without finding the exact stretch of DNA they claimed . was so important and without rigorously defining what they meant by *social values*. If people living in polluted areas was the problem, then gene jockeys could claim that those people must just lack the genes needed for social mobility.

As we will see in the chapters that follow, this tunnel vision has limited genetics research in a myriad of ways: It has precluded improvements in methods that would provide better data but would come at the cost of the gene-centric thinking; it has prevented genetics from divorcing itself from its eugenics past. Perhaps most damaging, it has prevented population genetics researchers from identifying the types of discoveries that their cancer or pharmacogenomic counterparts have identified—discoveries that meaningfully alleviate suffering from diseases and improve health.

CHAPTER 6

Blank Slates, Bombastic Scientists, and other BS

BOTTOM LINE UP FRONT: *Thinking that genetics play a minor role is not the same as believing in a blank slate.*

At some point in debating with any genetic-focused person, they will invariably accuse you of making one of two arguments: They will claim that you are saying genes have no impact at all and/or that you are claiming that group differences in genetics have no impact at all. The second claim flies under the banner of *Lewontin's Fallacy*, which we will dissect in the next chapters. The first example, the false claim that people who don't believe genetic differences are actionable must think genetics are totally irrelevant, is a claim referred to as *believing in a blank slate*.

If you ever hear anyone call someone a *blank slate'ist*, you can be assured of two things about them: First, they have a much higher opinion of their own intellect than is warranted, and second, they don't understand what a strawman argument is. You can guess that they are also likely a bigot, but that isn't as assured as them being the type to put their IQ score in their dating profile.

The blank slate—or *tabula rasa*—was an idea that was born in ancient Greece. The claim was that humans were not born with any innate knowledge. The main point is that humans needed to learn different behaviors and beliefs. For the most part, this was a meaningless exercise by philosophers, since whether you thought your child was born with the ability to pee in a toilet or not, you still had to practice enough to bring out the inner nature. However, like most ideas that stress the role of environment and nurture, it collided with the fragile ego of otherwise mediocre men. If it were true that any child could be taught logic, math, philosophy, leadership, or science, then women would have the potential to be mental equals to men, provided the laws allowed it. If anyone could learn if properly taught, then discriminating against people from other nations made little sense. As you can imagine, not everyone was on board with a philosophy that implied that women, non-whites, or even southern Europeans could achieve the same things as northern European men.

More modern opponents to *tabula rasa* will point to things like the existence of the motor or visual cortex. Their point is actually a sound one; the brain truly is born with innate *knowledge* of how to move the baby's arms and legs. The brain stem is born with the *knowledge* of how to breathe and pump the heart on beat. This, of course, is a poor counter argument. First, anyone who has been near a baby knows that the physical act of moving the arms or legs may be pre-programmed, but the child needs to learn the *knowledge* of coordinating those movements to walk, run, jump, and so on. Second, the muscle fibers also are born with the *knowledge* of how to contract the baby's muscles. Does that mean the muscles have innate *knowledge*, too? Would we say that hemoglobin has innate knowledge of how to carry oxygen, or might we more fairly say that hemoglobin has a set of biochemical reactions that are dictated by its structural and molecular makeup?

Maybe the failure here is best outlined by those quotes contrasting knowledge to wisdom. Knowledge is usually portrayed as learning the raw facts, while wisdom is framed as learning the application of those facts. So, maybe the brain is born with the knowledge of making sounds, but still needs to learn the wisdom of how to formulate words. So, in a sense, perhaps one can contend that the body (and brain) is in fact born with innate knowledge, but I'm not sure the original philosophers intended to include *basic biologic functions* in their definition of *knowledge*.

Twin studies researchers weighed in to defend the idea of innate, pre-programmed behaviors. We have already addressed the absurdity of the heritability calculation in Chapter 3. Even if we were to overlook that, they did not examine the *heritability* of behaviors such as crying when hungry or smiling when the baby saw her mother. Instead, they argued that the brains were born with a pre-programmed love of sailing, acceptance of Jesus Christ as the baby's personal Lord and savior, and the innate leadership skills needed to run the entire British Empire.

Let us run another thought experiment: Say we took an English prince at birth and transported him to the moon. As of this writing, Charles Windsor is King of England, and Prince William is next in line. William's brother, Harry, has split from the royal family and thus taken himself out of consideration, making the children of Prince William and Princess Katherine the next in line (George, then Charlotte, then Louis). So, what if they would have sent Prince George to the moon to live with a robot that fed him and changed his diapers but nothing more? No one else was there. No pictures or information of any kind was provided. Instead, he remained up there with nothing more than the feeding robot for 40 years and came back down to the U.K. What do you realistically expect him to do?

All his basic body functions would be intact, we presume. He would be able to move his arms, so that *knowledge* would be in place, for sure. He would be able to walk and probably run. Perhaps (but not assured), he might have his own version of spoken language; for example, he may have a specific set of grunts that to him communicate the same idea as when his fellow Englishmen say *porridge*. We know he certainly wouldn't be able to read anything. He might draw a picture of his feeding robot but wouldn't be able to understand nor communicate. But what else? Do you think he would love swimming? Do you think if you handed him a tennis racket, he would immediately be good at it since it was encoded knowledge? Would he have spontaneously developed a religion? How do you think he would fare on an IQ test? Importantly, do you think he should be the ruler of England? Because by the rules, he would be the *rightful heir* over (presumably) his sister, Charlotte, who would have been serving as Queen. Do we think George's leadership would be innately activated upon return from space? We do know, his legal rights would state he should be king.

Fair enough, many might push back against this analogy claiming that genes would still need some level of *nurture* to activate. But if the argument is that culture, political preferences, religion, intellect, and leadership are all as hardwired into the brain as lifting an arm is, then does this claim make sense? Maybe we should update the litany of quotes contrasting knowledge and wisdom to make one more apt for innate biology: Bodily knowledge is the ability to defecate, wisdom is knowing to do so in the toilet.

So, most of the political and social traits that the original critics of *tabula rasa* focused on don't really hold up to the idea of being innately wired into the brain. However, these *tabula sordida* (dirty slate) thinkers might have a better argument for disease states. Here, they could argue that if Prince George were to have the *genes for* Schizophrenia, then regardless of whether he was raised on the moon or in the U.K., his disease would manifest at some point. The twin studies researchers might spend time calculating sailing genes, but to the credit of population genetics, they more often looked to meaningful diagnoses like those found in mental and behavioral health. For support of this idea, they often point to diseases for which the causal mutation has been proven.

In Chapter 4, I outlined clear instances in which a single mutation in a single gene all but assured the patients of developing a medical issue. This *tabula sordida* claim is that since a mutation can destroy lung or brain function, it therefore means that genes must contribute to more subtle differences between lung and brain function. The idea that the existence of mutations in a given gene can harm prove the existence of mutations in the same gene can help was debunked earlier, but would certainly require some level of evidence to support it. Unlike in the cases of cystic fibrosis or cancer, the population geneticists rarely—if ever—connect those dots between mutation and function, and it does not appear that the field is moving towards doing so more frequently anytime soon.

Even if we overlook this lack of functional validation, there is still one limiting conceptual issue: If we are going to calculate that the lack of a mutation counts towards the importance of genetics, can't we do the same for environmental risk factors?

The genetic reductionist argument is that Michael Jordan wouldn't have achieved his greatness if he had been born with sickle cell disease, and thus, he owes some portion of his achievement to a normal hemoglobin gene. But it would be equally true to say that had Michael Jordan been in a debilitating car accident as a child, he wouldn't have made the NBA, and therefore, automobile safety is partly responsible for his career. Had his team been in a bus accident, had he eaten a tainted hamburger, had his mother taken medications that cause fetal harm, had he been exposed to toxic levels of lead, had he contracted Polio, so on and so on. Had any of these things happened, they too, would have derailed his career. So, do we also need to calculate how he owes his career to government officials inspecting the restaurants that he ate in growing up? Every time his plane safely landed in a new city, shouldn't we count that as a contributor to his eventual hall of fame status? If you want to consider the absence of the worst-case scenario for the 30,000 genes Jordan had as evidence that the genes *matter* to his career, that makes sense. But there is no end to running that same logic for all the myriad of environmental factors that might have also derailed him along his path.

Finally, as it relates to physical traits, the *tabula sordida* crowd has a stronger point. At the start of the population genetics era, people expected genetics to be able to *explain* 90% of someone's height—but the actual genetics are (at best) closer to 30-40%. But 30-40% is far from zero. No one would claim that both Shaquille O'Neal (a Hall of Fame basketball player who is over seven feet tall) and Muggsy Bogues (an All-Star Basketball player under six feet tall) have the same chance of having a seven-foot-tall kid if they were to mate with the same woman. If a couple with only brunettes in their family trees had a blonde baby, then a few pointed questions would be sent the way of the mother. Without a doubt, certain *physical* characteristics are encoded into the genome; however, as we will discuss in the chapters that follow, accurately calculating how much of those characteristics are owed to genetics is math that the population geneticists have yet to perform.

CHAPTER 7

The Fallacy of Believing in Lewontin's Fallacy

BOTTOM LINE UP FRONT: *Since there is more genetic variation within racial groups than between them, there is no biologic value to sorting people along racial lines.*

One of the first modern people to suffer being called a blank slate'ist was Richard Lewontin. Lewontin was a professor of genetics and evolutionary biologist at Harvard University starting in the 1960s and lasting through the early 2000s. To put a very accomplished career into a single sentence, Lewontin wrote many scientific papers mathematically evaluating how genes change with evolutionary pressure. He was also an outspoken critic of the kind of hereditarianism found in his field at the time. In 1972, Lewontin described that there is far more genetic diversity within racial categories than between them. Said another way, it would be normal for a Black man to be more genetically similar to a white man than a second Black man. Lewontin argued that since there was more diversity within than between racial categories, there was no value in sorting people by race (or ethnicity or ancestry).

A paper in 2003, however, claimed that this was a *fallacy* because although you cannot classify people by race using a single gene, you could make a good guess when you considered all the genomes in total. The nuts and bolts of why this paper was so flawed were well dissected in the journal titled *BioEssays* by the University of Illinois associate professor of evolution, Charles Roseman. Roseman outlined that it was precisely because Lewontin looked at the genetic profile of racial groups that he was able to tell that sorting along such lines was without value. Lewontin did not say it was impossible to sort people into sub-groups; he said that if the genetic difference within a sub-group is larger than the genetic differences between two sub-groups, then separating people has no biological value.

This does not mean that looking at the sub-groups separately never has any value, only that there is no biological or genetic relevance for those trying to find the real cause of a disease. For example, at this point in history, everyone knows smoking cigarettes increases your risk of lung cancer. The first two formal studies showing that cigarettes were dangerous were performed in the U.K. and then the U.S. When other nations heard about the data—which clearly demonstrated that smoking increased the risk for many diseases, including lung cancer—what do you think they did? Do you think the German health officials presumed that this finding must only apply to Brits and Americans and that Germans were probably fine smoking? No, of course not.

Since the biologic variation within Germany is larger than the differences between Germans and Brits, there is no reason to suspect a biologic finding in Germans wouldn't apply to Brits, and thus, there is no value in sorting people before running the analysis. The distribution of certain factors can still be different between the two groups. If more British people than Germans smoked, then it can be valuable to look at them separately when trying to assess why cancer rates might be different. However, that is accounting for the distribution of a risk factor, not trying to establish one. This error in logic is more evident in the genetics of lung cancer than smoking. Remember the discoveries of Dr. Winn demonstrating that lung cells need alterations in gene pathways related to *Wnt* to turn into a cancer. So, if you knew that mutations in *Wnt* might cause cancer in Norwegians, do you think they would also cause cancer in Italians? If a mutation in hemoglobin causes thalassemia in Greeks, would that same mutation cause thalassemia in Polynesians?

It isn't that every individual with a given mutation will manifest the same disease—not every mutation works that way. But on a population scale, saying a mutation in *Wnt* is cancer causing in Spaniards but totally benign in Koreans is to claim that Koreans are impervious to the harms from a mutation that you think damages the *Wnt* protein's function so significantly that it causes cancer in Spaniards. If you made this claim for an environmental risk factor, you might claim that "smoking is bad for your health, unless you are Irish. Irish people can smoke all they want and not have any problems." Anyone claiming that would be quickly recognized as *a fecking eejit*, and yet claims of ethnicity-specific immunity to mutations are made every day in population genetics circles.

This isn't to say that Spaniards and Irish have no genetic differences between them, nor that there is no value in looking at rates of genetic risk factors. Well-established genetic diseases have different distributions in populations. Cystic fibrosis is most commonly found in people of European dissent, while sickle cell trait and anemia are more common in people of African ethnicity. So, if you wanted to know why Black Americans have more sickle cell disease than white Americans, it would be valuable to look at the rates of the genetic mutations in one group versus the other. In such a case, you have established the risk factor and are asking how the risk factor is distributed in society. However, when researchers performed such an evaluation, they found that the group with the highest rate of sickle cell trait was actually a small section of Greece. Since sickle cell is a human genetic response to malaria, places with historically high rates of malaria were equally affected. Thai populations also have high sickle cell allele frequencies. So, while the disease is typically thought of as *Black*, the highest rates are found in people who would be defined as white Europeans and Asians.

It is safe to assume that anyone with a sickle cell mutation will be at risk for sickle cell symptoms; it may be that far fewer white Americans have these mutations, but it does not mean that white people are immune to the effects of the mutation. What Lewontin was actually saying was this: If the mutation is bad for one group, it will be bad for a similar group. What Lewontin was not saying was: There are no such things as groups.

Eugenicists (and eugenic-adjacent researchers) will often pull a bait-and-switch on this subject using a specific type of actual fallacy called a *motte-and-bailey*. A motte-and-bailey fallacy is named after a style of castle fortification where a raised enclosure is surrounded by a walled off courtyard. The term applies to the practice of arguing one controversial opinion, but when challenged, retreating to defend a non-controversial opinion or even an obvious fact. They use the defense of the reasonable position to defend their less-reasonable claims.

Here is an example. Say I want to make a new racial classification system. Instead of the Black, white, brown, red, and yellow put forth by the racists of old, I would like to sort people by their favorite sport. Doing so, I will also rank these new races so that people who favor American football are the master race (this is true, by the way), then basketball, then hockey, then soccer, then baseball. In this set up, I'm arguing that football fans are biologically superior to others. Any difference in characteristics between football fans and baseball fans are a result of their different genetics. And indeed, with a large enough sample size, I could find genetic differences in certain gene variations between the groups. I would presume that those genetic differences were the cause of why football fans were biologically superior.

Eventually, people would push back against my classification system (obviously, the baseball fans would). The critics of the innate superiority of football fans would point out that there is no biologic definition for *football fan*; there are no genes that are truly unique to football fans, and no evidence that any genetic differences were biologically impactful. But when these critics scoff, I could fall back by saying, "But there are genetic differences between football fans and baseball fans!" So, my controversial opinion is that football fans have genetic differences from baseball fans that make football fans superior; my *retreat position* is that football fans have genetic differences from baseball fans.

For centuries, people have claimed the idea of "racial diseases," which are claims that different races either experience different diseases or have different symptoms from the same disease, often because of a perceived genetically encoded inferiority of minority groups. The book, *Fatal Invention*, by Dorothy Roberts outlines several examples:

- During the civil war, it was assumed that Black Americans had smaller lungs, which caused their increased rate of Tuberculosis infection. When a microbial cause was proposed in 1869, it was initially dismissed under the assumption that Tuberculosis risk was racial. Not until 1882 was the idea widely accepted that Tuberculosis came from outside the body.

- The Tuskegee experiments were, in part, inspired because researchers wanted to see if Black Americans experienced the disease differently. In the notorious experiments, Black sharecroppers who had become infected with syphilis were lied to by the U.S. Public Health Service. Those enrolling in the study were given provisions but never told that the treatment for syphilis was already known (a simple set of injections with penicillin). The researchers hypothesized that Blacks would suffer more cardiac disease but less neurologic because of the assumed *inferior intellect* in Black people.

- Schizophrenia diagnoses during the civil rights era were edited so that defiance against segregation or believing in the Black power movement was included in the diagnosis for Black Americans. This seems like an obvious ploy to set up an extrajudicial system for locking up *militant negroes* without the need for a formal trial but physician groups failed to push back against framing schizophrenia as a racial disease.

- People have long believed that the same level of injury or the same pain of childbirth is experienced differently between Black and white people. This has been shown to result in substandard care in Black patients even though the rates of complaining about pain are the same.

With each *racial disease* that was proposed, some version of genetic causation was floated. Lewontin was arguing against this kind of thinking. If the biology of humans is the same, there is no reason to think that people will experience diseases differently (or at least, the variation in how people experience diseases will be similarly more variable within any group than between groups). Lewontin would argue that while there may be variation in asthma symptoms between two Black children, the range of that variation would be larger than the difference between the average white child and average Black child. Lewontin was correct.

If you take any two groups of people that are large enough, you will be able to see small genetic differences between them. No one—not Lewontin nor any researcher today—would deny that genetic differences could be found between football fans and baseball fans. The crux of the argument revolves around whether these differences hold biological significance or if they are primarily products of social constructs rather than genuine biological realities. Lewontin's work demonstrated that sorting on race has no more biologic value than sorting by sports fandom.

Roberts detailed a study that sub-divided the clinical results by the patient's zodiac sign and found significant differences between zodiac groups. If we still lived in a world that believed the zodiac was a true measure of difference, these kinds of studies could be used as *proof* of the weakness of Sagittarius. In fact, the computer algorithms that sort populations into *ancestry* groups can make a fit for any number of groupings the researchers request. When asked to sort a population into five groups, the computer abides; when asked to sort into 12, the computer abides. Here is one example from Ruth Johnson and colleagues published in the British journal *Genome Medicine* (image taken from their preprint on *medRxiv*).

Panel A shows the genetics of different people sorted when the computer is instructed to make 5-6 categories using a plot known as principal component analysis, or PCA. On PCA plots, each dot is a single person and dots that are closer together are more similar to each other than dots that are farther apart. Panel B is how people will sort themselves along self-identified race. Panel C shows how genetics differ if people are sorted by languages. In *Scientific Reports* in 2022, Eran Elhaik outlined several flaws in PCA, not the least of which is that the perception of distinct groups and their relatedness are entirely dependent on the number of people evaluated. And yet, the clustering on a PCA plot is often held up as evidence of genetic distinction between groups. Although the coloring above may feel distinct, when the image is set to *black and white*, the reality of a continuum with no true distinctions becomes clear.

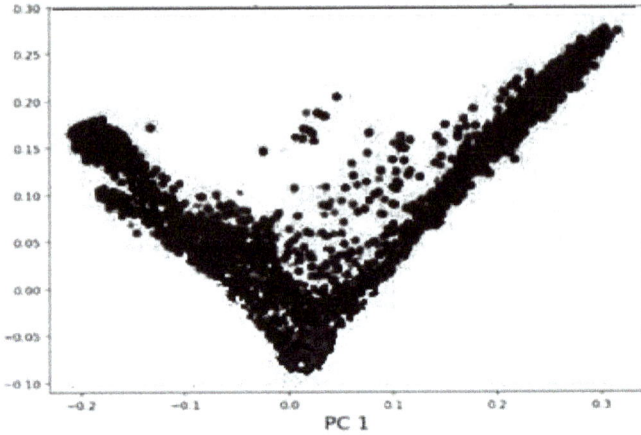

Yet for anyone who thinks that they could still sort people based on the plot above, the groupings remain arbitrarily limited by the study design. For example, here is the same style of plot sorting only people from Africa (image from WikiCommons).

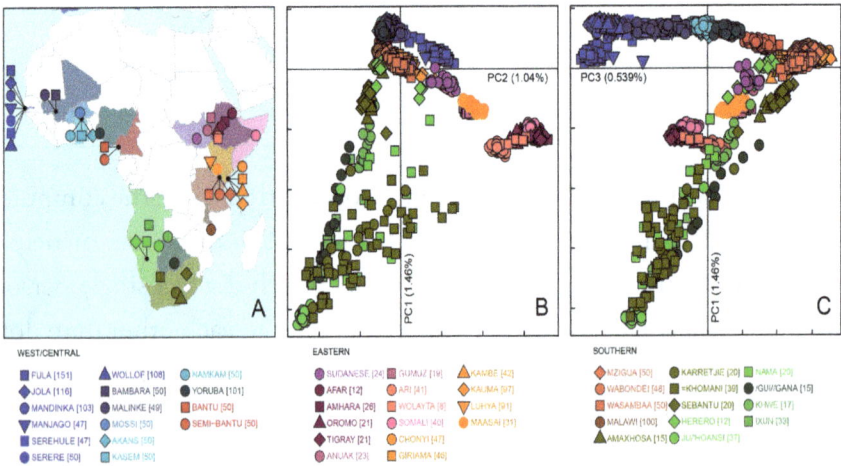

The same V-shaped splay and coloring groups can be seen. So, zooming in to treat Africa as a diverse continent of over a billion people, who all have different but overlapping family trees, suggests an equal amount of *genetic diversity* as seen in the global population. But treating all Africans as one monolith (as history has been known to do), gives the false impression that genetic boundaries are real.

The same illusion of genetic distinction can even be recreated by looking at only Norwegians, as was performed by Morten Mattingsdal and group in *The European Journal of Human Genetics* in 2021 (image taken from preprint).

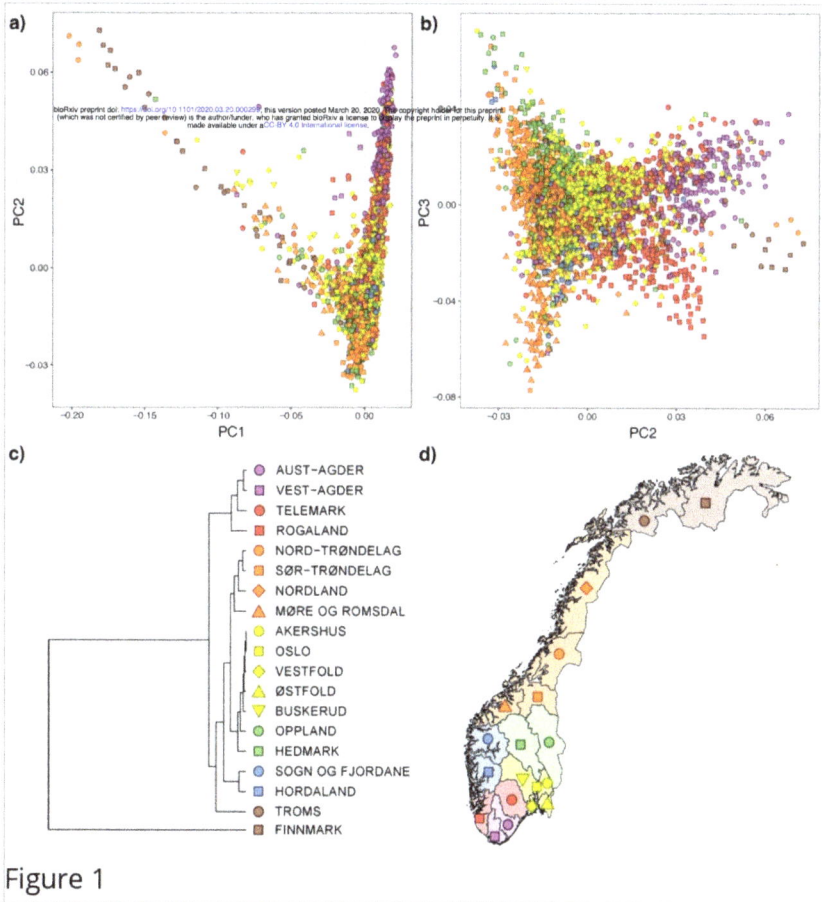

Figure 1

If dividing people on a PCA was all you needed to justify claims of being a unique ancestry group, the number of groups would be far greater than the reflex categories used by most population geneticists today. Central and southern Norwegians would be viewed as two different genetic populations rather than being lumped into "European" like the original PCA plots shown.

But if we want to divide people on genetics, we can use Johnson and colleagues' plots to deduce that if you sorted people by language instead of race, you would also get genetic variation between your groups. You would certainly also identify differences in disease rates and behaviors. People who speak Chinese would probably eat more Chinese food than people who speak Arabic. But would anyone be foolish enough to claim that people have genes for liking lo mein because the people you might put into the arbitrary *Chinese speaking ancestry group* also eat more lo mein than others? Well, actually, plenty of geneticists would be silly enough to claim that. One illustrative example is from *Nature Human Behaviour* where the authors led by Nana Matoba claimed that Japanese people have genes for liking green tea based on the logic that because Japanese people had both differences on a GWAS and in green tea consumption, the genes must play some role.

Here is an example of the motte-and-bailey fallacy in the wild. It came from Professor King Jordan, a geneticist at Georgia Tech. He posted in response to Professor Graham Coop. Coop was outlining his findings visualizing how remarkably similar the genetics are in a small cohort of people. His diagram is meant to outline the entire potential variation in humans based on all 2.9 billion base pair sites (this is the largest circle). The inner circle then depicts the 39 million sites that varied within his 609 people sample. Within that are the variations that were common, not just a one-time base pair change in one person. The take home point is that even though there may be at least one person in the world with a unique base pair at every location, for the most part, the vast majority of human DNA is shared between people and groups.

Graham Coop @Graham_Coop · May 16   ...
We can also place this Euler diagram back in the broader context of the whole genome. 8/n

Measurable locations in the human genome (~2.9 billion sites)

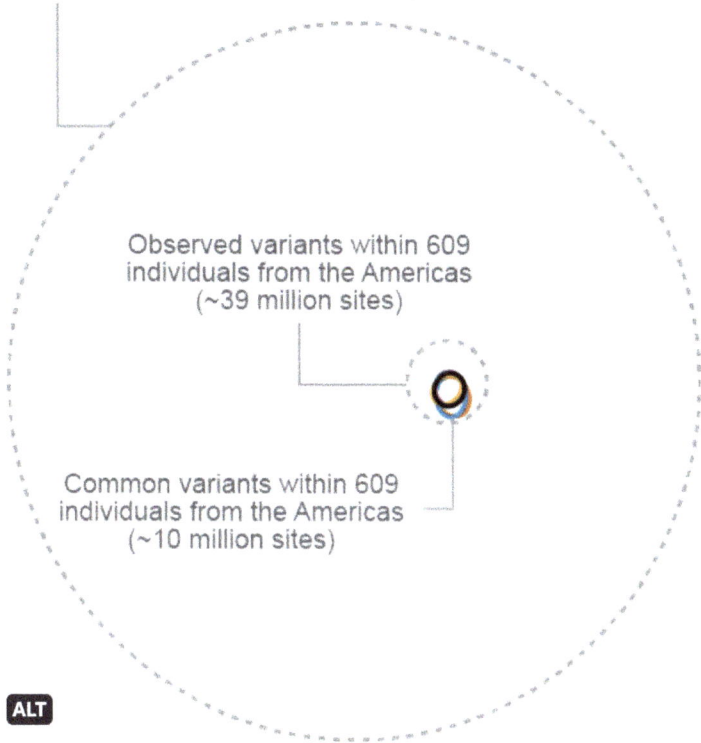

Observed variants within 609
individuals from the Americas
(~39 million sites)

Common variants within 609
individuals from the Americas
(~10 million sites)

ALT

Jordan disagreed. He felt that the representation was obscuring the large *biologic* differences between groups of people. His three-tweet thread states:

King Jordan @ikingjordan · 18h

The overlapping circles produced by your method, a direct result of eliding the underlying data structure (like Lewontin), show these populations as essentially genetically indistinguishable, when in reality they are highly differentiated. This is misleading.

2 100

King Jordan @ikingjordan · 18h

I see from the blog that your pedagogical motivation is to reduce racial bias among students by downplaying the extent of human genetic variation (Donovan et al. 2019, Biddanda et al. 2020). Lewontin was similarly motivated.

1 89

King Jordan @ikingjordan · 18h

These are noble intentions. But this is a genetic version of the moralistic fallacy (reverse naturalistic fallacy) – the leap from ought to is. Because groups ought to be treated equally, there are no meaningful genetic differences between groups.

Notice how the first tweet set up the claim that Black and white Americans are "*highly* differentiated" Then, a mere two tweets later, he accuses people of claiming that there are "no meaningful genetic differences between groups." The latter could be easily disproved by sickle cell mutations, which as discussed in earlier chapters, are protective against infection with malaria. The only reason there are differences in sickle cell hemoglobin rates between groups is the pressure of a large-scale environmental challenge like a life-threatening infection. The rates of sickle cell are thus a "meaningful genetic difference between groups," and would meet Jordan's motte threshold. But in no way does that mean the groups are as "highly differentiated" as his bailey claim implies. He changed the target from a controversial claim that "racial groups are highly different" to a non-controversial claim that "there are SOME meaningful differences between groups" and didn't even need to be pushed into doing so. For population geneticists obsessed with hyping genetic variations without proving their biologic meaning, it must be reflexive.

To bring in one final example on this subject, think of all the medications you have ever taken in your life. Have you ever had a headache and decided to take Advil (ibuprofen) or Tylenol (acetaminophen)? Or maybe you don't like pills and so when you had a sprained ankle, you decided to put an ice pack on it? Why did you take these steps to help your pain? Why not tape a flower to your forehead or soak your ankle in syrup instead of ice water? My guess is that you took Advil or the ice pack because you know those are proven interventions. You know that people have studied the best way to get over a headache or a sprain and (among many options) came away confident that both interventions have value. But here is the next question: Do you care where those studies were conducted?

What if you live in Los Angeles and have for generations but I told you the studies validating Advil was better than a placebo in controlling headaches were conducted only in Texas and Florida. Do you care? Do you worry that it is possible that Texas-style headaches can be treated with Advil but Californians' headaches will actually get worse with this medication? Do you fear that icing a sprained ankle is good in Florida, but those from Los Angeles might have ankle swelling that is impervious to cold?

When clinical studies for new drugs are performed in the U.S., typically a large drug company will enlist several clinics around the country to participate in the trial. Small scale studies may only have two to three sites while the largest ones may have 30-50 sites. What that would mean is that patients taking part in the study would go to one of 50 different clinics or hospitals in the U.S. and receive either the active drug or a placebo. Then, the trial would monitor for improvement and test whether those on the *real* drug did better than those on the *fake* one. While clinical trial designs can have a good degree of nuance between them, there are very few drugs approved in the U.S. in the last 50 years for which this basic setup has not been performed to some degree. I bring this up to say that even though there are over 6,000 hospitals in the U.S. (as of 2023), medications are—on average—tested in only around 30 or so locations.

So, I ask this: What should the FDA do with a drug that cured breast cancer but was only tested at sites in these 10 cities: Los Angeles, Fresno, Santa Barbara, San Jose, Dallas, San Antonio, Seattle, Portland, and Miami? Should the FDA approve the drug for the entire U.S.? If so, why? Can we assume that a drug that is safe and effective in a Floridian would be the same in Minnesota? Sure, we can guess that the drug won't be safe and effective in every single person who takes it; there will be some people in both states who would have bad reactions to the drug, some for whom the drug made no difference, and some who were cured. But on a population level, should the FDA only approve the drug for use in California, Washington, Oregon, Texas, and Florida?

The obvious answer is that the drug should be approved for all Americans, but this does not mean that there are no differences between Minnesotans and Floridians. It does not mean that evaluating the pooled genetics of Minnesota versus Florida would suggest the two states are completely identical. There can be plenty of measurable differences between the people in those states—some genetic, some cultural, some behavioral, and some environmental. However, if you are asking the biologic question of whether Tylenol can improve headaches through blocking the effects of inflammatory lipids and you get a positive answer in Floridians, you can assume an overall yes in every American. It may very well be that the differences between Floridians and Minnesotans could influence how *well* Tylenol worked, but you can assume the drug will operate in the same molecular way in both groups.

Now, rerun my thought experiment and swap Floridian for Cuban-American and Minnesotan for German-American. Any difference? Imagine you are an American of German ancestry living in Minnesota and you just received the terrible news that you have advanced stage breast cancer. However, the doctor tells you there is reason for hope because the FDA just announced approval of a drug that was 100% safe and effective against breast cancer in a study evaluating Cuban-Americans in Florida. Do you take the drug, or do you demand that millions of dollars and more years be invested in testing the drug in white Minnesotans first?

And why stop there? Why not test it in German-ancestry Minnesotans, just in case? And just to be confident, maybe they should run a trial for just German-ancestry Minnesotans living in St. Paul who went to college in Wisconsin and are named Janet? At what point does it become absurd to assume your biological response to a drug will be different from that of the people in the study? This isn't to say any drug is assured success or safety in any individual, but how similar do you have to be to the study population before you can feel comfortable that a drug that worked for others might also work for you? Even if there is pharmacogenomic evidence proving Floridians are at higher risk of reacting than Minnesotans due to allelic frequencies, the drug should be approved nationwide and *everyone* should be screened for safety.

Let us ask the question from the opposite direction: What if you found out that a particular exposure was poisonous to Cuban-ancestry Floridians; would you be concerned about exposing your German-ancestry child in Minnesota to it? If you heard that a town of predominantly Black Americans had seen its water poisoned with toxic levels of lead through a combination of political incompetence and indifference, would you want someone to study the effects in your drinking water *before* concluding that it would be harmful to you? If your mayor claimed, "We don't need to worry about filtering the lead out of our drinking water in our mostly white town because lead has only been proven toxic to Black people," would you endorse their stance and drink up? Again, this isn't to say any toxin will be equally harmful to all people, nor that all toxins will affect all groups equally. But the question remains: How similar do you have to be to the study population before you can feel comfortable that a toxin should be avoided?

Lewontin answered by claiming that you need only be human. You cannot assume a drug that works in mice or horses will work in people without first testing it in humans. A drug that works in mice might be more likely to work in humans than a drug that fails to help mice, but you can't assume that. However, a drug can be presumed to work in *humans* even if it does not work in every human. The population geneticists along with every online troll that quotes *Lewontin's fallacy* to defend racial purity would disagree. I can only presume that they would refuse a medication unless it were tested in a population as genetically homogenous as they are.

The genetics obsessed seem to want to argue that any differences between groups can be assumed to be meaningful until proven otherwise (they, of course, have no interest in proving that one way or the other, but they will make that assertion nonetheless). At best, the population geneticists would tell you that if Tylenol worked slightly better in Florida than Minnesota, it was safe to assume the reason for the difference would be most likely found in the innate genetics of each group rather than differences in the weather, food, pollution, or any other environmental factor. Is it possible that a gene variation might influence your response to Tylenol, that the gene variation may be more common in Floridians than Minnesotans, and therefore, differences in Tylenol response between the states might be partly due to genetics? Sure, it could. But, it would still be expected that: a) the gene variation is proven to actually cause differences in Tylenol response and was not just a measure of Cuban heritage that you mistakenly assumed to matter; and b) perform some contextualization in case the Tylenol-gene is only active in humid hot weather (in which case, the weather would govern if the gene matters, which would mean the weather would be more important). Yet the population genetics crowd never performs either of these assessments. As long as the genetic differences between the states could be put on the same graph as Tylenol use, they claim one can conclude that genetics *explain* the differences.

There is one other dimension to Lewontin's assessment: time. Since there is more variation within humans today than there is between humans today and our distant relatives, Lewontin argued that any finding that was *either* limited to one place *or time* would be on shaky ground. This is not to say that Lewontin claimed that genetics do not change over time. He was well aware of such changes and noted cases like the lactase gene mutations I mentioned in an earlier chapter. But Lewontin's point here was that since human genetics have not undergone enough change between antiquity and today, a claim that a gene *mattered* only in 16th century France would be unlikely to be true. This seems a near truism in fields outside genetics. If a drug worked to lower fever in humans today, do you think it would have worked in antiquity? If you are not sure, you could look up when aspirin was first used. Infection exposures and treatments have changed substantially since ancient times, but inhaling Tuberculosis would be as much a problem today as in the time of the Pharaohs.

In *The Genetic Lottery*, the author Harden attempted to place time limits on genetics by detailing a study that took children in an overcrowded and under-resourced Romanian orphanage and randomized them into either going into foster care or staying in the orphanage. The study subsequently found that the children who were placed in foster care had higher IQ scores than those left behind. Dr. Harden first posits that we don't know why the children receiving care did better than the ones living in neglect. She says this to set the bar for claiming environmental causation is knowing the exact molecular signaling going on in the child's neurons in response to starvation and abuse. Yet Harden takes this further by asking, "What if the study had been done in New Jersey in 2016? What if it had been done in 16th century France?"

If I were to try to answer her question, I would need to admit that I've not visited any New Jersey orphanages. Yet, I will give them the benefit of the doubt that they offer better living conditions than Romanian orphanages under an era of dictatorship. So, I suppose I would be forced to agree that randomly assigning children from a New Jersey orphanage into the foster system in 2016 would not have the same impact as seen in the Romanian study. But the analogy is flawed in a fairly major way: All orphanages are not 99.9% identical. If the orphanages in New Jersey, Romania, and pre-Calvinism France were always at least 85% the same, I *would* expect the results of randomly placing children into foster care to be the same.

But what if lives could be saved if someone could discover why Tylenol worked better in Florida than Minnesota? Where would you invest your research dollars? Would you bet on the weather, the cultural norms, or the genetic differences between the groups? All the above, maybe? If you decided to prioritize spending money on the potential genetic causes, how many failed attempts would be enough? If you ran 100 studies and never found a single gene, nor a set of genes that explained the difference, would that be enough to call it quits? While many people that still believe there is such a thing as Lewontin's fallacy are just racist online trolls, the next section of this book will take aim at those who claim that treating humans as different genetic groups will lead to better treatments for common disease.

PART II

Potential Rain Yields No Crops

This next section of the book aims to detail the flaws in the modern approach to population genetics. We will discuss the flaws in study design and interpretation that have led to the failing paradigm of genetic determinism. A quick note about the title of this section is in order: The title was a saying from the grandmother of the former NFL star, Hall of Famer, and current TV sports pundit, Shannon Sharpe. Sharpe's use of the phrase was to say that a sports team that is good *on paper* does not get any extra points for their potential; they have to go out and win. As it applies to science, a discovery that might become a treatment does not get points for potential; it counts only when it reduces patient suffering. To be a population geneticist is to be in a constant state of explaining that the payoff of your discoveries is just over the horizon. For the eugenicists, the so-called proof of the genetic basis for intelligence, and the so-called proof that Black people have less of it is also just a few years away.

PubMed is the search engine used to look for published papers in biomedicine—kind of like Google but only looking for scientific papers instead of some random guy's blog post. If you search PubMed for the phrase "potential therapeutic target," you get the following graph:

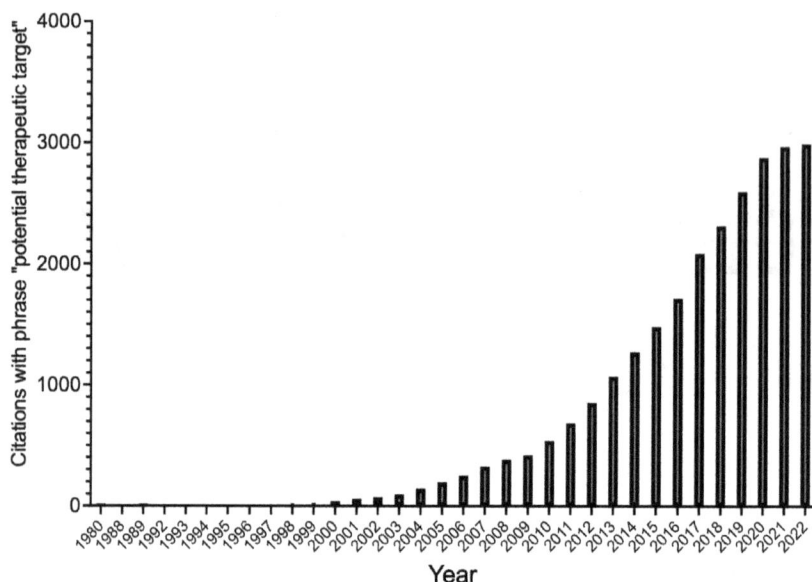

This isn't a graph of the total papers; this is a graph of the papers per year that are published using this phrase anywhere in the text. In 2022, just short of 3,000 papers claimed that they had found a potential discovery that will one day be turned into a new drug therapy. Not all these are genetically focused papers; in fact, the majority are not. Additionally, it takes years—on average, 12.5 years, in fact—for a molecular discovery to make it from a laboratory discovery to a prescription at your pharmacy. The point here isn't to decry people looking for basic science discoveries in hopes that they will one day become a treatment. The point is *potential rain grows no crops*. At some point, you hope to find a *confirmed* therapeutic target so you can make an actual therapy. This portion of the book will outline how modern population genetics has fallen short of its therapeutic goals with such consistency that much of the field has given up altogether.

CHAPTER 8
Putting Funding Over Function

BOTTOM LINE UP FRONT: *Refusing to test the functional consequences of genetic differences is inconsistent with the understanding of biology.*

As stated earlier, the original purpose of population genetic databases was to aid in identification of genes of interest in family-style genetics. Much like with the identification of *STAT3* as the causal gene for the related immune deficiency, researchers used these databases to sort out common versus uncommon alleles to focus the search for the ones that mattered. However, the current flavor of population genetics databases don't collect the full sequence; instead, GWAS is an assay of "tags" spaced throughout the genome.

GWAS studies look for variations not in individual letters, not in specific recipes, but within a collection of recipes. In the example of looking at paprika versus papryka before, the specific mutation was being evaluated in a specific word in a specific recipe. For a GWAS, you instead must imagine that you took the entire recipe book (all the words in the 30,000 recipes) and collated them into sets of roughly 100,000 *sections*. The sections (called loci) contain many genes within them, but the exact number varies. So, what you do is take people who, let's say, have had a heart attack and plot out how much variation there is in each of the loci. You then do the same for people who have not had a heart attack. So, let's say that at loci number 4578, there is a big difference between how variable the letters are for people with a heart attack versus controls.

You then calculate the statistical variation between them to see if it was statistically significant; that is to say, you ask, "If there was no real variation at loci 4578, how is what I am actually finding different from what I would have expected to see? And if what I'm seeing is different than what I would have expected, what are the chances that this result is by chance alone?" That *by chance alone* is captured by the p-value, which will be discussed in a moment. But for now, I will state, without judgment, that if the difference is above the GWAS adjusted threshold line of 5%, the difference is significant.

So, having found the statistical significance of your loci 4578, you plot it on what is called a *Manhattan plot*, like the one below (produced by an account that goes by GWASBot (@SbotGwa) on Twitter and posted for the public):

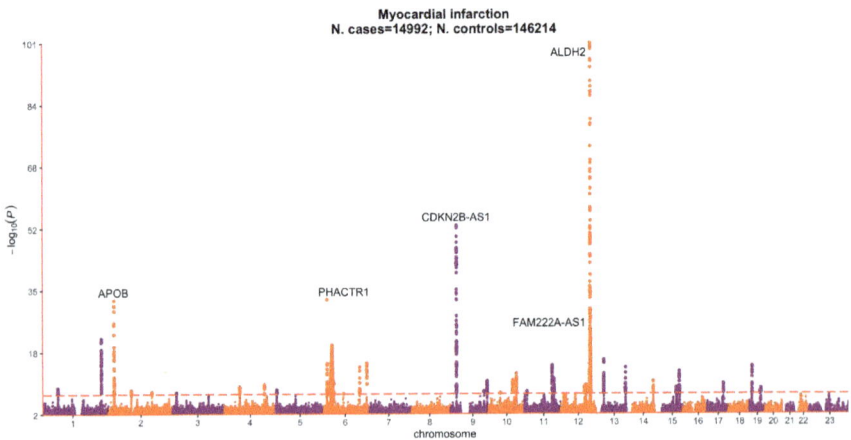

Myocardial infarction
N. cases=14992; N. controls=146214

Supposedly, these plots resemble the skyline of Manhattan. Personally, I think the fact that geneticists were willing to read patterns in their data that don't exist should have been a sign of how their field would play out, but here we are.

Note the dotted line across. That is the threshold that the GWAS crowd had decided upon for significance. The taller the *building* on the plot, the more *mathematically* significant the result. I stress mathematics here because it does not mean that this is clinically relevant. I don't want to get bogged down in how the significance is *adjusted*, but briefly, I'll remind you that the point of the p-value is *how likely the difference was caused by chance alone* even if there is no real effect. So, in statistics, you need to account for the number of times you run an assay, since each time you run it, you might stumble upon a positive result by chance. In GWAS, they run the same assay hundreds of thousands of times, so the analysis adjusts for all the tests to make the new target line as 0.00000005 or 5 times ten to the negative 8th power. The exact number matters for researchers but to understand the concept, you just need to know the chance of lucking-your-way into a positive result goes up with each time you check something. The GWAS analysis does indeed take this into account.

So, from the above, you note that at a loci on chromosome 12, there is a strong signal, meaning that the variation in that region is reliably different in people who have had heart attacks versus those who have not (yet) had one. Thus, you can look and ask, "Which genes are at loci 4578?" You might find none that make sense; you might find a clear candidate but know this: You can't actually know which specific gene in the loci was abnormal without far more detailed sequencing to look at the specific letters.

Going back to the recipe analogy, let's say we sort restaurants that have shut down from those that are thriving. We take their full menus and compare but, in this analogy, we can't actually read each ingredient or word. Instead, we can only look at the variation within sets of three recipes. Even though GWAS is looking at hundreds of thousands of loci, for our restaurant analogy, let's keep it simple and presume each menu has 30 dishes. So, that means we can sort recipes into groups of three and check 10 *loci* for variation. We run our analysis and it looks as if restaurants that have shut down have a potential abnormality in the third set of recipes than those that are thriving. We would then clearly ask, "Which recipes are in that set that might be predicting whether the restaurant failed?" We look and find the first recipe is chicken cacciatore, and the second vegetable lasagna. But the third is split because our *loci* are not perfect collations of genes; so, we instead get the first half of a recipe for tiramisu. So, we have two and a half recipes that could be the culprit as to why the restaurants shut down. In this case, what do you think would be the next step to take?

Hopefully, you would like to look at the details of the specific recipes. You have two and a half "candidates" for which the recipe caused the restaurant to shut down, but you don't know any further details. In fact, you don't even know if anything is truly amiss with the recipes. All you know is the variation is different but recall it could be "paprika" versus "papryka" or it could be "brine" versus "urine." Until you evaluate the signal on the individual letter detail, you can't determine whether what you found on GWAS matters, let alone how it matters.

In my discussions with population geneticists, the definition of "causal" is one that seems to create the biggest gap in understanding between well-meaning geneticists and other biologists. To population geneticists (who tend to be mathematicians), something can be defined as causal without ever having to leave a spreadsheet. If you have 100 variables in your dataset, and you change only one, then any shift in the outcome can be read as the one variable *causing* the shift in results. But humans don't live on spreadsheets, so such definitions fail in the real world. Take the following example: Say you want to make a model that can predict how well a group of people can speak Spanish. You measure several variables about them, such as where they grew up, who they married, and so on. Included in your data are their first names. For the group you are measuring, all of them are men with names like Todd, Steve, and Jim.

Since those names are not associated with speaking Spanish, those names would count against your group and predict that they would speak less Spanish than if you ignored their names altogether.

However, if you kept everything else about the group the same, but changed their names to Juan, Carlos and Bruno Madrigal, then *in your computer model*, you would see an uptick in Spanish fluency because such names would be positively associated with Spanish language skills. To a statistician, they would say that someone's first name "causes" them to speak Spanish because *their model* reliably measures a change in Spanish fluency following a name change in the database. In the real world, of course, if you literally changed a group of men's names to be of Latin origin, it would not impact language skills in any way. In the purely numerical world of the mathematician, changing your child's name from Michael to Miguel *causes* your child to speak more Spanish *assuming nothing else changes*; in the real world, *assuming nothing else changes*, naming your kid Miguel instead of Michael causes *solamente nada*. Finding causes for diseases on a spreadsheet is often a great first step, but medical research is meant to find causes in the real world so that those factors can be dealt with.

The early studies that made use of GWAS followed the model of trying to translate database *causes* into real world causes. The loci of interest were identified, then they drilled down to investigate for the specific sequences of the specific genes in those loci, and then looked to see whether the sequences in those genes were abnormal. They might hope to find the stop codon (recipe deletion) changes or some major harm like the brine-to-urine change. In effect, this was similar to what the family-style researchers were doing. The analysis involved identifying an allele (or set of alleles) linked to the disease. Using software, one could predict the impact of this genetic change on protein function. Subsequently, the allele could be replicated in a laboratory setting by introducing it into a cell culture or mouse. Finally, tests could be conducted to determine if the model exhibited abnormalities that resembled those observed in the patients.

The first sign of flawed logic came when drilling down from loci to sequence didn't corroborate the results of the GWAS. Let's say you ran that study on failed versus successful restaurants again and the loci containing chicken cacciatore came up again. You drilled down to look at the specifics and found that the failed restaurants had a misprint in which "add 1 tablespoon salt" was changed to "add 10 tablespoon salt." Eureka! The mystery of why the food was bad was solved! You excitedly repeated the study—this time a block over—and there, again, the chicken cacciatore loci pops up and again, you dive deeper. But this time, the only misprint for the failed restaurants is that "thime" is written instead of "thyme." Now, that is a little odd because that type of change would not be expected to have had any impact on the recipe at all. So, you move one block down and run your analysis again, and this time, the chicken cacciatore loci isn't even associated at all. But thankfully, when you move one more block away and run your failed restaurant analysis, the chicken cacciatore loci is significant again! You drill down but see that this misprint is in the opposite direction: "1T salt" is now "1t salt". Would this suggest that either too much or too little salt could ruin the dish? Plausible, for sure, but it would then be that much more important to check the actual flavor. If you ran a taste test and could show that both 10-times the salt and one third as much salt would both have equal impact on the flavor, then you might have something.

That being said, not every genetic difference would be easy to test in a lab setting. So, if a model was lacking or too difficult to generate researchers might instead jump right to a clinical trial. For instance, let's say you discovered that a gene related to vitamin D exhibited variations between the individuals affected by your specific disease and the control group. In such a case, you could consider trying out a vitamin D supplement to determine if it could offer any benefits to the patients.

Even with a model that might support your vitamin D genetic finding, you would still ultimately want to turn it into a treatment; so, you might as well try the treatment up front (as long as it seems safe).

For diseases like asthma, *the cycle of GWAS failures* played out as it had for the psychiatric diseases. One study had a loci association that contained the IL-4 receptor (a receptor that helps control allergic responses). But when researchers sequenced that loci, they didn't find any changes that would explain the difference in IL-4 function, nor were any changes consistent between groups. One GWAS would be associated with a loci on chromosome 3 but the next would be associated with a loci on chromosome 6, and then next on chromosome 10, and so on. Anytime people might bother to sequence the loci and find a sequence they thought might explain asthma, they would find that the same sequences were present in other populations but not associated with disease. This would be like running your restaurant evaluation in southern Norway and finding that the chicken cacciatore had 10T of salt instead of 1T; but when you ran the same GWAS in central Norway, they, too, had chicken cacciatore with 10-fold the salt in their recipe, but there was no association with failure. It would be akin to saying that overly salting your food in southern Norway matters but not in central Norway.

Here, you might be tempted (as the geneticists always are) to claim that the unique environments cause the recipe differences to matter. The *regional cuisine* aspects of my food analogy do lend themselves to a comparison with the gene-environment interaction framing used in genetics. I will discuss the details of the failed environmental evaluations in Chapter 14, but for now, there are two holes in the *regional cuisine* argument. First, if you can get away with an overly salted chicken cacciatore in southern but not central Norway, then it is the environment that is determining the impact of the misprint, not some inherent property of the misprint itself. Second, the existence of *regional cuisine* can be experimentally tested; if you think that southern Norwegians are intolerant of overly salted chicken cacciatore while central Norwegians are intolerant of too little salt, you can test that!

In eczema, the earliest GWAS analysis suggested the barrier protein known as filaggrin was different, but the models were either too difficult to create or did not show any downstream effect of a mutation. So, researchers performed a trial using a topical lotion supplemented with filaggrin. The hope was that if the gene caused the production or function of filaggrin to be lost, then supplementing with filaggrin might help. The trial failed, but the idea was no less valid. In psychiatry, there were numerous examples of the same basic pattern as seen with early eczema work: Researchers had carried the idea from genome to treatment; they came up short, but the approach at least made sense. However, while one could certainly defend the approach early on, the problem that began to form is that the trials *never* bore fruit.

GWAS studies were run in eczema again and again and again. As of this writing, there are over 120 different GWAS for eczema around the globe. 60% of those studies show an association with the loci that contains filaggrin; of course, that means that 40% don't, but let's take the *glass is 60% full* attitude on that one. The next most consistent association is the IL-4 receptor at only 10% of the studies linking the disease and the gene. Beyond that, each presumed loci association doesn't even show up in more than low-single-digit percentages of the studies. Maybe the GWAS results are inconsistent across environments because the environment is interacting with the specific loci you think are important, but it could also mean that their findings are not reproducible because the claims are invalid. If you never bother to test further, why should we assume you to be correct?

During the course of their use, GWAS study results developed a pattern of sorting into three overlapping groups: The first would fail at the stage of going from loci to specific sequence, the second would fail to make any meaningful model, and the third would fail to hold up to its claim of being a *potential therapeutic target*. Both the scientific journals and the public would only be able to feign excitement for headlines reading, "Researchers fail to validate gene as meaningful in disease" for so long. The field at large also had a little soul searching to do as it came up short of the intended goals of improving patients' lives again and again. So, at this point, there was a crossroads: either the field could change course to de-emphasize the loci hits of GWAS since the workflow of GWAS-to-treatment was clearly not all they had hoped for or they could simply move the goal post closer so the bar for success was no longer clinical relevance, but mathematical correlation. I suspect the reader can guess which choice they selected.

Ultimate success in genetic research (and in all biomedical science) is limited to discovering something that can either help with diagnosis, prognosis, or guide treatment. Genetic information might provide the basis for giving a patient a diagnosis. For example, someone with a clinical immune deficiency syndrome consistent with Job's Syndrome would be formally diagnosed after verification that their *STAT3* gene had a mutation. Alternatively, genetic information might help give patients a better sense of what was to come as might be the case in some cancers where certain mutations predicted better survival than others. Finally, genetic tests might guide treatment with examples again coming from the cancer world demonstrating that different mutations might suggest different chemotherapy cocktails would work better, even in otherwise similar cancers. This is not to say that every discovery needs to have an immediate impact, or that knowledge alone does not have inherent value. However, the guiding principal of biomedical research is to improve health and treat illness.

Having failed to achieve many, if any, examples of these stories, the GWAS acolytes redefined success into three new categories.

New goal post success category 1: Our GWAS is valid if it tells us something we already knew.

Suppose you tasted a restaurant's chicken cacciatore and determined it was too salty. You then looked at their ingredients and noticed that during prep, the cook added too much salt. You confirmed that adding less salt could fix the overall problem. Lastly, you performed direct tests to demonstrate a causal link between adding too much salt and the dish tasting too salty. If you were curious whether the recipe was the cause of the over-salting, what might you do? You would probably look at the specific recipe and look at the part related to salt to see if there was an error, right?

Suppose instead you just look at a recipe loci analysis like a GWAS would do; it tells you that the loci associated for overly-salted chicken cacciatore contains the recipe for chicken cacciatore (among many other recipes). How does this information help? Does knowing that the salty chicken cacciatore is linked with the area of the cookbook that contains the recipe for chicken cacciatore verify for you that the recipe was the cause of too much salt? No, for all you know, the misprint was "1 teaspoon salt" spelled correctly versus "1 teespoon salt." There is no assurance that the loci difference was your specific recipe of interest, and even if it is in the recipe you are interested in, it does not tell you that the difference is meaningful. Plus, you already know that the problem was too much salt.

Say one researcher had data from cell culture studies, mouse models, and an investigation of proteins in humans indicating that the vitamin D receptor played a crucial role in asthma. Next, a second research group runs a GWAS and it reveals an association between the loci containing the vitamin D receptor and asthma. Would the second study be considered a successful outcome? You already knew that the vitamin D receptor was important before you ran the GWAS. And if you wanted to check for the gene linked to your protein of interest, you could just do so directly. Yet, even if you found a sequence difference between asthmatics and controls, that difference does not prove there is a functional defect. People used to spend their entire PhD training time— usually 3-6 years—trying to link a sequence change to a change in protein function, structure, and/or expression. But now, the finding that a GWAS *hits* in an area of the genome that contains a gene you already know matters is considered validating in and of itself.

A shining example of this was published in 2017 by lead author Peter Vissher in *The American Journal of Human Genetics* entitled "10 years of GWAS: Biology, Function, and Translation". The paper was meant to be an outline of all the great things GWAS had provided medical science. The culmination of this argument was in a figure that claimed to summarize the times where a GWAS hit *directly* led to a new drug for a disease. The table is partially reproduced below:

Disease	GWAS Gene *Hit*	Known/Possible Drug
Type 2 Diabetes	*SLC30A8/KCNJ11*	ZnT-8 antagonists/Glyburide
Rheumatoid disease - Rheumatoid arthritis - Ank. Spondylitis - Psoriasis - Psoriatic arthritis	- *PAD14/IL6R* - *TNFR1/PTGER4/TYK2* - *IL23A* - *IL12B*	- Anti-IL6 - TNF-inhibitors - NSAIDs - Fostamatinib - Anti-IL23
Osteoporosis	*RANKL/ESR1*	- Denosumab - Raloxifene - Hormone replacement
Schizophrenia	*DRD2*	Anti-psychotics
High cholesterol	*HMGCR*	Pravastatin

The claim for Type II diabetes made in this table is that the GWAS signal in the loci containing *KCNJ11* is valuable because the drug glyburide targets the pathway the protein encoded by *KCNJ11* is involved in. But glyburide was approved five years before the GWAS timeline began, not to mention how long before that it was in development. So, the idea that the GWAS was helpful to the fight against diabetes is flatly incorrect. Rather, it is the success against diabetes that is being used to justify running the GWAS. Going through the other *examples of GWAS success since 2007* reveals TNF-inhibitors were first approved in 2003 and were in development for a decade prior. NSAIDs (the class of drugs that ibuprofen and naproxen belong to) came on the market in 1964, 13 years before DNA sequencing of any kind was invented. Antipsychotics for Schizophrenia were used even before that.

I had thought that this figure would be the sorriest example of a scientific celebration one could imagine. That was until 2023, when Vissher, along with Abdel Abdellaoui, provided an update on "15 years of GWAS discovery: realizing the promise" in the same journal. Amazingly, the list of discoveries that had (just five years earlier) been framed as the shining examples of GWAS success had been deleted rather than updated. Apparently, these *exemplar stories* were to be forgotten. A new claim took their place, in which it states that GWAS had been a valuable tool in looking for pathways involved in the response to SARS-CoV2 in the COVID-19 pandemic.

The argument they made was that *one* of the numerous (I mean, wildly numerous) GWAS studies run on COVID-19 patients came back with an association at the loci that contains the gene ACE2. The thing is, science already knew that the ACE pathway was involved in coronavirus responses. In fact, early in the pandemic, when clinicians were desperate for anything that might help, a trial of ACE-inhibitors was conducted to try to treat severe cases. The treatment did not provide benefit, but the idea had a firm basis in biologic understanding and was certainly worth a try. The doctors knew ACE *might* be important years before one GWAS had a hit on the loci near ACE. To be clear, I am 100% in favor of having run some of the GWAS tests for COVID-19—you can't totally know what you might find. But the important distinction is that the GWAS is only step one of a long process; loci need to be sequenced and models need to be built. It might end up telling you things you already know; it might offer something new. But the loci-association alone has no inherent value and can't be framed as some shining example of why we need to keep giving money to geneticists to conduct the association-only work.

The authors also point to an example of gene editing treatments used in a rare family-style genetic disorder; again, the disorder was not identified using GWAS; it was identified using actual sequencing like the monogenic disorder examples in Chapter 4.

Although not mentioned by the review of "greatest hits", there is one example of a discovery that was initiated by GWAS and carried all the way through to a therapy. The gene *BCL11A* is one that normally helps to turn off fetal hemoglobin production as a newborn grows into a toddler. As most people might guess, a fetus needs a slightly different oxygen carrying system while in the womb compared to after birth. So *BCL11A* helps assure that transition. However, if a patient has sickle cell disease, it would be better for the patient to have as much fetal hemoglobin as possible since their "adult" hemoglobin has the sickle defect. None of the functions of *BCL11A* were known until a team led by Dr. Erica Esrick of Boston Children's Hospital ran a GWAS on sickle cell severity. In that study, they found *BCL11A* was associated despite no prior reports linking the gene to the disease. They modeled the effects in cell culture and mice to discover why quirks in the gene might help those with sickle cell. In brief, a version of *BCL11A* that was less efficient at turning off fetal hemoglobin is an advantage in the setting of sickle cell.

The team not only followed the science all the way through the needed functional assays, but they also even implemented a gene treatment trial with the idea that breaking this gene in patients with sickle cell might improve their symptoms by increasing the amount of fetal hemoglobin made. Not to be overly negative here, but if the healthcare system is going to go through all the work of putting a patient through gene therapy, they would be better served fixing the hemoglobin gene itself rather than breaking *BCL11A*. But, credit where credit is due - this gene is an example of a GWAS hit that actually led to a therapy and I would be remiss to ignore it.

Overall, the question for these *successes* is, what would the world look like if they had not happened? What if no one ran a GWAS for COVID-19, and no one found a one-time association with *ACE2* in one population? Would anything have been changed if the money given to those studies would have been allocated to something else?

Taking confidence from findings that are consistent with prior work is an essential part of valid and reproducible science. However, there is a major double standard for the population genetic field. If a GWAS finds a hit in a gene encoding a protein that is already known to be involved in a disease, then the GWAS is seen as validated because it is consistent with the known evidence. But if a GWAS hit is for a protein that has never been implicated before, this is not viewed as evidence the GWAS might be wrong. Instead, it is viewed that the gene must be involved in some unknown way despite the results being incompatible with every study performed previously.

However, to make matters worse, even if the gene/protein has never been implicated before, the population genetics can map their associations onto physiology in a way more akin to astrology than biology. Instead of landing on a gene we directly know to be important, the researcher will celebrate finding a GWAS hit in a gene that has a known function that could, hypothetically, be involved in the disease process. One example of this is from the *Journal of Strength and Conditioning Research* from September 2019. The authors, led by Craig Pickering, tried to find genes that were associated with faster sprinting times. They claimed to have found 12 that were of "suggestive significance." To explain why some of the genes they found fit with their hypothesis that these genes partially govern sprinting times, they stated the following:

> *Of the other genetic variants associated with sprint performance in this study, the potential mechanisms are varied. NFATC2 ... forms part of the calcineurin-NFAT pathway, which has been implicated in both cardiac and skeletal hypertrophy—factors which are positively correlated with sprinting performance. TERT ... encodes telomerase reverse transcriptase, which acts to maintain telomere end length, likely protecting against exercise-induced DNA damage. rs12401573 is located in the SEMA4A gene (encoding semaphorin 4A) and appears to play a role in immune function, potentially mediating the postexercise immune response. rs8093502 is located in the CBLN2 gene (encoding cerebellin 2 precursor), which plays a role in synapse formation. CBLN1-null mice are severely ataxic, walk with an irregular gait, and do not maintain their balance. rs62247016 is located in the CNTN4 gene (encodes contactin 4), which plays a role in the formation of axon connections in the developing nervous system.*

To me, the idea that sprinting would be influenced by genes involved in dealing with inflammation from damaged muscles and development of how skeletal muscle and nerves interact makes sense. The problem is that genes rarely have just one function. Typically, they are involved in numerous processes. For example, if I wanted to write an explanation linking the exact genes listed above to asthma, I could write:

> Of the other genetic variants associated with asthma in this study, the potential mechanisms are varied. NFATC2 and rs12401573 located in the SEMA4A gene (encoding semaphorin 4A) are both involved in T-cell function, suggesting a role for immune cell infiltration into the lung tissue. TERT encodes telomerase reverse transcriptase, which

acts to maintain telomere end length, likely protecting against the pollution induced DNA damage associated with respiratory diseases. rs62247016 is located in the CNTN4 gene (encodes contactin 4), which plays a role in the production of antibodies. rs8093502 is located in the CBLN2 gene (encoding cerebellin 2 precursor) and is associated with a monogenic syndromic disorder which also includes respiratory symptoms.

Or what if I wanted to write these same genes for osteoporosis? I could write that NFATC2 is involved in calcium metabolism. SEMA4A being linked to T-cells suggests a role for immune breakdown of the bones. TERT may imply age related cellular decline is involved, while CNTN4 and CBLN2 are involved with neurologic innervation that may provide early signals to spinal compression symptoms.

For that matter, I could even use these same genes if I had run a GWAS for loci associated with believing GWAS is a valuable tool. I could write that NFATC2, SEMA4A, and CNTN4 are involved in neurologic function and neurodevelopment. The involvement of TERT may imply an inability to process age related cellular decline in the brain. And CBLN2 is associated with a monogenic syndromic disorder which has been linked to a reduction in cognitive function.

In theory, this practice of finding a gene, combing through its many functions, and then reporting why it might make sense could be done in a biologically valid way. The argument the authors are trying to make is that the gene, NFATC2, is so altered that it creates a measurable difference in sprinting times through changes in the *calcineurin-NFAT* pathway. Given that NFATC2 influences calcium metabolism, T-cell function, and neurodevelopment, if there were a genuine effort to connect the research findings to biology, researchers could investigate whether sprinters exhibit variations in blood calcium levels, bone density, T-cell function, infection rates, and whether they perform better on standardized tests or have lower rates of Alzheimer's as they age. If the claim is that a set of genes are so different that they meaningfully contribute to anything from asthma to cognition and 40-yard dash time, then the defects in those genes should be detectable in *at least some* of the other functions for which the genes are involved. Flour is used as an ingredient in an incredibly varied number of dishes. So, if you were blaming a batch of rotten flour for why a cake tasted bad, then you could check and see if all the other dishes you made with that flour have a similar problem.

One final version of the *post hoc ergo propter hoc* fallacy made by the population genetics is to couple their findings to the monogenic diseases discussed in Chapter 4 by claiming finding a GWAS hit the common version of a rare disorder is validating. If you recall the example of *STAT3* mutations leading to hyper IgE syndrome and eczema, the *STAT3* activity in these patients was about 75% less active than healthy people. So, one might guess that if a 75% loss of *STAT3* activity all but assures eczema, then maybe a 25% loss would increase the risk of eczema? Or perhaps a 2% loss might cause really mild eczema? This thinking is flawed for three reasons. First is the observation for a threshold effect in most disease states. Let us look at type 1 diabetes mellitus (T1DM) as an example. In this disease, the body's immune system attacks the beta cells of the pancreas. When enough cells have been killed, the ability to make enough of the hormone, insulin, is lost. Insulin is needed to regulate blood sugar levels and allow cells to take in the glucose they need to conduct normal metabolism. So, when insulin levels fall low enough, glucose levels rise, and cells cannot conduct normal metabolism. Eventually, a life-threatening condition known as diabetic ketoacidosis, or DKA, will develop. Without treatment with injected insulin, patients will die. A 2008 review in *PLoS One* by David Klinke found that patients with T1DM will develop DKA symptoms after about 40% of the beta cells have been killed by the autoimmune attacks. The disease may progress to kill off all the beta cells, but by that time, the patient will be sustained by insulin therapy. So, if you need to kill off 40% of the beta cells to generate harm, does having a loss of 20% of the beta cells mean you will have a moderate form of DKA? Would only losing 2% cause you to have a mild form?

The answer is, "It sounds like it might, but it doesn't." As T1DM begins to kill off the beta cells, the remaining cells are capable of increasing insulin production to make up for the loss. If 2% of the beta cells die, the remaining 98% will work 2% harder to make up for the difference. As the percent of cells killed climbs, however, the body reaches a point where the remaining cells can no longer compensate, and the patient's more emergent symptoms begin. So, if a 75% loss of STAT3 assures severe eczema, does a 2% loss assure mild eczema? It might, but this is where the types of follow up studies performed by Dr. Hsu are so valuable. In cases where the patient has milder symptoms due to a milder mutation, the term is called *hypomorphic disease*. It can be very challenging to validate a suspected hypomorphic mutation as the cause of a patient's milder presentation of disease because of the second limitation in this thinking: redundancy.

Proteins or genes rarely have only one function in the body. Furthermore, it is rare for one gene to be responsible for an entire function alone. Hemoglobin is one of a handful of exceptions given that no other molecule in the body can truly exchange carbon dioxide and oxygen. More often, however, multiple pathways run in parallel to regulate various biologic processes. If a gene is completely wiped out, other gene products might increase production to compensate. Some of our own work is among the numerous examples from mice exist in which a gene is deleted that would normally assist in fighting infection (for example, the immune signaling molecule IL-22), but the mouse does not seem to have worse illness because a different immune signaling molecule (such as IL-17) increases in activity to cover the gaps.

The final flaw in the claim that if big changes in a gene can create big harms, then small changes must be able to create mild harms, is that it ignores the need for an external trigger. Failure to properly assess the environment will be detailed in later chapters, but for now, look to the example of alpha-1-antitrypsin deficiency. In this disease, a single-gene mutation in alpha-1-antitrypsin causes an imbalance in the protein metabolizing enzymes in the lung, which ultimately results in the destruction of the proteins needed to keep the lung elastic. As the tissue is destroyed, a version of chronic obstructive pulmonary disease, or COPD, is all but assured. However, does this mean that someone with an extremely mild change in their alpha-1-antitrypsin gene would be assured extremely mild COPD? What would such a mild version even look like or feel like from the patient's perspective? Hypomorphic changes in alpha-1-antitrypsin may influence one's risk for COPD, but since they cannot cause the disease on their own, one would need cigarette smoke to make such hypomorphic changes relevant. So here, running a GWAS on COPD and finding a locus that contains alpha-1-antitrypsin wouldn't add any new information to the genetics of COPD and would not offer any guidance beyond advising people not to smoke. But a GWAS on COPD that landed on alpha-1-antirypsin would be celebrated in the genomics world as a shining example of success.

One of the favorite citations among those that want to claim that population genetics has improved medical care comes from Matthew Nelson and colleagues in *Nature Genetics*. The paper evaluated drugs that had been approved by regulators versus those that were not approved. The authors found that drugs that were approved were more likely to have a genetic signal in the literature than non-approved drugs. Genetic signature in this context was defined as either a GWAS hit or a monogenic target. To the author's credit, they separated out proven genetic targets (like *STAT3* or *BCR-ABL*) from those with only GWAS hits. They then reported that approved drugs were 1.8 times more likely to have a GWAS hit somewhere in the literature in a gene related to the disease target than non-approved drugs.

Despite how validating the population geneticists see this paper to be, it has two major flaws. First, the authors only began their search with drugs that reached Phase I clinical trials in humans. Phase I is the level to assess safety but could not be legally conducted without biologic proof the drug has effects in either animals or cell culture models. Thus, the authors are ignoring the years' worth of work that is put into drug development prior to Phase I trials. The second flaw is that, even accounting for Phase I enrollment, the drugs assessed had meaningfully different focuses. The authors' own data show that drugs for musculoskeletal diseases were more likely to have a genetic signal than those for autoimmunity or eye disease. This might impact the results because drugs are approved based on their disease of interest (i.e. drugs for eczema are vetted by FDA staff that are trained in dermatology or allergy, rather than cancer). So, if it were easier to get a drug approved for skeletal disease than eye disease, and there are differences in the GWAS hits between skeletal disease and eye disease, that alone could make it appear that approved drugs were more likely to have GWAS hits. Furthermore, it costs a lot of money to get a drug approved. Thus, differences in the number of large pharmaceutical companies investing in different fields would influence the results since common diseases would attract more pharma investment and more scientific attention for running the GWAS in the first place.

This paper, and others like it, are held up to imply that GWAS hits are valuable because they might guide new drug development but only show data for drugs we already know work well enough to test in humans. If you really wanted to test this hypothesis you would need to take GWAS hits and see how many could become drug targets from scratch. However, finding a gene association and then trying to make a drug out of it was exactly what the failed "candidate gene" era was predicated on.

New goal post success category 2: Our GWAS is valid if the loci associated contains a gene expressed in our organ of interest.

The favorite approach for any GWAS related to a neurologic or behavioral trait is to state that the GWAS found a loci association that contains a gene *expressed in the brain*. Every cell in your body (except red blood cells and platelets) contains your entire genome end to end. We will discuss the epigenetic mechanisms that govern how your liver cells know to stay liver cells and not become bone in a later chapter. For now, just know that cells only express a select number of genes related to their specific function in life. The cells of the pancreas, for example, are the ones that use the insulin production genes to make insulin and regulate your blood sugar. So, while any gene in your body could be expressed in any cell of any organs, epigenetics prevents that from happening.

So, how many genes are expressed in a given organ? This would be important to know if you were claiming that finding a gene expressed in the brain was important for brain function, correct? If only 1% of genes were expressed in the kidney, and you found a gene associated with a kidney disease that was also expressed in the kidney, that would be supportive of (but not confirmation of) thinking that the gene was important in kidney disease. But if 99% of genes were expressed in the kidney and you found a gene associated with kidney disease that was expressed in the kidney, that would be far less impressive.

In 2017, Drs. Negi and Guda from Nebraska University demonstrated that at minimum, 84% of all genes are expressed somewhere in the brain. So, if you run a GWAS for Autism and find any association at all, there is no less than an 84% chance the gene you find will be expressed in the brain somewhere. If you find 10 loci associated, the chances that *none* of those genes would be among the 84% in the brain would be 0.00000001%, or about one in 10 million. The other problem is that since you are studying a disease, if you found the gene was not normally expressed in the brain, that wouldn't exclude it from consideration, either. For example, maybe you found a gene associated with Alzheimer's, but that gene isn't normally expressed in the brain. Well, what if in patients with Alzheimer's, that gene is inappropriately expressed in the brain? The only way to know that would be to progress into the modeling and expression testing mentioned before. Overall, no one should be impressed by claims that a study in mental health has value because it flagged a gene that is expressed in the brain or that a study in asthma found a gene expressed in the lungs.

New goal post success category 3: Our GWAS is valid if the loci associated with our trait/disease are also associated with related diseases or traits.

What if I told you there was a gene for loving football? What if I told you there was a gene for loving chicken wings? Oh, and what if I also told you there is a gene for drinking beer? Do you think those genes would be more likely to be all found in the same person or less likely? If I told you the gene for drinking beer often hangs out with the genes for eating chicken wings and watching football, would you be surprised? Of course not, because those behaviors are so linked you would expect them to show up together in any analysis. What if I also told you that the genes for football, beer drinking, and chicken wings *did not* associate with the genes for watching field hockey while driving a Geo Metro and drinking kale smoothies? Would that shock you? I'm guessing still no.

But now, let's say I tell you that the genes for depression are also associated with the genes for having been abused. Also, what if I tell you that the genes for graduating college are not linked to genes for poverty? What if the genes for anxiety and depression are linked to the genes for despair? And the genes for despair are linked to the genes for drug use? Are any of these shockers? Again, I'm assuming no because we would all expect that depression would be found common in people with poverty or drug use. It could just as easily (and honestly more likely) to be that the outside factor causing these emotions causes them in sets. Despair is linked with depression because those emotions are linked. Drug use is not linked with graduating college because drug use may impair people's chances of success.

In 2020, in the U.K.-based journal, *Nature Communications*, Xu and colleagues claimed that they discovered 18 loci that governed smoking behavior. What's the number one piece of evidence that these genes really did cause smoking? That the same loci were also associated with traits like age of starting smoking, how often they tried to quit, whether their mother smoked, lung cancer, and overall health. These smoking genes were also linked to what time of day you have your first cigarette and how well your lungs functioned (meaning the more you smoked, the earlier the time you first lit up and the worse your lungs worked). Might one be able to come up with an alternate reason why smoking would be linked to worse lung function than thinking they share the same genetic profiles?

A 2022 paper in *Translational Psych* made the discovery that genes associated with happiness were also associated with satisfaction in family, health, and finances, but were negatively linked with depression and loneliness. Contain your shock that factors linked to happiness would also be linked to satisfaction but inversely linked to depression.

One of the more laughable connections was made by a group based in Sweden and published in 2014's *Psychological Science*. The authors claimed that practicing music had zero impact on music ability. Just in case you think I might be overstating their claims, the title of their paper was "Practice does not make perfect: no causal effect of music practice on music ability." Their logic? Since the genes correlated with practicing music are just as strongly associated as the genes for being good at music, they conclude that rather than practicing music leading to better abilities, musical practice *and* ability are *both* determined by genes. That's right, rather than practicing causing improvement, genes cause both! I presume these authors do not bother sending their children to school since one would assume the genes for studying language or math would also be linked to the genes for being good at those subjects. Just leave the kid at home and they would naturally start practicing math and eventually be good at it without the need to pay for any schooling.

My mom got me a violin for Christmas one year, but I never got around to trying to learn how to play it. I did, however, make ample use of the Nintendo she once got me. So, I suppose the fact that I can't play violin but could excel at Mario Kart is that I lacked the genes for talent and practice at music but have both the genes for playing and excelling at deploying the Blue Shell. Another obvious follow up would be to wonder whether the gene for playing and practicing the harp appeared 4000 years prior to the gene for playing and practicing the ukulele. Or is there a gene for string instruments and a separate one for percussion? Was there a spontaneous mutation in the 1980s that created the genes for playing and being good at synthesizer?

These three new goal posts were constructed so that geneticists never had to admit that the databases have not yielded any meaningful insights into biology nor any actual therapies. In fact, many of the current population geneticists today never set foot in a lab. They never collect a patient's sample nor run a single model that the founders (and successful scientists) of genetics used to run routinely. Instead, they sit combing through databases and write papers that are in effect:

— We found an association between loci X and our disease of interest.

— That loci contains a gene that might make sense for our disease based on work other people have already done, so we will assume our loci association is due to that gene and not any of the dozen other possibilities.

- We will not run expression analysis, will not test to see if the gene's pathway is abnormal, and won't perform a single functional study of any kind.

- We also won't be running full sequencing of the gene we think is important; we will just assume it has a mutation that harms protein function/expression because the p-value is above the dotted line on our plot that we pretend looks like the Manhattan skyline.

Since they know that the association only approach cannot suggest cause-and-effect, they will add one line in the discussion saying they "can't make causal claims." However, they will then immediately start using the term "contributes to" in references to the gene of interest. This is because they apparently believe the English language has no meaning and the word *contributes* does not imply a causal link despite its definition being "help to cause or bring about." As we will see in the next chapters, if you are going to go fishing for associations, the most important protection against finding a spurious link is having a negative control.

CHAPTER 9
Never Model What You Can Measure

BOTTOM LINE UP FRONT: *If you can experimentally determine what normal is, you should do so instead of just guessing at it.*

Having grown up in Colorado, college afforded me the opportunity to take a brief road trip to Las Vegas with my friends. On one such trip, we were all playing the roulette wheel. Most probably know this, but the roulette table has a big wheel with 38 total slots on it. In the U.S., two of those slots are green and numbered 0 and 00; the remaining are numbered 1-36 with half being black and the other half red. For each roll, a dealer spins the wheel one direction and then spins a ball in a track just over it in the opposite direction. Players then place their bets on which number or which color they think the ball will ultimately land on with no limit on how many numbers they can bet on. As the ball's momentum wanes, it will slide down into the wheel and bounce until it finally comes to a stop at only one number. If the ball lands on the color or number you put your money on, you win; if it lands somewhere you did not bet on, you lose.

One time, the dealer started the spin and a table full of people started placing their bets. By the close of bets, I would say about 75% of the numbers had at least one person who had money on that space. In addition, both black and red had betters, so any number between 1-36 would yield a winner. When the spin finished, the ball was on green 00, meaning that the entire table lost everything that had been wagered. Everyone let out a collective groan, but one lady at the table was furious. She screamed at the dealer for taking everyone's money and shouted aloud "What are the odds of that?!" I looked at her and said bluntly, "1 in 38." She walked off in a huff while the dealer chuckled. I did not think it was very funny at the time, since some of the money the dealer was scooping up used to be mine. But today, of course, the joke was worth the few bucks I lost. While the table seemed to have a laugh at the woman for her suggested misunderstanding of probability, in fairness, that was not what she was really asking about. I never knew her name but for the purposes of this story, let's just call her Kathryn.

Kathryn was not asking about what the probability was of hitting double zero; she was asking what the chances were that the table would have such bad luck. In a moment of anger, Kathryn was insinuating that the table had lost because the casino had cheated in a way that the loss was not due to bad luck but foul play. She didn't care about the chance that 00 would come up; she cared what the chance of getting 00 was by *chance alone*. Kathryn was debating the probability that the wheel hit 00 due to something other than chance, not the probability of landing on 00.

Science has decided that a 5% risk that something happened by chance alone is acceptable when drawing conclusions from the results. Whereas anything with a greater than five but less than 10% probability of happening by chance can be viewed as suspect (called a statistical trend) and any event with greater than 10% probability of occurring by chance is considered *insignificant*. If this sounds somewhat arbitrary, that is because it is. Statisticians routinely argue over how p-values could be improved or whether they should be used at all. Having a sharp cut off means that, for example, a drug that appeared to significantly improve symptoms of diabetes compared to the placebo would be approved if the probability of improvement occurring by chance was 4.99%, but the same drug might be denied approval if the probability was 5.01%. This type of outcome isn't common, but the point is that it exists and is used by scientists to determine whether someone's claims (like in this example, the claim that the drug helps improve diabetes) are valid or not. Officially, this is termed as either accepting or rejecting the null hypothesis. The null hypothesis is kind of the opposite of whatever you think is going on. So, if you think that your drug will improve symptoms better than a placebo, the null is that any difference between the results of those on the drug and those on the placebo have occurred by chance alone. A p-value of less than 0.05 means we reject the null hypothesis.

So, let's follow up on Kathryn's logic for a moment. Kathryn was implying that because so many of the numbers were covered, someone should have won. Let's pretend she stormed off to demand to see the manager or even the Las Vegas gaming commission that oversees casinos to assure fair play. Say the gaming commission decided to investigate. Would the fact that no one won on that one spin be enough to take action? If there are 38 numbers with an even chance of having the ball land on them, you would expect one out of every 38 spins to land on 00. Say they spin the wheel 38 times and find that 00 came up twice. Is that a sign that the wheel is rigged? Math says no (and so would the commission). One test for whether an observed phenomenon happens more frequently than expected is called the *Chi-Squared* test.

If you expected one 00 out of 38 but had two out of 38, the Chi-Squared test would calculate the p-value of 0.47, meaning there is a 47% chance that such a finding happened by chance alone. Scientists would consider this to be *statistically insignificant* and the functional conclusion would be that the wheel is fair, and Kathryn was out of line. However, here, we have a limit of how the p-value is used in hypothesis testing versus how such language would be translated to more common speech. The statement, "The probability that this happened by chance alone is 47%" could be just as easily presented in a different way, such as, "It is more likely than not that the wheel's results were caused by something *other* than chance." In normal speech, the line at which we make judgments is closer to 50% than the 5% used by scientists. Thus, the p-value here does not provide much help because, other than Kathryn, no one would really think a wheel returning two 00s instead of one during a 38-spin test would truly be indicative of the cheating. So, you could try to smear the casino by saying "there is a 53% chance their roulette wheel results are not due to chance," but no scientist would agree such phrasing would be valid.

To reach a *statistically significant* difference in 00 hits on only 38 spins, you would need to get 00 six times. So, if the gaming commission convicted casino owners for cheating based on p-values alone, they would convict a casino where the wheel hit 00 on six of 38 spins but exonerate a casino with a wheel that hit five 00s on 38 spins. Already, you can see some limitation on using p-values as the sole metric of validity. The gaming commission would never convict on that alone, but it might be something worthy of further investigation (i.e., they would not decide that five 00s out of 38 would be something to ignore and they might search the wheel for signs that it had been doctored or modified).

But now, let's say that the commission decided that only spinning the wheel 38 times isn't a great sample size. They really want to see if the wheel has a smaller bias in it, so they spin it 380 times. In that case, we would expect the ball to land on 00 ten times. So, the new question is: How many times would the ball need to land on 00 out of 380 for it to be a significant difference from the expectations? Remember, for only 38 spins, it took a 6-fold increase in 00 hits to indicate statistical significance. So, what would it be for 380? The number is 19 (significance could also be seen with only one 00 out of 380 because significance goes both ways on Chi-Squared tests). So, if you expected 10 out of 380 but got either one or fewer or 19 or more 00s out of 380, that would be significant. With the larger sample size, our ability to detect the wheel being potentially biased in favor of 00 went from needing a 6-fold difference to only a 1.9-fold difference.

At 38,000 spins you expect 00 to land 1000 times. Here, significantly more 00s are defined by seeing 1,064 hits, so we are down to a 1.06-fold difference. At 380,000 spins we would expect to see the ball land on 00 10,000 times but would call it *significantly more* at 10,196. This 1.02-fold variation from the expected could also be phrased as saying that if 00 came up 2% more than expected, we would say it did so by something other than chance. At 3.8 million spins, mathematical *significance* could be reached at only a 1.007-fold difference from the expected number. To be fair, swinging the odds of a casino game a fraction of a percent might be the difference between the house having the advantage or the player. So, it is not the intent of this analogy to suggest such small differences could be ignored. The point is that p-values alone cannot provide definitive conclusions about a given observation.

The second take home point is that larger and larger sample sizes mean that smaller and smaller differences can reach statistical significance. Those differences might not be something to ignore, but when you get to small enough variations, you need to know how much *noise* is in the system. Seeing a 0.7% variation in 00 hits might be outside of what mathematicians would expect, but it does not answer whether it is outside what casinos would expect. Suppose you manufactured 1,000,000 roulette wheels in perfect compliance with manufacturing regulations. If you spun each of them 380,000 times, do you think that *all of them* would land on 00 exactly 10,000 times? Wouldn't you expect there to be some range of 00 hits across that many different wheels?

What the gaming commission would really be looking for is whether the number of times the wheel landed on 00 was outside the range of normal considering the innate variability of roulette wheels, not whether it was outside the definition of *chance*. If many roulette wheels were tested,

380,000 spins each, and the range was 9,800 to 10,200, some of those values would be statistically significant deviations from the expected and some would be insignificant. But it would also mean that while a wheel that hit 00 on 10,197 times out of 380,000 spins would be *uncommon*, it would not indicate that the result was *unexpected*. Conversely, if testing thousands of roulette wheels 380,000 times revealed the variation between them to be only between 9,900 and 10,100, then any wheel that landed on 00 between 10,101 and 10,195 times would be considered unexpected, even while not considered statistically significant. This is the difference between modeling *by chance alone* and measuring *within expected variation*. The second one—knowing whether this falls outside the norms—is what Kathryn was really hinting at. If a result falls within the expected norms, even if on the extremes, then it is seen as a valid result.

However, the most important realization around interpreting p-values is one that lots of scientists get wrong all the time. Remember, I said that a p-value of less than 0.05 between a drug and placebo means we reject the null hypothesis. Notice I did not say that it means there is a 95% chance the drug is better than the placebo. It might effectively play out that way with drug approval agencies, but in statistics, that is not how the p-value should be read. To be fair, for a drug trial—as long as the trial was well randomized—the only difference between the group getting the drug and the group getting the placebo should be whether they had an active treatment or the *pretend* one (like a real medicine tablet versus a sugar pill).

But in the analogy of the roulette wheels, there are numerous other reasons a wheel spun 380,000 times might have landed on 10,200 00s that do not involve cheating. Maybe some have microscopically small dents in the ball, or extremely small differences in the depth of the wells, or maybe the casinos have small fluctuations in air circulation that makes a trivial difference on any one spin but can add up over a huge number of spins.

Again, many scientists get this wrong but only the modern population geneticists seem to have built their entire field upon this misinterpretation of what a p-value means. GWAS papers are completely enamored by statistical significance on the Manhattan plot. So much so that the obsession for larger and larger sample sizes has become akin to demanding the casino test with 3.8 million spins. The ability to distinguish between *significant difference* and *non-significant differences* increases with sample size but the ability to distinguish between *relevant differences* and *irrelevant differences* declines sharply. When the differences are so subtle as to require millions of evaluations to identify, then it becomes essential that statisticians calculate whether the results are within the range of expectations based on previous observations and not just whether they are outside math of chance. Every other field in biomedicine has used negative controls to help prevent misreading the p-values as an indication that *you are right* instead of *the null is wrong*. The next chapter will detail how population geneticists' refusal to use a negative control to establish the range of expected observations has turned their field into a giant Kathryn.

CHAPTER 10
Positive Conclusions Require Negative Controls

BOTTOM LINE UP FRONT: *Failure to use a negative control has led population geneticists to believe that all noise is signal.*

I know this next section will become dated as the world moves towards electric automobiles but try to imagine the sounds from inside of a racing car (with a combustion engine)—let's say a well-tuned 1969 Ford Mustang with metallic gray paint and black racing stripes. Imagine the sounds as the car takes off and accelerates down the road. If someone played that sound for you, do you think you could make a rough guess at how fast the car was going? Maybe you couldn't accurately guess when it hit 80 miles per hour, but you could easily guess if the speed was increasing and maybe even hear the gears shifting. If someone played two sounds from two different Mustangs racing, do you think you could guess which was moving faster? You probably could, based on how loud the engines roared.

But now, I want you to think about the sound of the engine idling. Anyone who has ever stood beside a car knows that once the engine is ignited, the car makes noise. If you were to measure the decibels coming off an idling car, they would be statistically significantly different from the level of noise made by a car that had not been started. So, the question is, when you hear the sound of a car idling, how fast do you perceive it to be traveling?

If you accurately identify that the car isn't moving, then you understand the concept of a negative control (defined as a test in which a negative result is expected). Most of the time you hear about a negative control, it is in the context of a clinical trial. Researchers will enroll a bunch of people, give about half of them a real drug, and give the other half a placebo. The idea here is that we know that just getting a sugar pill will have some benefit for people, so we want to see that the drug can be even better than that. We don't really care if the drug is only better than truly nothing at all. In fact, the placebo group in most clinical trials gets *significantly better than baseline*, meaning that sugar pills can make someone mathematically better than if they had received nothing at all. By a *chance alone* metric, sugar pills can improve just about every disease they have ever been tested against, up to and including a *significant* reduction in deaths from cancers or heart attacks. Just being better than *nothing* isn't the bar that medicine sets.

It is important that the placebo look exactly like the real drug, otherwise people will know it is fake and its effect won't be as important. One humorous example of this from the world of eczema is in relation to bleach baths. One research group suggested that putting a small amount of bleach into a bathtub could improve eczema symptoms versus giving people nothing at all. Someone wanted to repeat the study using a placebo. So, they made a new study and gave half the people a jar of bleach and half a jar of water and told the patients to pour the contents into the bathtub. Do you think you might be able to tell the difference between a cup of bleach and a cup of water? Well, the patients could, too, and the placebo group saw no benefit at all because they knew they were getting just water. So, the study was invalid because even though the researchers tried to get a placebo control group, they used a poorly designed one.

Properly used scientific assay always have a negative control. These controls are meant to perform at least two different assessments if not both. First, they are meant to isolate the background noise in the detector, and second, they are meant to provide a measurement of how impactful the most powerful confounder is. The most well-known negative control is a placebo pill from a drug trial. For any disease, there will be some small variations from day to day, so having a placebo group that can be monitored over time allows researchers to evaluate the natural ebb and flow of the patients' symptoms. However, a placebo also isolates the mental aspects of drug response—called the placebo effect. People believing they are on a drug that will help can oftentimes generate benefit even without any biologic medication included in the pill. Having a placebo allows us to isolate the influence of positive thinking on the response to the treatment. For some diseases like headaches, the placebo effect can be very large and end up *curing* a lot of people of their headache symptoms with only a sugar pill. For other diseases, such as fatal cancers, the placebo effect will be very small (but not imperceptible).

Other examples of negative controls are found throughout basic biomedical science research. Some assays use antibodies to detect whether a certain substance is in the sample (such as those used to detect whether a person had any SARS-COV2 in their blood). One limit of antibody tests is that if the antibody binds incorrectly, or if something cross reacts with the antibody, it could provide a false positive. To get around this, researchers will use what is called an *isotype control*, which is just an antibody that is otherwise identical to the one used in the test except that the researcher knows it won't stick to the target in the normal way. So, the negative control in the COVID test might be a similar antibody that we know ahead of time does not bind to the virus. If something is wrong with the test, then both antibodies might bind and create a signal. In that case, the evaluator would know that the test was an error because it showed a positive signal even when it was not supposed to.

In another example, research might use viruses to alter organisms, such as putting a gene for *glowing in the dark* into a plant. In that case, the negative control is called *an empty vector*—a situation where the virus is present, but it does not have any glow-gene inserted. By using an empty vector, researchers can evaluate how much of an impact just exposing something to the virus is. In the case of the glowing plants, it seems a little excessive since the plant isn't going to glow from interacting with a virus alone. But in cases where someone might want to test gene therapy in a lab, such as that type discussed for bone marrow treatment of patients with deficiencies in *STAT3*, it makes sense to be sure that the reason your patients' cells appear to have immune activation is because of the gene you inserted and not just because they were exposed to a virus. RNA sequencing uses a sample where the primers are present but no input RNA is found, just to ensure there is no contamination. Microbiome research will perform an *air swab* where someone will literally wave a swab in the air then test for microbes. If the swab has lots of signal, then you might worry your *real* samples could be contaminated.

So, what might be a good negative control for population genetics? To answer this, first ask what the most notable confounder in the assay would be. If your assay is meant to identify when differences in gene pools create differences in disease rates between populations, then the confounder would be the reverse situation when differences in the populations create meaningless differences in the gene pools. This would be an example referred to as *reverse causation*. An easier example to understand is the following: Say a researcher tells you that they studied the habits of billionaires and noticed that "a significantly higher percentage of billionaires have personal chefs making their meals. Therefore, we conclude that if people want to become billionaires, they should hire a personal chef." Or how about claiming that since more people carry umbrellas in Seattle than Phoenix, and it rains more in Seattle than it does in Phoenix, then carrying umbrellas must cause it to rain? You spot the flaws here.

Obviously, no one would be surprised that being a billionaire and having a personal chef are mathematically linked. But the direction of that link is obviously that being a billionaire allows you to hire a personal chef, not that hiring a personal chef makes it more likely that you will become a billionaire. As it would thus relate to genetics, a negative control would be a situation where you expect the differences seen were not due to genetics.

Here is just one excellent example, collected from the U.K. Biobank by GWASbot again:

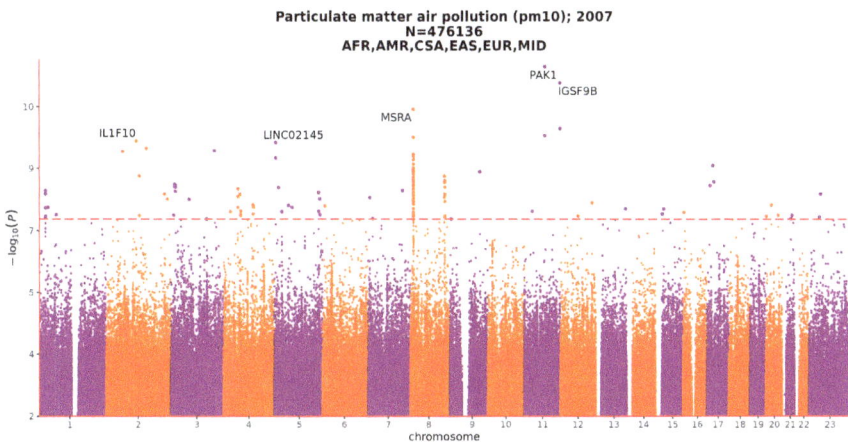

Particulate matter air pollution (pm10); 2007
N=476136
AFR,AMR,CSA,EAS,EUR,MID

This shows the genes *associated with* the pollution in the air around someone's house. Not asthma, not coughing, not lung function—just the pollution measured in the air. Don't be fooled by the small number annotations for the loci. Each dot is meant to suggest that that segment of the genome is linked to living near air pollution. Thus, if we viewed this as a sign of causal links, this GWAS would claim that nearly 50 loci somehow control the pollution in the air around you. Now, no one in their right mind would argue that genes *control* pollution in the air, so any of these associations are a result of what is called *population stratification* or *population structure*. These terms mean, effectively, that different groups of people have *sorted themselves* into gene pools. This is in contrast to when genes sort people into groups (for example, a mutation in the cystic fibrosis transmembrane receptor gene can sort people in those with clinical manifestations of cystic fibrosis and those without such symptoms). People tend to live in areas with other people that are more similar to them than others living in different areas. As a result, you will tend to meet, interact with, and then maybe mate with people in those areas that are slightly more similar to you than others. In the GWAS above, these findings suggest that people living in high pollution areas are slightly more genetically alike than compared to people living in low pollution areas. Anyone who understands the realities of how wealth is inherited will understand that this is a graph of how separate the U.K. gene pools might be if comparing rich white people in the county-side and poor white people in the industrial zones of big cities.

When Paige Harden was shown this example of *genes for living near air pollution* GWAS online, she attempted to defend it by pointing people to a paper by Xu and colleagues and discussing how GWAS can teach us about environmental factors for a disease. The paper by Xu, like many others on the subject, argues that GWAS can provide insights into the environment. It specifically focuses on the connection between GWAS hits that don't directly cause lung cancer but might influence behaviors, such as smoking. One 2016 example from *PLoS Genetics* by Gage and colleagues discussed that variations in the gene region *CHRNA5-A3-B4* are more common among people who have or currently smoke, compared to those who have never smoked. To their credit, the authors point out that these findings mean that one needs to be careful when a GWAS hit is found for a disease that has a clear environmental contributor. However, they only warn researchers not to assume the

GWAS hit is *directly* linked to the disease rather than being open to the possibility that the gene is spuriously associated.

But let's look at some other examples:

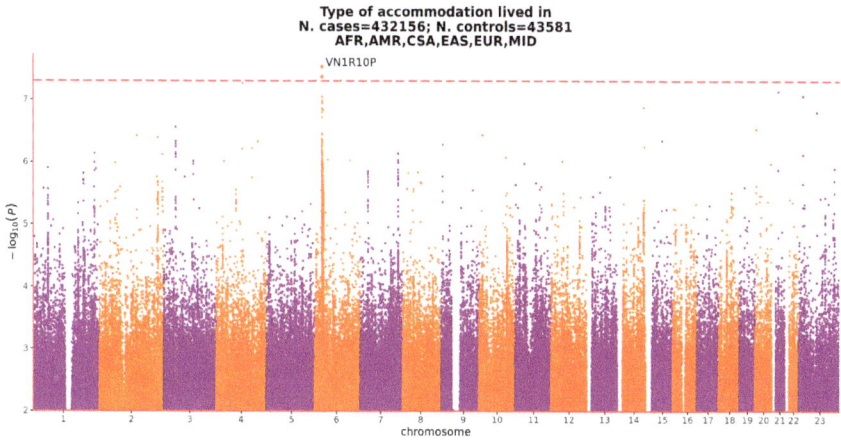

Type of accommodation lived in
N. cases=432156; N. controls=43581
AFR,AMR,CSA,EAS,EUR,MID

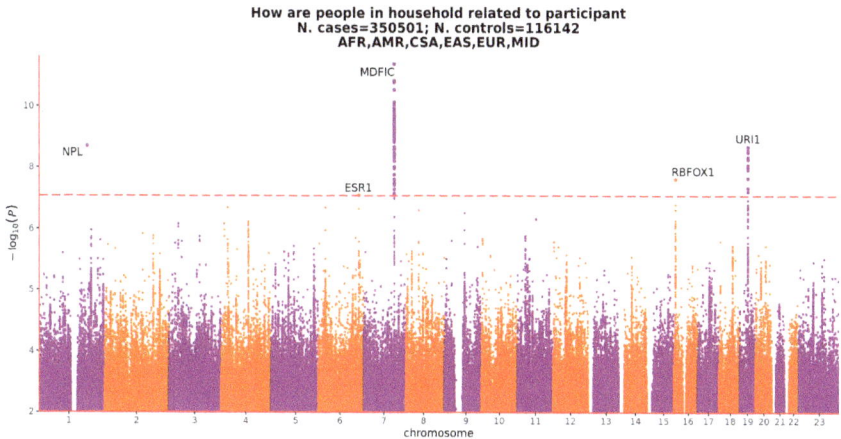

How are people in household related to participant
N. cases=350501; N. controls=116142
AFR,AMR,CSA,EAS,EUR,MID

If we interpreted these as causal based only on statistical significance, it would lead you to believe that the gene *VN1R10P*–a gene with no known function and that isn't thought to be made into RNA–is a significant determinant of whether you live in a house or apartment. Then, a completely different set of genes, led by *MDFIC*, contribute to how many people live in that house or apartment with you even though this gene isn't even human and is instead one of the many virus genes floating around the human genome.

The example below implies that *FOXP1* would *contribute to* whether your normal commute in England is via a bus, train, or car.

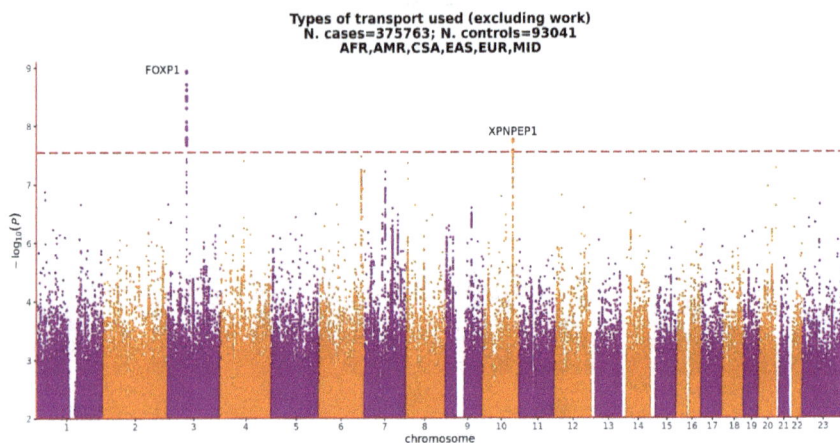

Types of transport used (excluding work)
N. cases=375763; N. controls=93041
AFR,AMR,CSA,EAS,EUR,MID

Perhaps the best example of a negative control is one that the geneticists have never run. Imagine that we collected genomes from Canadians and sorted them by their ability to speak French. What would you expect?

Remember, genes are not inherited randomly; you have to get half your chromosomes from Mom and half from Dad. All the ancestry sites are predicated on the fact that certain genetic sequences are more common in certain groups. So, one can guess how much of your genetic background comes from Iceland or China or elsewhere. These sites all overstate their findings (for example, the idea that you can tell that someone is 22.7% Irish is pseudoscience) but in broad strokes, it is correct that you can make guesses about the part of the world where most of someone's family tree would be rooted.

French-Canadians are concentrated in the province of Quebec. On the population level, people of French-Canadian descent will be more likely to marry and mate with someone who is also of French-Canadian descent for many reasons, not the least of which would be the probability that a French-Canadian may exclusively speak French. I'm sure someone out there has some wild story to tell, but most people do not even have so much as a one-night stand let alone a relationship with someone with whom they lack a shared language. So, if we ran our GWAS *for speaking French* in Canada, we would expect to find genetic links that are nothing more than a reflection of being from a family that was predominantly French-Canadian versus English-Canadian.

One reason that using the ability to read a specific language would make for such a great negative control is that the brain is not innately wired to understand the written word. During neurodevelopment, parts of the brain appear destined to assist in speaking or understanding spoken words but, if left to its own, there is no part of the brain that will naturally develop the ability to read or write. This is why every parent of elementary school-aged children has had moments where their kid could hold a full conversation about a story but might struggle to read that same story. Obviously, practice and good teaching overcome this programming, but there is no innate wiring for reading or writing. Genetics could play a role in the process of the brain attaining reading skills overall. But in terms of *specific* language, or *specific* religion, or *specific* sports team fandom, these are entirely dictated by environment. If you are never exposed to the French language, there is a 0% chance you will ever learn to speak French, and no one would spontaneously become Episcopalian without exposure. Rather than rely on an arbitrary p-value, researchers could use this to calculate how big of a gap in gene pools might be possible from just population stratification and mating preferences (these are officially termed *assortative mating*).

A study from *Survey Center on American Life* asked people about what types of behaviors in a potential partner would make it less likely the respondent would date the person. 77% said smoking was a dealbreaker. 69% said unemployment would prohibit dating. 68% said that people living in another state would make it less likely they would start a relationship, even in today's digital age. Other examples included whether the person had either too strong (42%) or too little religious beliefs (49%), veganism (40%), feminism (42%), political affiliations (25% said they would not date a democrat, 33% said they would not date a republican), being too tall (20%) or too short (34%), and 25% would not consider dating someone who never attended college. The survey did not ask whether inability to speak the same language would make someone less likely to date a person but I'm willing to bet it would at least outperform smoking.

Let's take an extreme example of how confounding by assortative mating might influencing trying to find genes that might cause asthma. We take one group of people who have asthma that are French-speaking, unemployed, smoking, atheist, vegan, feminist, libertarians who were under five feet tall. The second group did not have asthma and were tall, German-speaking, non-smoking Church accountants who ate meat and voted authoritarian. If you ran your study sorting people only by asthma, any gene differences you find could be related to asthma, or could be due to the fact that such groups would rarely mate with each other.

So, revisiting that study about GWAS and smoking that Harden claimed was somehow validating for a GWAS on air pollution, we realize that the GWAS for smoking would have been an excellent example of a negative control. We would expect that people who smoke would have sorted themselves into a slightly separate gene pool if the non-smokers would not even give them a date let alone give them a child. Rather than pointing to a meaningful contributor to smoking, that study indicates that simple mating preference can create enough of a gap in the gene pools to manifest associations with 22 loci with p-values as low as 2e-14.

After abandoning the need to assess if a GWAS hit was real or spurious, geneticists began to make endless claims of genetic causation. To shed light on this era, I spoke with David Curtis, Honorary Professor of Genetics, Evolution, and Environment from University College London. Dr. Curtis himself researches the potential genetic contributions to Schizophrenia.

Dr. Curtis wrote a blog post entitled *"Why Every GWAS is Evil,"* which included several criticisms. First, Dr. Curtis notes that any effect of a gene encouraging Schizophrenia on a population level would need to be tiny. Any person with a gene that carried a large risk of having a serious mental health disorder like Schizophrenia would be less likely to have children. Added up over generations, such alleles would become uncommon as they would be swapped out by variants that did not carry such harms. People would avoid mating with someone experiencing Schizophrenia, especially in eras before treatments were available and if symptoms were severe. Curtis argued that running a GWAS for Schizophrenia was destined for failure since it would be very unlikely that a meaningful allele could be found with such a tool. When a loci association was found, Curtis remained skeptical but awaited the types of functional models that were a staple of familial and cancer genetics. However, such studies either never came or consistently failed to translate to humans. Dr. Curtis' frustrations with the approach mounted when his colleagues continued to run GWAS in other populations rather than nail down whether their initial findings were legitimate. Despite having failed to translate any GWAS hit from loci-association to meaningful drug target in white Europeans, the field wanted to check for loci in Asian and African cohorts. "They kept looking for hits when they hadn't followed up on the hits they already claimed to have found," Curtis outlined. Eventually, Dr. Curtis resigned from the Psychiatric Genetics Consortium out of frustration with how the field had prioritized data mining rather than searching for biologically valid discoveries.

However, Dr. Curtis did not yet anticipate just how far afoul of biology the field of population genetics was headed. As the expectations for a *success* in GWAS shifted from identifying a treatment target to just a p-value above the Manhattan plot cut off, more and more absurd claims would flow into the literature. The effect was a perfect example of what is known as *Goodhart's Law*, which states that when a measurement becomes a target, it ceases to be a good measurement. At its purest form, the p-value is supposed to be a measure of confidence in a result of a comparison or intervention. However, the second that the goal became finding a gene association with a p-value less than 0.05, the p-value went from measurement to target. The incentive in GWAS research went from finding a drug to improve patients' lives to finding a gene link that had a significant p-value and publishing it.

The psychologist and writer, Randy Cima, kept a tally of failed candidate genes (up until 2018) that he called *The Gene Fool*. Many of the ones that made the list were for mental health disorders like ADHD and Autism. However, some of the more humorous gems include:

- A claim that one gene variant could explain 30% of bad driving.

- A claim that a gene caused suicide, which one would really need to explain how such an allele would have survived evolution. Unless, as was pointed out by Dr. Curtis, the effect size was so small it would be negligible on a population level.

- A gene that made women more likely to want a divorce (no comment was made about whether the men had a gene that made women more likely to want to divorce them).

- A gene that determines whether women orgasm. *The Guardian* wrote this one up in 2005 claiming that having an orgasm gene would be an evolutionary disadvantage since "women who orgasm very easily may be more likely to be satisfied with poor quality men."

—   Another recent one was a claim to have a "gene for income." Numerous researchers have claimed to have found genes *associated with* high income. Here would be the classic example of the need for a negative control to sort out population stratification. Entire books, for example, *Capital in the Twenty-First Century*, have outlined how rich people tend to have rich kids who marry the rich kids of other rich people. Thus, any gene linked with income would be just as (if not more) likely to be related to a separation in gene pools than a true cause of income differences.

One of the most disgusting *"gene for"* claims has to be the claim that there are genes for being abused as a child. In 2021, the *Nature*-family journal, *Translational Psych*, published a paper claiming that variations in *FOXP1* were associated with being abused as a child. You might recall *FOXP1* from the random GWAS bot examples before that *predicted* what mode of transportation people prefer. The listed (and biochemically proven) functions for *FOXP1* include assuring proper structural development of the heart and arteries, suppressing cancers, and regulating the growth of stem cells. According to the authors of the *Translational Psych* paper, however, the allele also controls how children behave. The authors did not know whether the child was abused by a biologic relative, since the databases did not distinguish between foster parents, distant relatives, or unrelated strangers. So, they did not claim, nor even mention, that the gene could govern abusive behavior by the biologic parent of that child. Instead, their hypothesis was that either the gene variation caused a child to behave in such a way that it would encourage an adult to abuse them or make it more likely that the child would report the abuse. The authors wrote:

Specifically, FOXP genes might influence externalizing traits and so be relevant to childhood maltreatment. Alternatively, these variants may be associated with a greater likelihood of reporting maltreatment. A clearer understanding of the genetic relationships of childhood maltreatment... may ultimately be useful in developing targeted treatment and prevention strategies.

Did you catch that? The only two mechanisms entertained by the authors is that either the gene makes someone more likely to tell an authority or the abuse suffered by these children is partly their fault due to behavior that causes adults to abuse them. Their work did not define abuse as "being spanked when acting out" or "being punished a lot." The work defined abuse and maltreatment in a way that child protective services and the legal system might. So, this isn't about a gene that might *cause* a kid to be more willing to stay out past curfew and thus be more likely to be grounded. The authors are talking about the type of abuse that, if discovered by authorities, could get a child removed from that abusive home and land the abuser in jail. Of course, these authors never detail what behaviors specifically they think would cause a caregiver to become abusive to the point of potential intervention by the police. If I felt their insights worth my time, I might ask the authors "What would your child have to do to get you to abuse them?"

As bad as this paper is, it was topped by one from *Lancet Psych*. This 2021 report claimed that it found 14 genes for being abused as a child. This paper used American data from veterans who had donated their DNA and medical records after their service. I have to question whether the soldiers were told that their confession of being abused would be used to imply they behaved in a way that brought that abuse on. Either way, the authors went on to suggest that the genes for being abused were also linked to the genes for being depressed but not linked to genes for heart disease. Can you think of a few non-genetic reasons that a history of serious maltreatment as a child would be linked to depression but not heart disease? Because these authors couldn't.

For all the animosity I carry for Richard Dawkins and the eugenics-endorsing paradigm he has pushed upon society, I do extend him grace and sympathy for his self-reported history of being abused by a priest as a child. My grace substantively wanes when I see Dawkins then claim the following: "Horrible as sexual abuse no doubt was, the damage was arguably less than the long-term psychological damage inflicted by bringing the child up Catholic in the first place." This implies that raising a child in a religion is worse than what he terms "mild pedophilia". While Dawkins has received plenty of pushback against these statements, I won't pile on here and will instead do my best to remind myself and the reader that the type of trauma inflicted on people like Dawkins can warp someone's views and self-esteem in ways that those who have not suffered those harms could ever really understand. Yet, I would like to ponder what Dawkins would think of someone claiming that his abuse was caused by a selfish gene that wanted to best assure its survival by inducing behavior in him that the gene knew would make Dawkins a target for abuse?

Even if we accept that, somehow, these interpretations are valid, what would we do with this? Should we screen children for these alleles? If they are found to have these alleles, should we put the kid on a drug that would target *FOXP1* and prevent their abuse-inducing behavior (a drug that would apparently also encourage them to take the bus instead of a train)? What if the authors' second hypothesis is correct and instead of encouraging abuse, the alleles encourage reporting the abuse. Then what? If we gave children an anti-*FOXP1* drug and the rate of reported abuse declined, how could we know that the drug was preventing abuse or preventing reporting?

One thing all the most absurd claims have in common is a lack of what is known as *falsifiability*. For a claim to be falsifiable, it has to be at least theoretically possible to design an experiment to prove that the claim is false. Not every laughable example lacks falsifiability. An entire sub-field of population genetics is dedicated to food preference, like the paper that claimed Japanese people have a gene variant for liking green tea, or another that claimed that one's favorite flavor of cheese is genetically encoded. In theory, you could identify the sequence in the *green tea gene*, mutate that gene in a mouse so that it shared the gene variant, then see if it would consume more green tea. Or you could mutate the *cheese gene* and see if you could get a mouse that would exclusively eat Havarti. But if no one can test your hypothesis, then no one can disprove it. How would you test if a mutation caused higher income? There is no way to make a cell or mouse model that shows you when the mouse gets rich. To test this on humans, you would need to mutate an embryo and follow it over a lifetime to see if it made more money than a control group; this type of study is completely unrealistic and could never be conducted (even if somehow you got ethics approval to perform it).

To be blunt, as commonly practiced, the entire field of *evolutionary psychology* (or evo psych) is one that lacks falsifiability. A claim that "men evolved to like women with big breasts because it signaled breast-feeding abilities" can't be tested. Short of accessing a multiverse where you can see what happens if women's nipples grew on their elbows, so you can see whether men then decided to dedicate magazines to elbows instead of breasts, there are no methods that can facilitate testing these claims. Basically, any claim of "the behavior evolved to..." can't be tested and thus it cannot be falsified.

As was true for the candidate genes claiming to have found a gene for divorce, the one major issue with this is that it makes the claims of *multiple genes combining to account for education* non-falsifiable. As was well pointed out by Dr. Joshua Dubnau, a researcher in neurodegeneration at Stony Brook University, lack of falsifiability is a defining feature of pseudoscience. He wrote on Twitter:

The fact that genes influence complex cognitive processes is an obvious truth. Brains are the product of selection after all. The issue is that Dr Harden and her field make a fallacious and harmful leap from this truth. The false intellectual leap is the idea that one can meaningfully study genetic underpinnings of sociological traits (general intelligence, socioeconomic status, educational attainment) using the methods of their field. There are statistically sound correlations between gene variants in human populations and societal measures of success. These correlations exist both because genes impact behavior and because society limits or enhances achievement based on many human characteristics.

What sort of genetic human characteristics impact societal achievements? skin color, hair texture, nose shape, facial features, height, weight, and cultural differences that correlate with hundreds of others and with hundreds of gene variants. And herein lies the fatal flaw. The flaw that undermines this subfield of genetics is that there is not one experiment that can be done, or conceptualized, that would distinguish whether a correlation between gene variant and societal achievement is causal in a biologically meaningful way.

Let me explain. If a gene variant causes differences in (e.g.) skin color (or is linked/travels with it), it will have a causal impact on societal achievement because of racism of the society, not because of impact on the biology of the individual to render them less capable of achievement. And there are no experiments that can distinguish such sociological explanations from the claims of the field.

Two things follow: first, this field produces zero biological insight. Second, the impossibility of disproving the claims is the definition of pseudoscience. Not all pseudoscience deserves to raise our intense loathing. Some is silly but benign. But this field is caustic to humanity in two ways. First, it sucks resources from more meaningful research. More importantly, it causes great harm by breathing air into an invigorated eugenics political movement. I don't impugn the motivation of the researchers, they are earnest but unwitting tools.

The number one feature of pseudoscience is the lack of falsifiability. You can't really test if someone's Zodiac sign or the position of the moon influences human behavior. There is no lab in which you can truly move the moon around and see if people's behavior changes. You can't disprove that ancient aliens visited Earth. In theory, you could prove them correct if you found evidence of such a visit, but you can't ever *disprove* such claims since someone can always claim to you that you just haven't found the evidence yet. You can't test how a gene could enhance income and can't test if a gene induces divorce. However, just because a study has falsifiability, that doesn't protect researchers from their lack of negative controls.

One prominent but far less humorous GWAS study included claims of finding a gene that caused criminal behavior. The study (also from King's College London, but not related to Dr. Curtis) claimed that if a boy were severely mistreated as a child and had a variant of the MAOA gene, he would be more likely to commit crimes. This time, it was published in the American journal *Science*. This gene is also one linked to crime and violence under the lay-press moniker *the warrior gene*. This got a lot of attention in the media and fueled the idea that criminal activity could one day be reduced by identifying people with this gene and either blocking its effects at best or eliminating them from the population at worst.

Because those with the MAOA gene were *less* likely to commit crimes in the absence of abuse, the researchers claimed that the gene must be triggered by childhood trauma. The authors claim that the Māori people of New Zealand were more violent than their white countrymen because of their variants in this gene.

Since the gene is a neurotransmitter expressed in the brain, that alone was seen as validating. The main problem with the study was that it could not be repeated; when people in other countries tried to recreate the findings, they simply couldn't. There was also an issue with how variable one might define *aggression*. Sure, if one person were super aggressive and one timid, you could tell the difference. But to do these kinds of studies, you have to truly put a number on how aggressive someone is. What would it mean for you to become 11.4% more aggressive at work? How would you measure that? Since the original researchers could define that anyway they wanted, the risk of bias becomes profound since a small tweak in the definition could make the difference between statistical significance and not. So, with no functional model, and only inconsistent and nebulous links at the loci-level, one could expect what would happen when a full sequencing was performed. Indeed, the full sequencing failed to show any real difference in the MAOA sequence and none that correlated with violent behavior (even if loosely defined).

To be overly fair to these researchers, years prior, a group in the Netherlands claimed that they had found a family where deletion of MAOA was linked to aggressive behavior. You should already realize why this isn't exactly a defense. Saying that total deletion of a gene might be linked to a certain outcome is not the same as saying a nebulous link to a loci-level association of the gene could do the same thing. This contrast also highlights Dr. Curtis' prior comment that genes with massive impact would be rare. If select variants in MAOA really did cause behavior problems, they couldn't be major since 30-56% of men have one of these variants. A behavior problem at the level of the family with the MAOA deletion would carry a substantial reduction in the chances those children would ever have children of their own, and as such, would not spread in the population far enough to be picked up on a population level analysis.

So, how could negative controls have prevented *the warrior gene* nonsense? Well, one example would have been to sort the same people in your study along characteristics you don't believe to be genetic. Recall the example given in Chapter 3 for a twin study that claimed that children's first initials were genetic. Had the researchers collected several types of data on their subject and then sorted along those control topics they could contrast how tightly linked their MAOA variants were to other factors before turning to assessing aggression. For example, was MAOA also linked to whether subjects preferred spaghetti versus fettuccine? Was MAOA linked to the men's favorite baseball team? Could MAOA levels predict boxers versus briefs preference? By only sorting on violence, and only assessing for violence, the researchers can't say whether the link between the MAOA variants, aggression, and history of maltreatment was unique or whether that same link could have been seen for other unrelated (and non-genetic) factors.

Next, a negative control might have protected against the incentive for chasing a significant p-value. The researchers sorted people into high and low-MAOA groups. However, it isn't as if high- versus low-MAOA levels were written on Moses' stone tablets brought down from Mount Sinai. What if the researchers set their separation point a little higher or a little lower? What if they instead looked at the exact numbers instead of binning people into high or low? For example, they could have taken the exact values of MAOA and contrasted it against their arbitrary score for aggression. Having a firm small, medium, and large concept of the world works for coffee shops, but in science, if a variable can occur over a continuous range of numbers, it should be treated as such.

The incentive to alter the MAOA cut off until it lands on a spot where the separation generates a significant difference would have been high for the researchers. Having a negative control could allow them to test how their modification changed other parameters. For example, if they moved the MAOA cut off up and then, suddenly, there was a significant difference in the subjects' favorite baseball team *and* aggression score, that might suggest that their change was just inflating links across the board.

Alternatively, they could have used something other than MAOA to see whether they could get that same separation with something unrelated. What if they separated people by low- or high- levels of calcium in the blood? Or hemoglobin levels? Or cholesterol? None of these would be related to their theory that some brain transmitter caused violence. So, if they could just as easily sort aggression in people based on nonsense as on MAOA, it would argue against their conclusions being valid.

The problems of lacking negative controls have only been magnified in the GWAS era as they moved away from candidate studies. As was outlined in the example with roulette, as sample sizes increase, smaller and smaller differences between groups suddenly become *significant* mathematically even while they become less relevant biologically. If you can pick up exceedingly small fluctuations, it becomes imperative that you can assure yourself that you are not just picking up noise. However, GWAS researchers often point to their increasing sample size as some sort of validation (i.e., the fact that they can find a link across an ever-expanding size population means that link must be real). This leads to claims like those made by Song and colleagues in *Nature Human Behaviour* that there are not only genes that govern processed meat consumption, time living at current address, or preference for ground coffee over instant, but that such genes have even stronger mathematical links to phenotypes like blood pressure and hip size (as measured by smaller p-values).

Instead of using a negative control, GWAS researchers, like Visscher and Abdellaoui mentioned before, attempt to dodge the issue in two ways. First, they claim to be able to adjust their model to look for assortative mating. However, their models for assortative mating were made after they already decided that intellect and education had enough genetic contribution to justify dedicating their careers to investigating it. The researchers would thus be biased against any model attempting to account for mating preferences that might erase the link between education and genetics.

Even with higher end mathematics, the larger point remains: You should never try to deduce what you could just measure. If you enroll people into your study and feel that their height and weight might be important variables, then just measure them. Don't guess; don't try to derive them from their genetics any more than you can guess their ear-to-nose ratio. Assortative mating can be directly measured. Sort people by language, sort them by their favorite sports team, sort them by zip code, or sort them by religious affiliation. Then, check for genetic links to know what the range of variability might be. Only after you get a sense of how well people can sort themselves into gene pools can you trust the results of your experiment trying to see how powerful the gene pool is on sorting people into groups.

Although not entirely akin to a negative control, one approach to better understand assortative mating was presented by the University of Colorado PhD student Tanya Horwitz in a 2023 edition of *Nature Human Behaviour* (I will admit that occasionally, *Nature* journals do publish anti-eugenics work). Horwitz wondered if people in the U.K. Biobank were more similar to their spouse than the general public. The top similarities were age and place of birth. Finding that people marry people of about the same age and from similar areas is to be expected. Other highly similar traits included political views, religion, education, time spent watching TV, time with friends and family, height, and smoking and, as a result, lung function. Importantly, similarities with partners varied across the population, suggesting that the likelihood of marrying someone of a different religion or educational background has changed over time and differs between cultures and hometowns (something most sociologists would have guessed).

Some traits had exceedingly low partner correlation including handedness, birthweight, and disorders like eczema. Only a few traits indicated "opposites attract" through a negative correlation in partners, including hearing ability and whether you are a morning person or night owl. If you have a partner who likes to sleep in and always tells you that you don't listen, this probably hits home for you.

Importantly, Horwitz points out that failure to account for the complexities of assortative mating can either inflate or deflate the calculated role of genetics depending on the trait, calculation method, or populations. However, her analysis indicated that 89% of the traits in the Biobank have evidence of assortative mating that is ignored by most genetic models.

Work led by Dr. Richard Border of UCLA published in *Science* in 2023 took things even further. Border found that assortative mating can compound over generations. Tall people marry tall people and then have tall children, who marry tall people. The compounding effects of this can make it seem as if genes are predicting height when Border's team calculated mating overlaps with 74% of the perceived genetic signal.

I will note that these works may still underestimate the impact of mating preferences because the participants in the Biobank are not truly a reflection of the full diversity of the U.K. population. And this work still does not answer the question of how genetically different Catholics may be from Protestants. However, they are an earnest attempt to calculate the overestimates of genetic claims that arise from a failure to account for how variable mating preferences may be across time, space, and culture. Plus, these examples were performed in a way that removed the incentive for researchers to tweak their model to maximize the genetic signal for any individual trait.

The second—and by far the most common—way the GWAS researchers try to dodge the need for a negative control is to just pretend no such thing exists. If you claim that everything has a genetic component, then there is no such thing as a negative control. Even the geneticists that push back against GWAS over-interpretation make this error.

The GWASbot account that put forth the various absurd examples of GWAS shown earlier is run by a group led by Andrea Ganna in Finland. You might think with all the unserious GWAS plots the bot spits out that their group would be critical of the approach and careful not to overstate mere associations. You would be wrong.

The group has a paper in *Nature Reviews Genetics* about the genes involved in the COVID-19 response. They state:

> *Compared with other common complex diseases, studying the human genetics of infectious disease poses additional challenges including uneven exposure to the virus within a population, the differential treatment of patients with severe disease under a pandemic emergency and the implementation and uptake of vaccination programmes. Nonetheless...*

They are admitting that you can't run a genetic study on infection control at a population level without accounting for exposure, vaccination, local mitigations, and access to care. But then they simply say "nonetheless" and go on to dedicate the next 10 pages claiming that anywhere from 19-55 alleles make COVID more severe despite not adjusting for any of the factors they previously stated may confound their results.

In their essay, "*Why Biology is Not Destiny*", Professors M.W. Feldman and Jessica Riskin critiqued *The Genetic Lottery* for the *New York Review of Books*. The scathing review referred to Dr. Harden as

> *...a dedicated frog boiler. She introduces many comfortably room-temperature premises: measurement is essential to science; people differ genetically; genes cause conditions such as deafness; a recipe for lemon chicken produces variable results but never leads to chocolate-chip cookies. Lulled to complacency by such anodyne and often homey observations, we soon find ourselves in a rolling boil of controversial claims: genes make you more or less intelligent, wealthier or poorer; every kind of inequality has a genetic basis.*

The authors deride Harden's claims that GWAS derivations are devoid of human bias by pointing out that the "estimating, measuring, counting, weighting, correlating" are all active processes. They go on to point out that:

> *...people are making interpretive decisions at every stage: how to define a phenotype and select people to represent it, how to count these people, which single-nucleotide polymorphisms to consider, how to weight and aggregate them. Interpretive decisions are of course essential to all science, but here there are a great many opinions dressed up in facts' clothing.*

Feldman and Riskins point out that GWAS studies have been better at gaining researchers "funding and professional advancement" than discovering any meaningful treatments. They mock Professor Harden's feigned ignorance as to how moving children from under resourced orphanages into foster care could increase IQ, and they even compare her approach to phrenology. But they, too, make the claim that genome and environment cannot be assessed independently when they state:

> *This is impossible, even as an ideal, because the environment is in the genome and the genome is in the environment. We can no more*

POSITIVE CONCLUSIONS REQUIRE NEGATIVE CONTROLS

> unbraid genetics and environment than we can unbraid history and
> culture, or climate and landscape, or language and thought.

Kudos should go to the authors for denouncing the neo-eugenics of *The Genetic Lottery*. However, while genes being essential for human life can't be debated, just because genes are involved in any biologic process does not mean that differences in DNA sequence necessarily govern biologic processes. Every single process in the human body is sensitive to the pH—the measurement of how acidic or basic the conditions are. Nearly every normal, healthy process in the body requires oxygen. Every cell is so dependent on the electron transport chain for survival that shutting off the electron transport with drugs like cyanide can cause violent death in mere minutes. So, the fact that genetics are involved in the response to cigarette smoke is no more impressive than the ubiquitous involvement of protons or electrons.

Knowing that DNA must exist for a biologic process, does that mean that the carbon, nitrogen, oxygen, and hydrogen atoms that comprise DNA also *contribute to* disease and cannot be *unbraided*? What would you think if you were watching the pregame show for your favorite sport, and someone asked the sportscaster, "What do you think will be the most important factor in deciding who wins today?" and they answered, "The existence of the ball"? It is technically true, but why not answer, "The subatomic forces that govern the universe in unknowable ways"? The claim being put forth by Harden and every population geneticist out there isn't merely that genetics are *involved* in development, but that *differences in genetic sequences contribute to* outcomes. Knowing DNA is involved in all biologic processes does not preclude the need to find a control comparison in which DNA's contribution would be expected to be negligible.

Overall, negative controls can't solve the lack of falsifiability of claiming that genetic differences cause college attendance, but they could help show that any genetic differences found were more associated with education than the myriad of social differences that might lead to separations in the gene pool. However, the entire field of population genetics is balanced on their ability to pay only lip service to the reality that population stratification is the second biggest risk to the validity of their field. Overall, the refusal to meaningfully engage with environmental science was, is, and will forever be the biggest failure of population geneticists, particularly once they started to aggregate their noisy data into claims of *polygenic* signal.

CHAPTER 11
The Plural of *Noise* is Not *Signal*

BOTTOM LINE UP FRONT: *Polygenic scores are what you get when you throw bad money after bad.*

Imagine you are sent in to investigate the exact cause of why so many people at a church potluck got sick after the event. Because you want to be exhaustive, you tally every single food ingredient everyone ate and then run an Ingredient-Wide Association Study for who got sick and who did not. You run your analysis and find something like the following:

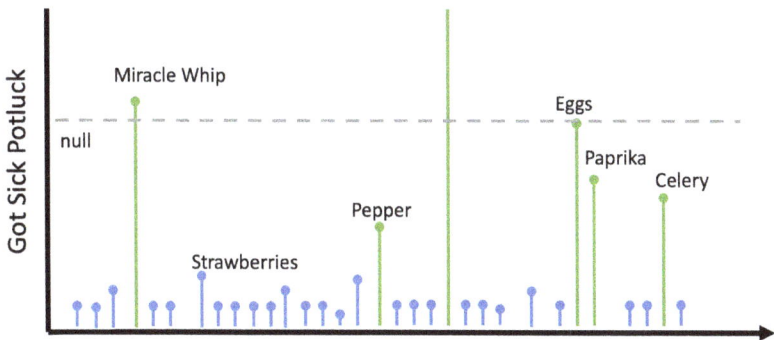

What is the culprit? Do you think people who ate French fries got sick? After all, potato is the most associated variable. Most people looking at the results would correctly identify that the people who ate potato salad were the ones who got sick. You naturally bundle the different variables into what scientists might call *pathways*. No one at the party randomly ate Miracle Whip on its own (well, at least it is very unlikely anyone would do that) but you can collate this information into a single dish. What looks like dozens of variables influencing getting sick is actually only one: whether the person ate the potato salad.

But what of the strawberries? Alone, they are non-significant, so you might argue to ignore them altogether. But the signal is stronger than the background level and would rank as the seventh strongest link. God willing, no one put strawberries in the potato salad. Obviously, given what we know about recipes, the interpretation should be that people who ate the potato salad were also slightly more likely to eat some strawberries. So, if you were tallying a *risk score* for who got sick, the wise approach would be to collate the variables into a sensible pathway, then ignore the variables that were *along for the ride*. What you should not do is make a risk score that includes strawberries or any of the other dozen ingredients with extremely low signal. If you did this, someone who ate a lot of strawberries might be given a risk score higher than someone who had only a bite of the rancid potato salad. The goal should be to sort the signal from the noise, not to combine them.

After the GWAS candidate gene era had failed to produce any meaningful treatments, population geneticists turned to collating the various weak genetic signals into a *polygenic score* (abbreviated PGS). This is sometimes called the *polygenic risk score* (PRS) or *polygenic index* (PGI), depending on the researcher. I will use the term polygenic score, but it does not matter what moniker you give it; the PGS is like potato salad with strawberries mixed into it.

Referring to a *pathway* in biology is meant to indicate related metabolism or signaling. Here is one open-source example for glycolysis from the Kyoto Encyclopedia of Genes and Genomics (abbreviated KEGG).

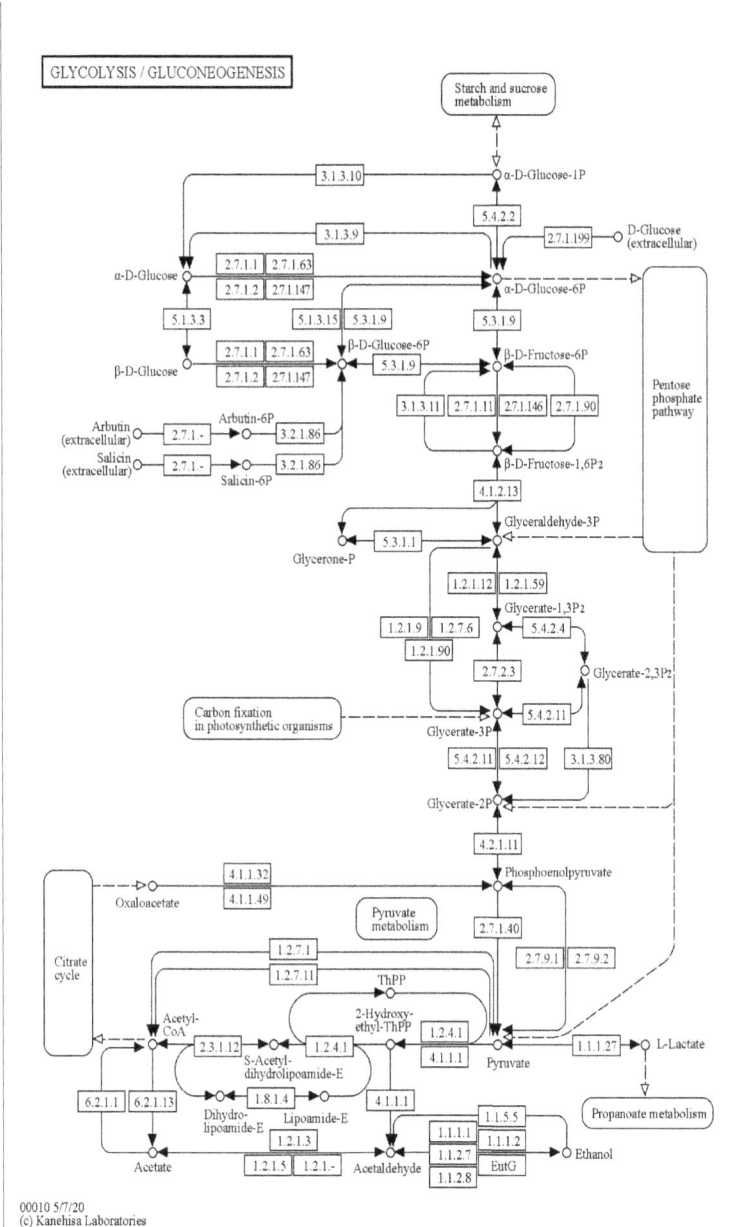

GLYCOLYSIS / GLUCONEOGENESIS

00010 5/7/20
(c) Kanehisa Laboratories

Each box with the numbers is an enzyme; each circle is a metabolite. Each enzyme is encoded by a specific gene, which would then be transcribed into an mRNA transcript before being made into a protein. At the top, you can see alpha-D-Glucose-1P as the start of this pathway.

Glucose-1P, in this case, is broken down to Glucose-6P by the enzyme given the number 5.4.2.2 (named phosphoglucomutase). As you travel down the pathway, glucose is ultimately broken down into pyruvate, which can then transit into other pathways from there. Notably, metabolism is an incredibly complex web of reactions. For example, KEGG depicts human metabolism as the following, with each dot indicating a different metabolite and each line indicating a different chemical reaction.

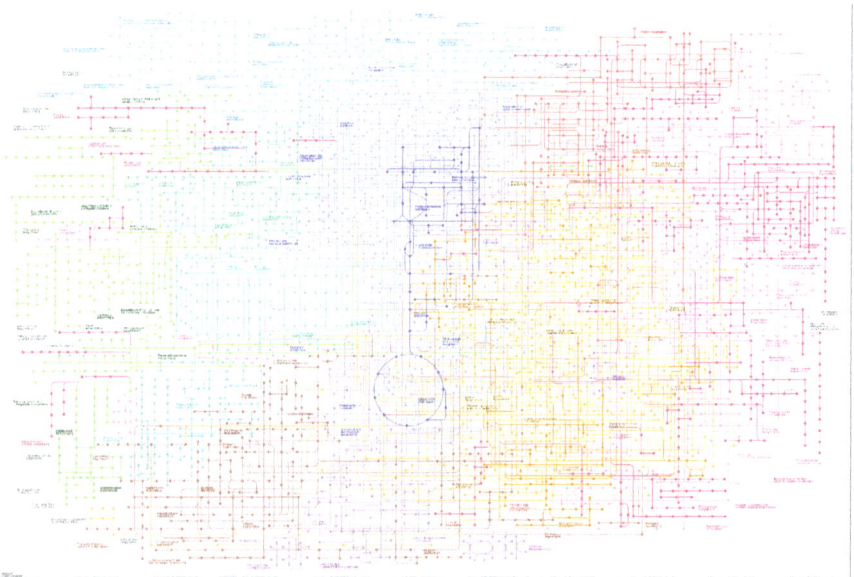

The interconnectedness of metabolism means that you cannot alter any one point on this map without impacting all the connections surrounding it, and all those connections surrounding those connections. However, sorting the tangled web of metabolism into pathways allows researchers to make more digestible experiments and design targeted therapies. Genetic signaling pathways operate in the same manner. For example, here is one of the many signaling pathways that involve the gene *STAT3*, which was discussed in Chapter 4.

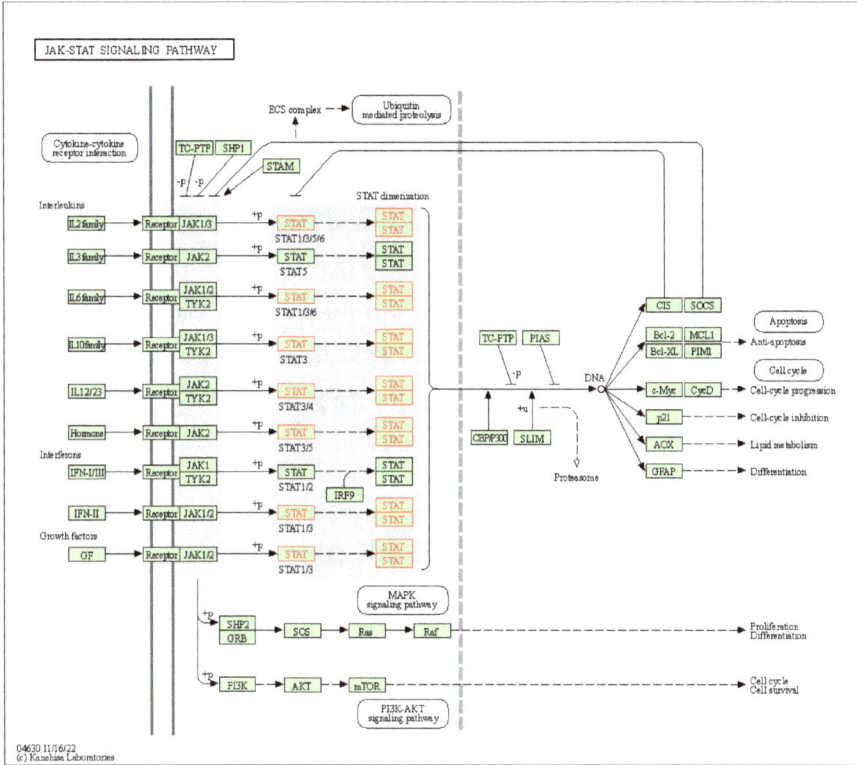

In this image, every box is a protein, and the arrows indicate interactions between them. As of this writing, KEGG indicates STAT3 is involved in at least 32 different signaling pathways, which makes figuring out exactly what might be wrong in patients with STAT3 deficiency so difficult. However, by simplifying into pathways, you can see that the JAK proteins have effects upstream of STAT3. This opens the possibility that using drugs that modify JAK proteins (called JAK inhibitors) might improve symptoms for patients with defects in the downstream STAT3. Although it remains uncertain whether JAK inhibitors ultimately benefit patients with STAT3 defects, exploring biologic pathways provides researchers with valuable insights into potential drugs worthy of investigation. Just as potatoes may have been a widely used ingredient at a potluck, collating results into *dishes* would allow someone to accurately evaluate the source of food poisoning.

On paper, the reliance on pathways was supposed to be the idea behind the PGS. In 2007, a group led by the same Dr. Visscher mentioned prior reported collating results from individual GWAS hits into a PGS. In their paper in *Genome Research*, they wrote:

> *Identification of causal variants and elucidating disease pathways through genetic and functional studies is difficult and time-consuming, particularly if there are many risk loci with small effects. However, knowledge of all risk loci or knowledge of causal variants at any one risk locus is not necessary for the prediction of the risk to disease of individuals in the population.*

It seems worth noting that sometimes, things are "difficult and time-consuming" because doing things correctly takes time and effort. There are an endless number of late-night infomercials claiming that losing weight is "difficult and time-consuming" and then claiming that squeezing a spring between your thighs for 20 minutes a day, three days a week will be able to supplant the hard work and time needed. Thus, the presentation of PGS was originally a replacement for the "difficult and time-consuming" traditional approaches. You will also note that the authors assume that some risk loci have *small effects* despite those loci not having functional studies to prove their contributions to the disease. Overall, however, Visscher is offering an easier pathway to disease identification. Rather than the time-consuming work of finding gene candidates and proving their validity, such as was discussed for *STAT3*, Visscher indicates that this process can be done quicker if we collate GWAS results. Ostensibly, the idea was correct.

If a defect in the glycolysis pathway was the cause of a disorder, then defects throughout the pathway might suggest a similar disorder. Or mutations in genes from overlapping pathways could combine to create greater impact than either of the genes in isolation. Combining independent risk factors is a widespread practice in medicine. One example is the Framingham Risk Score for heart disease. The Framingham score combines the various risks of heart disease posed by smoking, age, sex, cholesterol, and blood pressure. Each of these factors present its own independent risk but combining them created a better predictor of whether a person might go on to have a heart attack.

Cancer researchers made a version of the PGS by combining genes validated using the "difficult and time-consuming" practice known as proper science. If mutations in both *BRCA* and *CHEK2* could increase the risk of breast cancer, then having both mutations would present a bigger risk than either alone. Of note, the cancer researchers used the term PGS first, and it was copied by the association-only collators. The problem with using the same term for the cancer-like PGS and a PGS for Schizophrenia is that the candidate gene era had not produced any validated results. Where cancer researchers could combine numerous validated risk alleles to add up small harms into larger values, those in Schizophrenia work were tallying associations that lacked any biologic validation. Furthermore, in cancer, the directionality of risk was consistent. Having a *BRCA* mutation may not lead to everyone getting breast cancer, but having a *BRCA* mutation was never associated with a reduced risk of disease. Meanwhile, diseases like Schizophrenia or asthma would often claim one gene sequence increased risk in one study, while that same sequence was associated with a decreased risk in another. Thus, as the Visscher paper implied, the thought became that combining all the results—the potatoes and strawberries alike—might reveal pathways that could create success out of the failed candidate gene trials.

Again, the idea had merit. For example, a defect in enzyme 5.4.2.2 would damage the ability of the patient to turn glucose-1P into glucose-6P. But a defect in 5.3.1.9 would create a similar disease, since the patient could not break glucose-6P into fructose-6P. If a given pathway were the source of the patients' disease, then it would improve detection under the expectation that risk alleles would concentrate in the sugar metabolism pathways as opposed to less-related pathways.

The pathway alteration is very evident in RNA analysis. RNA sequencing (shorthand referred to as *RNAseq*) looks at all the transcripts in a sample and combines them to look for pathways of interest. Although the results may vary slightly between patients with uncontrolled diabetes, in all cases, *RNAseq* reveals abnormalities in the transcripts related to glucose metabolism. Pathway alterations are also evident in metabolomics, which aims to evaluate all the metabolites that are different between samples. In diabetes, doctors may measure glucose for the diagnosis, but metabolomics reveals abnormalities in all the metabolites related to glucose, such as fructose, pyruvate, and acetyl-CoA. For RNA and metabolites, it isn't just that researchers value pathways over individual transcripts or metabolites; they *demand* pathway analysis to validate findings. It is impossible for only one metabolite to be abnormal. If you were to inject someone with loads of glucose, the body will begin to metabolize that glucose to either use or store the energy. Activation of the *JAK* proteins could not happen in isolation either since you would expect to see downstream impacts on *STAT3*. This is why no drug is without side effects; you can't block COX2 (the target for drugs like Ibuprofen) without disrupting every molecule in the pathways that COX2 is active in. The idea that only one spot in an entire pathway would be skewed without throwing off everything else is considered invalidating in the fields of transcriptomics and metabolomics; for population genetics however, it has become the norm.

Yet, despite what they may have hoped for, the genomic signatures in diabetes didn't fall into any discernible pathways. It wasn't as if small changes in each of the glucose metabolizing enzymes were spread out across an entire population and pointing to glucose as the key to the disease. Although GWAS may occasionally stumble upon a result that had been previously validated in transcriptomics or metabolomics, combining GWAS results did not provide the same clarity as other modalities. As the failures of PGS to create any new treatments mounted, a shift occurred in the definition of success similar to the moving goal posts of GWAS successes. Whereas the original claim was that PGS would speed up the process of identifying causal genes that might make for good treatment targets, the new definition of success was how well the PGS could mathematically overlap with variation. Now, instead of looking at the ingredient hits and trying to find the culprit dish that genuinely explained who got sick and who didn't in the real world, they combined everything into one giant stew in hopes of *explaining* variation in a theoretical world.

Even this, however, was fruitless for behavioral geneticists. A paper entitled "Polygenic Risk Scores and Physical Activity" reported a maximal difference in activity between those in their study population equal to 17 metabolic equivalents of training, or METs. One MET would be the energy used to sit and watch TV for an hour. Hiking for an hour might be seven METs while 17 METs would be competing in a cycling race. The total amount of exercise difference *explained* by the PGS? 0.4 METs, which is less energy than it took you to read this chapter. For eczema, one American report in the June 2022 edition of the *Journal of Allergy and Clinical Immunology (JACI)* claimed that PGS could explain 13% of eczema rates but only if they limited the analysis to white Americans. 13% may not sound bad, except when you realize that the PGS they derived had zero predictive capacity for an individual. If you had Group 1 with a low PGS for eczema and Group 2 with a high score, then the researchers could predict that Group 2 would have about twice as many patients with eczema as Group 1. However, if you had Person 1 with a low PGS and Person 2 with a high score, you could not in any way reliably guess whether either would have eczema, let alone guess which one. Once again, the lack of negative controls in the population genetics means that even predicting eczema in Group 1 versus 2 does not mean the genes cause eczema; the alleles might merely reflect mating practices that overlap with eczema risk factors (like income, pet ownership, nationality, etc.).

A given PGS could be calculated for one group in one country, but it would have no value in a second group in a neighboring country. One defense of these inconsistencies included the idea of *gene-gene interactions*. That is, the reason a PGS wouldn't work in different groups may be that a gene that was more common in Group 1 was interacting with the important genes in different ways than genes more common in Group 2. However, as shown in the *STAT3* signaling example, genes interact with each other in somewhat predictable ways. So, if *STAT3* were detected as important in one group but not the other, then the other genes that might be interacting with *STAT3* should be somewhat identifiable. Yes, there are still interactions that remain to be discovered, but the point is that if science has already revealed dozens of genes that interact with *STAT3* across 32 pathways, then genuine *gene-gene interactions* should include at least *some* of the genes in those pathways.

A more legitimate excuse for PGS inconsistency would be allele frequency. Perhaps variations in *STAT3* would be too rare in Spaniards to generate a link with eczema, but common enough in Norwegians to be picked up. So, a PGS that relied on *STAT3* for tallying the eczema risk might undercount people in Spain. However, there is no way for *STAT3* to contribute to eczema without all of the pathways which include *STAT3* also being important. Recall that the original expectation for PGS was that it would have the same expectations of pathway detection as *RNAseq* and metabolomics. If this had been true, then the fact that *STAT3*, *JAK*, and *TYK2* all interact would still have identified the pathway as important. For example, if Spaniards only had meaningful variants for *JAK*, while Portuguese had *STAT3* variants and the French had *TYK2* variants, the combined PGS would still have implicated the *JAK-STAT* pathway. Thus, if the PGS for eczema correctly captured the biologic pathways involved in disease, then even if different populations might have alterations in various parts of the pathway, the pathway's overall importance would be preserved.

To be fair, some researchers understand this. 2023 work from Demela and colleagues in *Nature Communications* looked at a sample of immune mediated diseases. The group compared GWAS results with *RNAseq* from the same disorders and combined them into pathways. By doing this, they were able to identify pathways that had a shared signal. Unfortunately, their work did not identify any new drug targets; for example, they found *CTLA-4*, which is already a target for drugs. However, their work is the first of its kind and, conceptually, it offers the chance that the days of just looking at genes without contextualizing the results in the rest of biology may be coming to an end. If, five years from now, people have repeated this type of analysis one hundred times and still failed to find anything new, we might need to revisit its value.

Another problem with PGS is that genes are not the only variables that can be summated for better prediction. Dr. Venk Murthy, a cardiologist at a university that my ties to Ohio State compel me to refer to only as "that school up north," wondered what value PGS could add on top of more traditional clinical metrics. His team collected only the patients' age, sex, and history of whether their parents were overweight or obese. This information alone could only *explain* 5.1% of the variation in the body mass index (BMI) of white patients 25 years later. I'll use the term *explain* since this is the standard use, but we will see that the more appropriate phrasing should be "mathematically overlaps with". If they added the PGS for obesity to that base information, the explanation went up to 13.6%. So, PGS added to base information appeared to *explain* an additional 8.5% of the variation. Seems great, right? They could nearly triple the ability to predict BMI 25 years later by including genetic information. However, Dr. Murthy then added the baseline BMI—that is, the BMI when the patient initially signed up for the study. Age, sex, parental obesity, and baseline BMI could then explain 52.3% of the variation in BMI 25 years later. Adding genetics to this information only pushed that number to 54%. But if they just added two more BMI measures—that is, they followed people for three total visits and saw how stable their BMI was—then age, sex, family obesity, and three measures could explain 81.3%. Adding PGS —with all its flaws in disconnection with biology or pathways—pushed the number up to 81.4%.

What is notable here is that the variables Dr. Murthy looked at were still pretty limited. They added self-reported fitness and activity levels, which did not add much, but what if you could add other variables? The ability of PGS to add anything to the patient's prognosis effectively evaporated with pretty basic insights. What would happen if you added data on their dietary choices? Or what about any of the other factors known to impact risk of obesity like living near fast food, living far from fresh fruits and vegetables, smoking, chronic illness, antibiotic exposure, poverty, and so on? For all the obsession of combining every perceived genetic link to make a stronger combined signal, those in the field never see the need to tally any environmental variables. Dr. Murthy's work demonstrated that if they had, the value of PGS would diminish.

Geneticists often favor height as one of the prominent traits to *explain* study outcomes through genetics. The polygenic crowd celebrated a major milestone of being able to explain over 50% in the variation in height if you added PGS to the average height of the two parents. But what if you used more than just parental averages the way Dr. Murthy did for BMI? My father is six feet tall whereas my mother is only five-foot-three-inches. So, the mid-parental height equation predicts that I should have been five-foot-nine-and-a-half inches. I, instead, ended up six-foot-six-inches (okay, six-foot-five-and-a-half inches but let a guy round up). There are only three possible explanations as to how I ended up eight inches taller than my parents' heights predicted. The first is that my parents each carried the genes for height, made no use of them for themselves, and then passed them on to me so I could grow taller. The second is that each of my parents had about half the genes needed to *make a tall baby* and needed to combine them just-so for this to work. The third would include the massive differences in my environmental exposures growing up: a better socioeconomic status leading to both better access to complete nutrition and a less stressful pregnancy for my mother, moving from inner city Cleveland to the suburbs around Denver, not smoking as a young adult, and not spending my time from age 18-21 dealing with the stresses of fighting in the Vietnam war.

Population geneticists love to defend their work by comparing a PGS for a trait or disease with just one single environmental variant. When people dismiss PGS for (let's say) obesity as being trivial, population geneticists will defend it by saying the statistical associations are about equal to those living in a food desert. A food desert is a term used to describe a community that has poor access to fresh fruits and vegetables, resulting in diets that are heavy in unhealthy processed foods. No one who works in public health, especially those who specialize in obesity, would claim that setting up a fruit stand in an impoverished area would make a massive impact on obesity (even assuming you made the food affordable). Therefore, recognizing that the influence of a PGS implies that even if one could achieve the impossible task of editing thousands of genes in a person all at once, the outcome would be as insignificant as simply giving them a free orange. This doesn't provide the defense that geneticists envision.

And yet, Dr. Murthy's work demonstrates that even this defense is wrong because the perceived impact of a PGS depends upon intentionally ignoring those other environmental variables. Environmental changes are hard and behavioral changes are even harder, especially if those behaviors become rooted in cultural norms. However, when you hear people talking about modifying our environments, you will often hear the terms "systemic change" or "systematic change," precisely because they know small, one-off changes won't cut it.

Let's break this down before anyone chimes in with "it could be all the above; gene-environment interactions might be at play." There are natural experiments that would clue us into whether nutrition or genetics are more important in height. A 2016 report in *Nutrition Reviews* found that high-income nations had gained over five centimeters (2 inches) in average height between 1930 and 1980. Medium-income nations gained an average of four centimeters, low-medium nations gained less than three centimeters, while poor nations did not get any taller over those same 50 years. So, clearly, interventions related to nutrition can have profound effects on populations, even though their genetics do not change over a similar timespan.

The gains in nutrition are entirely attributable to the insistence of metabolomics to bundle findings into pathways as a means of validation. Nutrition factors related to height are actionable in ways that PGS are not. If my parents had been told, "Your son has a PGS that predicts his height will be 6-foot-6 inches," what would they do with that information? If they had been told I had the genes of a future 5-foot-3 man, would they give me "more genes"? Had PGS been able to live up to their initial promise and predict pathways, then perhaps a low PGS for height might be based on genes for protein digestion. Then, my parents would have known to assure that I eat more protein, or maybe use the PGS to select the optimal source of protein (such as beans versus chicken). In contrast, when a dietitian performs an assessment and cautions that your child is at risk for short stature based on nutritional factors, they can also tell you exactly what to change.

A 2016 project in *Economics and Human Biology* led by Dr. Tomas Kalina identified that protein and calorie consumption alone could explain 75% of the global variation in average height. If you were to add in chicken and rice consumption, child mortality, fertility, and urbanization, that number jumped to over 87%. To be fair, Kalina's group also identified a role for the lactase gene mutations mentioned in the early chapters. Populations with more of the mutations allowing for adult activation of lactase were taller due to increased ability to consume milk protein. However, national income could still easily trump genetics because rich nations with high rates of lactose intolerance could afford to cultivate protein from other sources. So, adding the potential effects of every single gene variant, including potentially spurious ones, can account for 50% of variance if you also know the parent's height. However, nutrition combined with basic economics could account for 87% before adding in antibiotic exposure or exposure to growth stunting toxins.

This is why the term *explains* is so flawed. When using bare minimum information for Dr. Murthy's study, the PGS *explained* an additional 8.5% of BMI variance, whereas with minimal information and a few clinic visits added, it only *explained* 0.2% of the variance. Since *explained* is just a contextual value, which can vary widely depending on how good the model is, I think the term *mathematically overlaps with* would be more apt, or the more common phrasing of *associates with*. Or, more succinctly, *correlates with given my assumptions and adjustments* might be apt, especially since such phrasing would trigger the truism that it *does not equal causation.*

There is a propensity for some people to believe that technological advances can supplant hard work and expertise. To be fair, this is true in lots of ways. The best navigator alive could not beat a well-updated GPS. But the idea of PGS was that one could replace the decades-long process of dozens of scientists working to identify and validate the role of *STAT3* in the patients with hyper IgE with nothing more than a weekend and a computer connected to the U.K. Biobank. Hubris barely describes such a literal jumping to conclusions. In the end, the only thing that PGS made faster wasn't the ability to correctly identify causal genes; instead, it was the ability to publish papers spuriously claiming to have found causal genes. Visscher was correct that collating results into pathways could be a superior method compared to looking at individual genes alone, but doing things faster only helps if you don't lose accuracy. And you can't know if you lost accuracy without using the similar validations methods of the previous approaches (in this case, through the functional studies or cross-validation with transcriptomics).

Over time, the gap grew between GWAS and PGS reports and the basic understanding of biology offered by other detection systems such as transcriptomics and metabolomics. Instead of identifying pathways across humans suffering from the same disease, PGS results were framed as increasingly contextual. Results in one population would only be applicable to that population and, as such, could not be the basis for any prospective drug development. This was termed *portability*, meaning that results in one group were not *portable* when going to look at other groups. Population geneticists are exceptional at manipulating semantics. Every other field would use the term *reproducibility* to describe the ability to reliably find similar results in different study populations. The findings that smoking was bad for Brits was reproduced in the studies on smoking in Americans. Population genetics use the term portable because such semantics allows them to construct a different standard of evidence without admitting it.

The important distinction is that environmental science focuses on sorting *independent* variables from *dependent* variables. For example, both smoking cigarettes and paying cigarette tax correlate with lung cancer risk. The more cigarettes you smoke, the higher your risk of cancer. The more cigarette tax you pay, the higher your risk of cancer. But the only reason cigarette taxes are associated is that they are dependent on smoking. Paying sin taxes does not *independently* predict risk. In contrast, no matter what else you do in life or how much sin tax you pay, smoking will increase your cancer risk. In an environmental model, cigarette taxes would commonly be excluded from the analysis because that variable only matters in the context of smoking. Thus, if PGS are as dependent on the environment as their lack of portability implies, they too would be excluded in the final models of most statistical models looking for the causes of disease.

To be fair, there is some justification for the lack of reproducibility between populations. *The Genetic Lottery* discusses it in lip service before referring to it as "complicated math" in the chapter dedicated to convincing racists not to be racists. Yet you will recall from earlier chapters that certain alleles may be common in one group but relatively rare in others. If the allele is rare enough in a population, then it is not likely to generate enough of an impact on that population to be detected. If you were running a population-scale analysis for what made for a good meal, the impact of sauerkraut would be larger in Germany than Italy. Recall that if you were running a GWAS for reactions to abacavir, HLA-B*57:01 would not be expected to be frequent enough among Asians to generate a detectable impact on abacavir reactions.

Yet even when allele frequency isn't the limiting factor, PGS may lack portability due to differences in which genes and alleles are near each of the tested loci. When genes tend to get passed on together due to physical proximity to each other, that is known as linkage disequilibrium. Two genes that occur right next to each other in the genome are more likely to be inherited together than two genes that are far apart. While every grouping of humans is at minimum 85% identical, the exact order of the genes may differ slightly. As populations moved and the genomes shuffled, the exact genes may not be in the exact order and linkage disequilibrium differs. Since GWAS does not look for the exact allele, a loci associated in one group may be near a subtly different set of alleles than the same loci associated in other groups. Think of it like shuffling a cookbook over time. The recipe for chicken cacciatore might be the fifth main dish listed in one cookbook and be right next to the recipe for chicken alfredo. But in a different cookbook, cacciatore might be the ninth main dish listed and be next to beef minestrone. So, a loci hit in a GWAS does not make a polygene-*ic* score, it makes a polygene-*ish* score.

Cancer and pharmacogenomics do not use polygene-*ish* scores and instead only tally alleles with verified risks. The polygene-*ish* scores used in most population genetics are made by using loci that may differ in the exact alleles they represent. So, a polygene-*ish* score in Europeans made from loci A on chromosome 1, loci B on chromosome 7, and loci C on chromosome 12 might not represent the same genes as the same loci on the same chromosomes in an Asian population. There may be some overlap on which alleles are linked to the given loci, but you should not expect the loci to be perfect matches for each nearby allele. Dr. Harden notes that because of difference in allele frequency and the polygene-*ish* score phenomenon, a PGS for one population should not be expected to be functional in a different population.

This is true but limited in two obvious ways. The first is that linkage disequilibrium only occurs within a given chromosome. So, if a GWAS found only a single hit on loci A of chromosome 1 in Europeans while a second GWAS found only a single hit on loci A of chromosome 8, linkage disequilibrium is not capable of explaining that away. Second, the linkage disequilibrium explanation evaporates as soon as you invoke a specific allele of a specific gene. If the hemoglobin gene is involved in your diseases of interest, then the exact loci association may differ between populations, but every population has that gene somewhere. So, if you are claiming that a given allele for hemoglobin matters for a disease, then allele frequency might still differ between groups, but there is no reason that the allele should be counted differently in one population versus another. Remember, this is how the successful polygene-*ic* scores were made for pharmacogenomics. Collecting truly causal alleles into a unified database meant that each patient's risk alleles could be tallied regardless of which exact loci they were nearest, or which unimportant genes were flanking them.

If *RNAseq* or metabolomics were to tolerate the same lack of *portability* as genetics, it would mean that someone could claim that diabetes could be defined by glucose levels in white people, pyruvate levels in Black people, and some completely unrelated metabolite in Asians. This practice was essentially researchers selecting their populations to fit their data rather than fitting their data to the population in what could best be called *statistical gerrymandering*. This practice was cautioned against well before PGS took hold. In 1909, Thomas Hunt Morgan warned about the possible implications of the Mendelian practice of coming up with more complicated genetic explanations to explain observations. He wrote:

> In the modern interpretation of Mendelism, facts are being transformed into factors at a rapid rate. If one factor will not explain the facts, then two are involved; if two prove insufficient, three will sometimes work out. The superior jugglery sometimes necessary to account for the results may blind us, if taken too naively, to the common-place that the results are often so excellently 'explained' because the explanation was invented to explain them. We work backwards from the facts to the factors, and then, presto! explain the facts by the very factors that we invented to account for them. I am not unappreciative of the distinct advantages that this method has in handling the facts. I realize how valuable it has been to us to be able to marshal our results under a few simple assumptions, yet I cannot but fear that we are rapidly developing a sort of Mendelian ritual by which to explain the extraordinary facts of alternative inheritance. So long as we do not lose sight of the purely arbitrary and formal nature of our formulae, little harm will be done; and it is only fair to state that those who are doing the actual work of progress along Mendelian lines are aware of the hypothetical nature of the factor-assumption.

Sadly, we will see in the next chapter that the results of geneticists losing *sight of the purely arbitrary and formal nature* of their claims led to substantial harm by turning the PGS into a potent tool of bigotry and a hindrance to improving patients' lives.

CHAPTER 12
The Jump Shot Gene

BOTTOM LINE UP FRONT: *Excusing the inconsistent results across populations ignores biology and emboldens bigotry.*

It was the final minute of the final game of the 2016 NBA Finals. The Golden State Warriors and Cleveland Cavaliers were battling for the title. The prodigal son and greatest player of all time, LeBron James, had returned from *having taken his talents to South Beach* to bring a title to his home state. LeBron had just made what is known as *The Block* to keep the game tied. Kyrie Irving took the ball on the right side of the court and made a key three pointer over the Warriors' Steph Curry. Curry, for his part, came back the other way and found himself guarded by Kevin Love. Love is a much taller, but less nimble, player than Curry and was forced to try to prevent the best shooter in NBA history from getting off a tying three. Considering the mismatch, the situation, and the opponent, Love put forth what should be considered 10 seconds worth of the greatest defense in basketball history. Curry missed, the Cavs rebounded the ball and Cleveland was (as the announcer proclaimed) "a city of champions once again."

The basic biochemical processes involved in each of the players' performances were identical. LeBron used the same actin and myosin in his muscles to make the game saving block as everyone else. The excitations in retinal cells needed to visualize the rim were the same for

Kyrie Irving as they were for Curry on his attempt. Love's motor cortex fired neurotransmitter signals through the cerebellum, which were carried down his nerves by sodium and potassium balance, until they ultimately found their way to the synapses on his muscles as they responded to Curry's rapid shifts. The same adrenaline was pumping through each of the player's veins. Each of their skin sweat glands were producing liquid in hopes of cooling their bodies. All these events were made possible by the energy provided by the molecule ATP that had been produced by breaking down the sugar from the sports drinks they consumed and glycogen from the meals they ate in preparation for the game.

All these processes are indeed genetically encoded. Myosin in the muscle tissue is encoded on several genes, including *MYH11* and *MLC*. The acetylcholine needed for muscle activation is encoded by the gene called *CHRNA1*. Oxygen was carried to the muscles and carbon dioxide returned to the lungs by hemoglobin encoded into DNA, as well. So, it might seem reasonable to believe that differences in these players' performances, and certainly the difference in skill between these players and your average person, might be caused by differences in the genes encoding the key molecular factors needed for elite athletics. A 2019 report in a *Nature* publishing group paper claimed to have done just that. The authors claimed that the genes *FOLH1* and *VNN1* were linked to elite performance. *FOLH1* breaks down polypteroyl-glutamate into 5-methyl tetrahydrofolate (related to folic acid). *VNN1* is involved in pantothenate metabolism. The implication here is that somehow, subtle differences in how athletes metabolize folate and vitamin B5 contributes to elite athletics. Although these results are almost certainly spurious nonsense, let's assume the findings are true. If every biochemical process in LeBron, Curry, Love, and Kyrie were identical during their finals climax, and if folate and B5 were truly a path to elite athletic performance, would you expect that eating avocados (a source of both folate and B5) would impact the athletes in a dramatically different way?

Let's assume for a moment that all four of these likely future Hall of Fame basketball players have the *FOLH1* and *VNN1* gene variants needed for elite athleticism. Then, let us pretend that we gave them all the same diets, enriched with folate and B5. If our vitamins successfully improved Love's performance but failed to enhance the remaining players, what might you think *went wrong*? You might go to *gene-environment interactions* but all four were in the same arena, in the same game, on the same 4700-square-foot plot of hardwood. You might say, "Gene-gene interaction," but which genes? If some other genes were canceling out the benefits for everyone but Love, shouldn't those genes have been negatively associated with elite athleticism on the analysis? Or would you guess that the plot of land that Love's great, great grandfather was born on made the difference?

If you generated a poly-metabolite score for athletic performance, you would find that the four athletes are identical. Every organ system uses the same cells, proteins, and small molecules to perform the same functions. There is obviously a difference in efficiency, and some people can have genetic defects that can cause breakage in those systems, but the basic aspects of athletic performance cuts across every human. In contrast, genetically speaking, LeBron, Curry, Kyrie, and Love would all be 99.9% identical but would also be assignable to different ancestry groups by the standard practice of population genetics. Even though each player was born and raised in America, LeBron and Kyrie would be assigned to the group with African ancestry, Love would be European, and Curry would be considered *admixed* given his multiracial background. So, the fact that the athletes can be sorted genetically in ways that they cannot be sorted biochemically begs the question, "So what?"

The same is true for diabetes, asthma, or any other disease you can name. A healthcare provider can look at a patient's lab results and know that they have kidney failure. A nephrologist can look at the salts in their blood and determine how to adjust their dialysis settings. But no one can look at the lab profile from a patient and determine the patient's heritage. So, how exactly do the GWAS jockeys propose that the genetic variations have an impact? How can the measurable and predictable differences in genetics between Kevin Love and LeBron James impact their athletic performance *without* generating any signal whatsoever on the biochemistry those genes encode for?

Like in the GWAS era before it, the PGS researchers failed to find any pathways or meaningful insights into biomedicine. Yes, they could mathematically *explain* variation, but even that was dependent on a purposeful refusal to include social, economic, and environmental variables in their models. Rather than ponder whether their PGS was incorrect, the field decided that it was the populations that were incorrect. PGS became increasingly population dependent. One potentially valid way this could have occurred might be if the PGS for the same trait was impacted by cultural differences. For example, a gene that aids in hand-eye coordination would be valuable in every elite athlete but would be more valuable in a culture that prefers basketball than one that prefers soccer. Another example might be that a gene that aids in long-term storage of glucose would again be beneficial to all but would mean slightly more for endurance athletes than sprinters.

Yet this was not how the PGS was contextualized. Even though everyone's muscles use the exact same building blocks encoded into the exact same genes, somehow the hypothesis formed that the constellation of genes for muscle growth could be completely different based on one's ancestry. And it wasn't just European versus African; the PGS became so specific that one calculated for Southern Norwegians could not be used in Central Norway. A 2023 paper by Ding and colleagues in *Nature* (contain your shock) argued that there are different polygenic scores for every possible trait for every ethnicity, and those further varied within ethnic groups. The ability of PGS to predict results for any of the traits Ding's group measured were poor, but if they first sorted people into the same racial categories that the founders of racial hierarchy selected (with different names, of course), then the PGS became slightly less terrible at predicting group-wise effects. Ding's group took things a step further, showing that the PGS was variable even within the same ancestry group, meaning there would be a set of *North Italian asthma genes*, *Central Italian asthma genes*, *Southern Italian asthma genes*, and then *Sicilian asthma genes*. But saying certain genes matter in one group but not another requires that you also claim that those proteins matter in one group but not another. Yet, it would be hard to claim that a PGS for eczema for Southern Norwegians shouldn't include the same skin-specific molecules as would be found in Central Norwegians. You would have to conclude that, somehow, Norwegians would be harmed by subtle changes in filaggrin, but Swedes would be impervious to the same exact changes.

To some extent, this population variation is a result of linkage disequilibrium and the polygene-*ish* score phenomenon I spoke about in an earlier chapter. However, once the specific alleles are being invoked, those caveats do not apply. Say, for example, the main gene governing three-point shooting percentage is the *SWISH* gene. Also presume that African American basketball players have variants of this *SWISH* gene that improve their shooting percentages. These variants may not be exclusive to African American players, but the allele frequency is much higher, and this confers a group advantage. One might run a GWAS for three-point shooting percentage and identify the association with *SWISH* gene variants, then calculate that some portion of the "disparities" in shooting percentage between groups is, at least in part, due to the increased presence of the advantageous *SWISH* alleles. This exact logic is what is performed for genetics of racial disparities in healthcare. Some researchers will run a GWAS for, say, eczema in African Americans and identify some particular set of alleles that increased the risk of eczema. Then, they will contrast the allele frequency for these deleterious variants against other groups and conclude that the disparities in eczema rates are, at least in part, due to the increased presence of alleles that increase the risk of eczema.

Since the functions of the brain are far less understood than the functions of the skin, these *population specific* claims are far more prevalent for mental health and behavioral genetics. Claiming that one set of genes governs muscle contractions in LeBron, but a different set of genes governs muscle contractions for Curry requires you to claim that different sets of proteins and small molecules govern muscle contractions in the two athletes. This claim is easily disproved. But what if someone asked about *basketball IQ* as a metric of performance? Just as for muscle movement, you cannot claim that one set of genes matters for intellect in LeBron but a different set matters for Kevin Love without suggesting that the two differ in which proteins and small molecules control cognition between them. To believe this, you would have to believe that every single process of shooting a basketball was biochemically identical for every NBA basketball player *except* the *decision* to shoot the ball.

Determining a phenotype to assess by GWAS can sometimes be simple. For example, it is easy to define people by their height on a tape measure and then sort by size. Kyrie Irving and Steph Curry are both listed as six feet and two inches, Kevin Love as six feet and eight inches, and LeBron as six feet and nine inches. But what about more complex concepts like *competitiveness?* Which player has the best decision-making skills or which player has the most heart? Barbershops across America might be able to spend an entire day arguing which player has the *highest basketball IQ* because you can't resolve the debate as simply as you might for asking which player weighs the most. Failure to find anything consistent should probably be seen as a sign that complex traits will not be explained by a handful of amino acid changes, especially when those changes do not align with any teaching in the field of biochemistry.

For those who truly believe that IQ is a valid measure for intelligence, answer this question: What would be the true test for athleticism? How would you define an athleticism quotient (AQ)? If we defined athleticism as the ability to play soccer, then every professional basketball player would be considered non-athletic (heck, kicking the ball is a penalty in their sport!). If we defined athleticism as the ability to sprint across land as fast as possible, then everyone in the National Hockey League would be non-athletic, or at least their athleticism would be considered less than their sport's performance might suggest. Should we define athleticism as the ability to perform gymnastics? In this case, most professional athletes would be less athletic than a random pre-teen taking lessons. How about we define athleticism as the ability to tackle NFL running back Marshawn "Beast Mode" Lynch? By that definition, only a couple hundred or so men in the history of sports would be allowed to be considered athletes. Or how about we use the ability to take a cylindrical bat and hit a fist-sized ball traveling at 100 miles per hour?

The reality is, there isn't an easy way to define or enumerate *athleticism* that doesn't get tangled up in the question of "who is more athletic, 8-time gold medal sprinter Usain Bolt or 23-time gold medal swimmer Michael Phelps?" If medals are the metric, then Phelps clearly wins, whereas if running speed is the metric, then Bolt is the answer. However, even if you tried to combine them and say both sprinting and swimming would count towards your *athleticism quotient*, then a person who barely made the decathlon team might be considered more athletic than both of these famed athletes. Serena Williams won her seventh Australian Open professional tennis championship without losing a set, *while at least two months pregnant*; if we included that as part of our definition of *athlete*, then who else would even come close?

Even if you tried to define athleticism by basic movements like running, lateral motion, or reaction times, that, too, would fail. This is shown every single year at the National Football League's Combine, where the top college players perform drills to demonstrate their abilities to teams that might draft them. Every year, some player will put up amazing numbers on the drills, be viewed as a super athlete, get drafted higher than their college performance would have suggested, but then end up being a bad-to-mediocre player in the NFL. Conversely, every year, someone will put up mediocre numbers that will be looked back on with laughter when they enter the Hall of Fame and/or celebrate their Superbowl victories.

An arbitrarily narrow definition of *athlete* would also encourage the same type of bigotry that we have seen with defining intellect strictly by IQ. If we defined *athleticism* by soccer ability, then those of Latin or European descent would be—on average—more *athletic* than any other socially defined race. If we used cricket skills as the definition of athletic, then people with backgrounds in Asia, India, and the West Indies would be *superior athletes*. If we use football as the only metric, then Americans would be the only athletic people on Earth. If we, instead, picked hockey, then only white people would be among the top athletes.

Might you make a *race-specific* measurement of athleticism? In such a scenario you might design a *black AQ* that was mostly based on basketball, but a *white AQ* based on hockey. You would then need to design an *East Coast white AQ* that focused on sailing and squash while *Midwest white AQ* measured football and baseball skills. You'd need an *Indiana white AQ* that is basketball heavy, a *French white AQ* that is mostly soccer related, and a *Canadian white AQ* that is hockey-centric. German born NBA Hall of Famer Dirk Nowitzki would thus be considered athletic if measured on the *Rucker Park Black athlete scale* but not the *German white AQ*.

Although it originated in France, the first widespread use of IQ testing was by the U.S. military as a way to determine whether someone could understand basic orders well enough for the infantry. It was meant as a "yes or no" test whereby if you scored below 85, you would be rejected from military service out of the fear that you could not follow or understand commands well enough to be trusted on a battlefield. On a mass scale, it was never intended to be the lone measure of intellect, nor was it intended to claim that scoring 105 was meaningfully better than scoring 100. It was originally intended on identifying kids who needed extra help, but was quickly embraced by those looking to look at overall behavior and support white supremacists views. So, anyone you hear holding IQ up as some infallible metric of intellect might as well be holding hockey skills up as an infallible metric of *athleticism*. It would be a metric that was not culturally neutral when it was created and would not be truly indicative of athletic skills.

Even if you did make an athleticism quotient that would be used as the sole metric, one major reality would immediately become obvious: Whatever score someone achieved would not be indicative of their innate athleticism nor the athleticism of their in-group; it would only comment on their score at that time and place. For example, if you decided your AQ was going to be just running and jumping, would you be able to compare the results from one group who took the test in freezing cold rain with those who had pristine weather? If an elite athlete scored poorly on the AQ test because they had an active injury at the time, would that score be a fair metric of their innate athleticism? Good nutrition would impact the AQ, a good night's rest would impact the AQ, and training and preparation for the test itself would impact the AQ—all of which would mean the score would only comment on the performance at that moment and would not be a true metric of someone's innate athletic skills.

Here one might argue that the weather, breakfast, and injuries would be noise on an individual level, but the group averages might be the same. Two populations might be reliably different because the effects of weather and nutrition would even out across the group. However, such large assumptions require experimental validation. If your methods assume that the entire environment of the U.S. or even all of Europe is homogenous at scale, you should probably double check your predictions are correct. Yet anyone who has traveled would know that the weather, nutrition, and numerous other environmental variables are profoundly different even within a single U.S. state. Rather than evening out, expanding your study population increases the environmental confounders in your results.

Furthermore, if you wanted to guess at the average AQ for a very large group, it would be important to make sure you had a representative sample. If you tested the jumping ability of NBA basketball players, it wouldn't tell you very much about the average vertical leap for all Americans. Similarly, you could not fairly compare the AQ of an average college student population with the AQ from the residents of a retirement home. And yet, that is exactly what the authors of one of race sciences most beloved studies did for IQ. In one example, they reported an average IQ for the entire population of Equatorial Guinea based only on a single publication that assessed the IQ for children diagnosed with mental disabilities in that country. This study is termed the "national IQ dataset" and despite its obvious flaws and the racist musings of the lead researcher, it still gets uncritically cited in academic journals focused on cognition.

Despite the inability to make one true AQ test, it would be fair to say that no matter the test you developed, elite athletes would outscore your average person. So, one might point out that the test wouldn't be totally arbitrary, even if it were imperfect. Yet, if the only thing your test can tell you is that LeBron James is more athletic than an overweight allergist, it wouldn't be a very valuable test. Certainly, the test could not pretend to have such fine-grain detail as to score one person's athleticism as *102* while the next person's was *101*. The test might not be totally arbitrary on the 50,000-foot level, but on the micro-level, *arbitrary* would be a fair term to throw at it.

This isn't to say that there are no arbitrary metrics in medicine. One of the many measurements of atopic dermatitis is called the Eczema Area and Severity Index (EASI for short and pronounced *easy*). This metric claims to be even more precise than IQ since it measures down to the fraction of a point. So, you could walk away claiming that Eve's eczema is "28.7 bad" but Jason's is only "28.6 bad." IQ apologists often point to these clinical metrics to claim that even if these tests are not perfect, they do measure *eczema* just as IQ could provide a general understanding of *intellect*.

However, what the defenders are missing is that these metrics were designed for clinical trials to measure change over time. The point of EASI is not to say Eve's eczema is 1.0035 times worse than Jason's even if the numbers work out that way; the point of EASI is to be able to state that when Eve was started on a new drug, her eczema got 75% better. Even if EASI isn't *the* perfect one for both Eve and Jason, even if EASI should be tweaked for one or both of them, so long as it has that *ballpark* level accuracy, it can be helpful to assess changes within an individual. The metric was designed because the researchers hoped it would be reactive to changes and dynamic, not because it was meant to say anything about Eve's innate biology. So, if IQ were being used in a clinical trial to show that nutrition programs were helpful or that lead poisoning is bad, it can have value. If it is being used to genuinely claim that Eve is 1% *innately* more intelligent than Jason, it is a failure since it could only truly measure that Eve scored 1% better on the test at a specific point in time, at a specific place.

One other important distinction between even flawed clinical metrics like EASI and polygenic scores is that clinical metrics can be used across populations because diseases are similar no matter where you go. Even though eczema is a skin disease, and thus would be more appropriate to have variation across racial categories, there isn't a *Koreans only EASI* or *EASI for people who are at least 85% Scandinavian*. The EASI metric accounts for the fact that eczema can appear different across different skin tones. For example, light skin tends to turn pink or red while dark skin turns darker brown. I will say that the photo books used to assess eczema severity across skin tones are not as widely available to clinicians as they should be. However, the National Eczema Association has a site dedicated to providing example pictures to help providers recognize and score severity in all people. Other metrics in eczema include a 0-10 scale for how bad the symptoms of itching are for the patient. That metric is used in the same way for every patient. Not all metrics in eczema are written for a global population, mind you. The Investigators Global Assessment (or IGA for short) for too many reasons to detail is possibly the worst clinical metric in all of medicine, and sadly, it is the only metric used by the FDA for assessing whether an eczema drug should be approved. To name one of the many reasons the IGA is poor is that it uses the presence of *pink* as one of its criteria and thus would overlook symptoms in skin of color.

Realizing that eczema symptoms and severity are race neutral, let us revisit the paper claiming to *explain* white Americans' eczema burden using the PGS. We can further add in a paper in the same journal a year prior that claimed that they could *explain* 11% of Canadian eczema burden but could improve that number to 37% if they evaluated a "diverse population". The authors defined *diverse* as including both English white Canadians and French white Canadians. By making an Anglo- PGS and Franco-PGS, the authors reported they could *explain* up to 37% of the eczema variance. To believe these reports, you would have to believe that the genes for white American eczema are distinct from the genes for white English Canadian eczema, which are also distinct from the genes for white French-Canadian eczema. You would have to believe that in only 400 years (a blink on the clock of evolution), the selective factors for skin health between white people sharing a border were different enough to select for different alleles but similar enough to still present a risk for the exact same disease manifestations.

For the differences to occur through nonadaptive genetic drift would require that random mutations happened to land every white North American in a place where they had a set of their own independently cultivated *eczema genes*. Neither of these reports included a single environmental variable that might have provided a better explanation. Neither of these reports included a negative control or any other approach to account for possible population stratification. And the researchers effectively hung a "whites only, French use the side door" sign on their study's recruitment desk. And still, they could not concoct an equation that was more predictive for eczema than just knowing the person's zip code of birth.

A group led by Dr. Eric Jorgenson out of San Francisco took an *eczema* PGS derived from European Americans and applied it to African Americans. They published their findings in 2020 in the *Journal of Allergy and Clinical Immunology*. The authors deserve credit for their report, but the results were not surprising. Despite African Americans having a two-fold increased risk for eczema, the increased burden could not be explained by the PGS calculated from white Americans, by a genetic ancestry PGS, or a PGS for skin tone. The authors concluded, "Our results suggest that social and environmental factors are likely to play an important role in atopic dermatitis disparities and warrant additional research." Yet, the writers of a summary editorial chaired by Esteban Burchard disagreed. They wrote:

> *The low yield of this prior GWAS and the lack of association for the current study's polygenic risk score might be due to a number of factors, including disease misclassification, the frequency of relevant genetic risk factors in populations of African origin (rare variants), lack of representation of these populations in prior GWAS studies used to inform polygenic risk scores, and lack of representation of population-specific risk variants on the gene chips used to study these populations.*

So, to reiterate, Jorgenson's group found that genetics cannot explain why Black Americans suffer from more eczema than white Americans. Jorgenson concludes that this must mean that non-genetic factors are the cause and effort should be put into looking into potential environmental factors. The invited summary postulates instead that it could be: that Black Americans are incorrectly diagnosed with eczema and instead might have some other disease; that Black Americans have a bunch of various eczema mutations that are individually too rare to detect but create the disparities; or that we just don't have enough data on what the *Black eczema genes* are to properly label all the Black Americans with eczema.

Burchard continues:

> *In addition, there might be biological differences in endotypes of disease that are population specific... The authors suggest that this finding might suggest that the social and environmental factors associated with African ancestry are more important in AD disparities than genetic factors. Although this might be possible, it is also plausible that local genetic ancestry or rare variants might be more relevant than global ancestry. Indeed, population-specific rare variants are underrepresented or missed on standard GWAS platforms and hence not necessarily correlated with global ancestry estimated by using these platforms. As such, it remains quite possible*

that some of the disparity is due to poorly identified local genetic factors.

So, the authors are proposing that Black eczema operates through a different biology than white eczema. Kevin Love's skin sweats the same as LeBron's, grows hair the same as LeBron's, and repairs wounds the same as LeBron's. But somehow, the claim is that he would suffer eczema through completely different means. In 2008, Burchard tested a number of people from different backgrounds to look for alleles linked to asthma. The number of people they tested was so small it could comfortably fit in any high school gymnasium, but Burchard concluded that they found a mutation that was *only* in people of African ancestry. He proposed that, one day, they would use this data to find a drug so that if LeBron were having an asthma attack, he could take a race-based inhaler that would be less useful to Kevin Love. Fifteen years on and no one other than the people in Burchard's lab have benefited from his publication.

One of the recurring themes for the failures of genetics is their tolerance for the types of thinking that would be laughed at in other fields. Would Burchard propose that there will one day be a Black person's sports drink? Would the Cleveland Cavs need to have their equipment manager swap out Gatorade for White Boy-Ade for Love, Mulatto-ade for Curry, Black guy-ade for LeBron, and 'Militant Negro'-ade for Kyrie? If Burchard believes in the idea of a unique Black biology for the skin and pulmonary systems, surely, he must believe this for the metabolic system, as well.

At least we can say eczema is a skin disease, and so, maybe you would think pigmentation could play a role. But not only did Jorgenson demonstrate that pigmentation PGS did not play a role, epidemiology clearly shows that the only areas of Africa with eczema rates even close to the U.S. are the highly industrialized cities on the continent. Furthermore, highly pigmented populations in India or Australia do not carry an increased risk of eczema, unlike for keloid scarring (where the scar grows out of control and can become burdensome in its size). The highest burden of eczema is found in the Swedish; the Swedes are known for many things, but a dark complexion is not among them. So, being dark skinned isn't a risk for eczema, being dark skinned in America is.

The only biochemical differences between dark skin and light skin are the balance between folate and vitamin D. In dark skin, the increased protection from the sunlight preserves folate but makes it harder to make vitamin D. In light skin, the increased penetration of the sun's rays generates more vitamin D but causes a greater loss of the skin's folate. Although keloid scars are not fully understood, what evidence is out there suggests that folate metabolism is central to their formation. So, a disease truly associated with pigmentation in the skin appears to be linked to the only metabolite that is elevated in pigmented skin. Makes sense. But the vitamin D and folate differences are literally only skin deep. Serum levels of the two hormones are not different between races, indicating that if you want to claim that racial differences are a matter of biology, you had better be prepared to explain how vitamin D and/or folate in the skin are responsible.

When population geneticists insinuate that African and European brains operate via different biology, liberal/scientific Twitter pushes back, even while the bigots rejoice. Yet when researchers like Burchard imply that African and European skin operates via different biology but can invoke neither vitamin D nor folate, no one seems to notice. I do not know the authors of the editorial, so will not assume their motives or biases, but their logic stems from an idea all too pervasive in medicine: the idea that the default hypothesis for an increased rate of disease in Black Americans should be that there must be an as-yet-discovered inherent defect in Black biology. Letting go of the idea that there is a racial defect for eczema seems just as hard, even if not nearly as controversial, as the idea that there exists an innate, racial defect for educational attainment.

Some of these *Black biology* ideas were encoded into medical records via *normal values* on lab reports. While many are thankfully being phased out, for far too long, medical systems reported racially-defined interpretations of kidney function. In essence, lab reports would have a line stating, "This is your patient's kidney function, unless they are Black, in which case, their kidney function is different."

Lung function tests, for a time, also had different interpretation values for *Black kid lung function* than non-Black. These test metrics were born from an intent to minimize the need for intervention and dismiss disease in the community. Upon the slightest scrutiny, the tests were revealed to be minimizing disease and have been removed from most (but sadly not all) healthcare systems in the U.S.

These Black biology lab values were derived by comparing the function of the lungs or kidneys in African Americans and European Americans. When differences were identified, the researchers could have predicted the environment may play a role in why the groups were different. If they had, they would have uncovered numerous environmental exposures that both worsen lung and kidney function and are more common in Black communities. Instead, they assumed the differences they found were innate and codified a claim akin to saying "Black organs function differently than everyone else's".

Apart from these notorious examples, however, no field of medicine acts as if ancestry should dictate their care. Imagine a dietitian saying, "You should limit how may donuts you eat, unless your grandfather was Scottish. Scots can eat the entire baker's dozen." Or imagine a dentist saying, "Oh, you are 1/8th Native American? Well, in that case, you don't need to floss!" You can't imagine it because those kinds of sweeping statements based on demographics would be laughable.

Here, these types of researchers make it painfully obvious that they have never actually cared for a patient, let alone cared about patients. It may be fair to have a patient in front of you who is describing symptoms that could indicate that they have HIV. It may be fair to be more concerned if that person has been engaging in unprotected sex with multiple same-sex partners. What is unfair, however, would be to ultimately decide to test them for HIV *only* because they identify as homosexual. Or to decide not to test your heterosexual patient for HIV despite their reporting a similar number of sexual partners. Or if a white man was in your clinic with all the symptoms of prostate cancer, would you decide not to test him because of his skin tone? As stated, for any medical finding to have meaning in the clinic, it must help guide decisions. Can anyone describe a clinical situation where, all other things being equal, a person's sexuality would be the deciding factor in whether to test a man for HIV? Can you think of an example that, all other things being equal, you would send a Black man for prostate cancer work up but not a white man?

Women experiencing heart attacks will tend to share left arm pain and indigestion symptoms with men, but are less likely to report chest pain. And yet, here, too, if you were caring for a patient who was short of breath and had left arm pain but no chest pain, failure to evaluate for a heart attack based only on their gender would be malpractice. So, the valuable medical information is that some patients can have heart attacks without chest pain and such patients are more likely to be women; however, since this finding isn't exclusive to women, clinicians cannot use the patient's gender as a proxy for the decision to evaluate for heart disease.

While the rest of medicine moves towards the realization that a patient's social group should not play an active role in their care, those in population genetics are tripping over themselves to recreate the *Black people kidney function* days of old. Worse still, they are doing it under the banner of bolstering diversity equity and inclusion (DEI) initiatives. These claims go beyond implying that there is a gene for a good three-point jumper, and it just so happens that Kevin Love is one of the few white men to inherit it. Rather, these claims imply that Kevin Love has the white guy three-pointer gene, Kyrie Irving and Lebron have the Black man three-pointer gene, while Steph Curry and his teammate, Klay Thompson, each have *distinct admixture race* three-pointer alleles.

Just as the editorial by Burchard insinuated, population geneticists believe that if databases keep sequencing more Black people, they will be able to find the genes that contribute *to* racial disparities in healthcare. The mental bias that more African representation in the genetic databases will uncover the innate causes of Black inferiority in disease outcomes is being draped in the sheep's clothing of wanting to improve those outcomes through the promise of deriving *race-specific algorithms* (as was promised in the *New England Journal of Medicine* by Oni-Orisan mentioned previously). There is no race specific biology—not in cognition, not in eczema, and not in three-point percentage.

In a way, this is the genomics form of white saviorism. Insisting that genetics databases diversify their enrollees is great. But doing so while implying that the reason we need more diverse genetic information is so that science can aid minorities by discovering the genes encoding their innate disadvantages isn't exactly an anti-racist stance. Claiming that we need to sequence more impoverished white people so that we can find a drug to reverse their inborn errors isn't an egalitarian position, either. Therefore, regardless of whether the belief in the ability to find a *Black eczema* PGS is held with a genuine intent to improve racial equity, overwhelming ignorance, or malice, the result is the same: to continue to feed the types of biologic essentialism that manifested in the Buffalo killings.

So, if someone tells you they are researching the genes for why people of African ancestry are more prone to asthma or why left-handed South Asian dairy farmers develop gout, ask them, "Do African Americans have genes that increase their three-point percentage?" Grant them that no gene *determines* if a given shot will be made, but if the genes truly act *probabilistically*, then over the course of a game or season, the effects should add up in terms of more made shots. So, ask them if Steph Curry has genes that increase his shooting percentage, how many of his over 3,000 three-pointers made can we *explain* with genetics? If Black people have genes for jump shots, why don't they seem to have proteins or metabolites for three-pointers? Do Hall of Fame white basketball players like Larry Bird have the same jump shot genes as their Black teammates and opponents? Or does Larry Legend have his own European *SWISH* alleles? And most importantly, how on Earth can someone evaluate the role of the jump shot genes *independently* of the thousands of hours of practice each of these players has dedicated to his trade alongside world class coaches and nutritionists? Kyrie Irving made his three pointer, Steph Curry missed his; to believe that genetic sequence differences between them had anything to do with that is modern day race science and incompatible with biology.

CHAPTER 13

The Equal Environment Assumption Continues to Make an Ass Out of Geneticists

BOTTOM LINE UP FRONT: *Inability to recognize their own implicit bias influences otherwise well-meaning research by geneticists and clouds their views on the role of the environment.*

Imagine you are strolling through an incredibly affluent neighborhood on a sunny afternoon. Every house is a sprawling mansion, and every lawn is perfectly manicured. You notice a young woman jogging down one of the wide and inviting sidewalks. She is young, blonde, white, and attractive. She is wearing designer jogging clothes and high-end running shoes. What assumptions might you make about her? You obviously assume she lives somewhere in the neighborhood since nothing seems out of place for a young white female running through an affluent neighborhood. But does your mind assume what she does for a living? Does your mind guess that she might be a lawyer, a doctor, or business mogul? Or might you wonder if she is a stay-at-home mom whose husband is the one who earns the money needed to afford their lifestyle? What kind of car does she drive? Is she nice or rude to service workers? Does she do her own lawn work and if she does not, does she know the names of the people who do the work for her? What are her political beliefs, what are her favorite TV shows, and does she prefer wine, beer, or (my guess) mimosas?

Now, imagine in the distance you see another person out jogging. He is also white but is a male covered with tattoos, with multiple piercings, and hair dyed to have a green stripe across half his head. Rather than high-end running gear, he is wearing baggy jean shorts and ratty looking skateboarding shoes. What assumptions are you making now? He seems more out of place than the young woman, and so, to *place* him, your mind might need to come up with far more complex connections. Unlike the young woman, you probably don't assume he is a stay-at-home dad going jogging while his breadwinning wife is at work. Maybe he still lives with his rich parents in the neighborhood but has not reached his potential? Maybe he is a musician or maybe he doesn't live there at all and is running from a house he just robbed?

Dr. Jabraan Pasha presented this scenario in his presentation titled "Destigmatizing Implicit Bias". Dr. Pasha is an internist with over a decade of clinical experience and host of the podcast, *Lean In with Dr. Jabraan Pasha*. Dr. Pasha in no way attempts to minimize the impact of implicit bias nor excuse the effects of acting on one's flawed assumptions, but he does aim to *destigmatize* the fact that making mental shortcuts is a part of the human experience. Humans are designed to recognize patterns. Thus, anyone with enough experience driving through rich neighborhoods will recognize the pattern of what people who live there tend to look like, tend to wear, and tend to do for a living. There are few affluent neighborhoods in which a slightly disheveled looking man with tattoos will be as common as a brunch loving blonde, and because of that, our minds will always make different assumptions about why different people might be present in the same space.

Dr. Pasha also makes it clear that the impact of explicit bias is a serious harm, such as overtly racist statements made from one person to another or overtly discriminatory laws. However, he argues that the impact of implicit bias can be even more damaging precisely because it goes unrecognized. The point of *destigmatizing* social bias *is to admit that pattern recognition is natural* and not something to feel guilty about. Acting on such bias, especially codifying it into policy, is where serious harms can arise. For example, there is no shame in noting that the tattooed man in clothing not designed for jogging is uncommon for a rich neighborhood; however, the decision to call the police on him based only on the assumption that he must be there with ill intent is a problem.

After the murder of George Floyd in 2020, America seemed to begin recognizing the role of race in society. Academic centers attempted to do their part with a proliferation of trainings in *diversity, equity, and inclusion* or DEI. As of this writing, new trainings and lectures on DEI are still popping up around America. Some of these programs have been effective in improving racial equity, while others have been nothing more than an attempt to be seen as *doing something*. Most of the pushback against DEI initiatives wase not made in good faith and was more a reflection of a desire to maintain established social hierarchies. I need to stress that the pushback against DEI programs does not fit into a *both sides* model that is the norm in major media coverage. Some states, which claimed to care about *local government*, have banned programs that attempt to address DEI and have banned books that discuss racism. Meanwhile, professors have been fired for teaching about Black history or women's studies. I say this to make sure it is clear that the legitimate criticisms of some DEI efforts made in an attempt to genuinely make the world a better place are vastly outnumbered by authoritarian nonsense.

Dr. Pasha outlines that for all the work poured into DEI programs both before and after the 2020 Black Lives Matter protests, the impact of most programs is underwhelming at best. The only interventions that appear to meaningfully improve scores on implicit tests are actually meeting people from other groups. If you grew up in a culture where it was more

common for women to be stay-at-home moms, you will be more likely to
initially assume that the woman jogging through a rich neighborhood
was a stay-at-home mom married to a man who serves as the home's
breadwinner. If you were raised by a lesbian couple who were both
lawyers, your initial guess might be different. If you personally know a
lot of successful punk rockers who jog in their skater gear, then the man
mentioned running through the same neighborhood would not strike
you as out of place. So, if actually meeting people is what matters, what
should we make of academics that work in a field that is so homogenous?

There is no shortage of diversity in the field of genetics compared to
other endeavors, but like all of science, the differences between who
works in niche sub-fields can be stark. The crowd working on *genes for
education* is nearly entirely white, mostly male, and all from upper middle
class or above. They trained in similar institutions with similar mentors.
They go to the same conferences with the same people every year. Even
when they construct panels to discuss their field's shortcomings, the
scientific *critics* who get invited are typically within the genetics field itself
rather than those actively researching environmental science or social
determinants of health. Of no fault of their own, most of them were
raised in societies with structures and beliefs that imply that people living
in poverty are *less than*. They live in societies that have both current
systems and long histories of implying that minorities are not only
distinct from white people but are of *lower stock*. Some of them grew up
in a British culture that taught them that nobility could be inherited
along with the rights to the Indian sub-continent. This is not to imply
that any of them harbor these views overtly, but if they have grown up in
a biased society and come into prominence in a sheltered and privileged
field, what kind of implicit bias might they bring to their work?

The Groundwater Approach is a model written by Bayard Love and
Deena Hayes-Greene of the Racial Equity Institute. The model states that

if you were to come upon a lake that had one dead fish floating in the water, it would make sense to research that fish. It would make sense to presume that there must have been something uniquely wrong with that one fish. But if you were to come upon a lake where 50% of the fish are floating dead, you will know not only that it makes more sense to study the lake, but you will know not to get into the water. But what if you come upon a lake and 25% of the black fish are floating dead while only 15% of the white fish are floating? Then, what should you presume? I would argue that the implicit bias that geneticists bring to the table is that they would assume that there must be something *inherently* flawed in the black fish that explains the differences in results. The field of population genetics was built by people who believed that non-white people were genetic *degenerates* and operated under that assumption for a very long time. While none of the current crop of prominent geneticists may believe that, they still carry the bias that *certain* people have an "increased rate of genetic predispositions."

From an intent standpoint, these ideas are extremely different. From an operational standpoint, they are the same thing. If one researcher states that "people of African ancestry have an increased burden of genetic predispositions to asthma" and the other says "Black people have genetically defective lungs," these statements are different in intent and scientific nuance but not in interpretation. If you were given a choice of whether your child would have a fatal reaction to eating a peanut or not, you would clearly opt against an allergy. So, you don't need to be overly politically correct in avoiding the claim that any gene that would increase the risk of dying from peanuts would be less desirable than a version of that same gene that did not increase such risks. Thus, if you believe that genes that cause food allergy are real, and you think that the average Black person has more of those variants than the average white person, then you are advancing the idea that Black people are genetically defective for food allergy. You might not be claiming that Black people are genetically defective *overall*, but this is the message that is sent when geneticists attempt to explain racial disparities by claiming that Black people must have *increased genetic predispositions* to allergy, diabetes, COVID, cancer, heart disease, kidney disease, and so on.

The implicit bias in genetics causes those in the field to be more open to the explanation that the dead Black fish must be innately more susceptible to a shared exposure rather than thinking that there may be a difference in exposure that may explain the results. This is called *biologizing*, which means that any difference observed between groups is assumed to be caused by biology and not the environment. The discussion of the PGS for eczema in the last chapter is a perfect example. Despite evidence that probing the genome could not explain the disproportionately higher rates of eczema in Black Americans compared to white Americans, Burchard insisted that people not lose sight of the possibility that Black Americans were innately inferior in some biological way. The terminology used was obviously more politically correct than that, but the concept is the same.

Underlying all these biases continues to be the original sin of population genetics—the equal environment assumption. Again, this error isn't limited to eugenics-friendly researchers. Despite being arguably the first truly anti-racist geneticist, Lewontin had a tendency to make this error. Despite being born of social justice initiatives, The Groundwater Approach also, in part, makes this error. The analogy of the lake comes from a skewed thinking that a *shared environment* even exists outside the controlled conditions of a laboratory. The real environment is far more dynamic than the analogy of a fish swimming in a lake can capture. If someone pours a toxin into one part of the lake, the laws of diffusion will assure that eventually, the entire lake will have a near-equal distribution of that toxin. But this is not true for the environments that humans traverse.

Lewontin was the first to note this error in hereditarian thinking and the conclusions about heritability that the genetic field was drawing from Mendel's work. Lewontin put forth a now well-known thought experiment using a drawing depicting two different pots of plants, such as the one below.

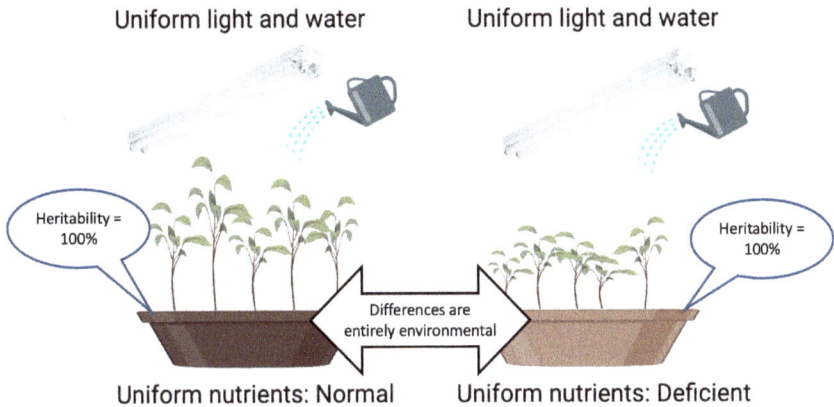

Lewontin put forth that one pot might have adequate soil, water, and light, but would still have differences in the heights of the different plants. In a second pot, the soil, water, and light may be deficient, but the plants would still have variations between their heights. In both pots, the differences between the heights of the plants would be calculated to be 100% heritable. The equal environment between the plants within each pot would mean that any remaining difference would be due to genetics. Yet, the differences *between* the two groups of plants would be entirely environmental. The height trait would be 100% heritable but the finding that the deprived plants were shorter would have nothing to do with genetics. Lewontin was clearly making this analogy for bigots who concluded that since the heritability of certain traits were high within Europeans, then the differences between Europeans and other races must also be genetic. However, even Lewontin's anti-racist analogy did not capture just how dynamic the environment truly is, even within the world of plant genetics.

I spoke with Dr. Kevin Bird, an expert in plant genetics at Michigan State University. If you were voting for who would best embody a modern-day version of Lewontin-style anti-racism, Dr. Bird would be a top pick. While there was much to discuss about his fondness for scrapping with race scientists online, my question for him was why population genetics for plants and animals has been so much more successful than it has been for humans. Generations of race scientists had pointed to the agricultural advances in crops as evidence that human eugenics should work in the same way. So, I wondered why the idea of taking lessons from plant genetics and applying them to humans was so wrong.

Dr. Bird relayed that the first key distinction is that no modern plant geneticist overlooks the importance of the environment. In a field of genetically identical corn plants, a well-known phenomenon is for the plants on the edges, and particularly the corners, to be slightly shorter than all the other stalks. This is because the plants on the edges take the brunt of the wind and other challenges while the plants in the middle of the field are protected by those surrounding them. For similar reasons, when a field has hills, the plants at the top of the hill will be subtly different from those at the bottom. Each time I drive to my in-laws, I pass a corn field planted with genetically identical plants that confirms Dr. Bird's observation that the plants closest to the road and most exposed are shorter.

Farmers exhaustively try to maximize the environment to keep their fields growing well. Yes, they might want to select the best seed, but they work to assure every plant has adequate nutrient delivery through complex irrigation. Another limitation on drawing human conclusions from plant genetics is that crops are bred for characteristics under strong selection. For example, crops are selected for drought or cold resistance. These are big phenotypes that would be expected to have a significant impact on survival. The plant either survives the drought or it does not. Crops are not bred based on subtle things like whether the leaf has 15 serrations per edge versus 16. Therefore, plant genetics does not use proxy measurements. The researchers who look at education would prefer to work on intellect or cognition, but they don't have a genuine measure of what that means, or don't have access to data from enough people who both donated a DNA sample and took an IQ test. Plant genetics is focused on growing more crops per a given area; thus, their measurement is clear, since you can count the exact number of bushels of strawberries collected at harvest. If plant height, leaf serration number, or berry redness happen to correlate with yield, so be it. But no one in plant genetics would focus their life on measurements that are not what they truly want to look at.

As such, failure of any genetically modified organisms (GMOs) won't be brushed off as a sign that a gene must work in Greek strawberry plants but not Italian ones. If Gene X is really a drought resistance gene, then moving it from one seed of corn to another should create drought resistance in the recipient crop. If this fails, it is a sign that some assumption must be incorrect and further work is needed; it is not an excuse to write a paper claiming that Gene X has unique impacts on drought resistance in only one set of Greek strawberries planted in one small field in Athens. The final advantage for plant geneticists that Dr. Bird outlined is that the ability to screen GMOs, or other crop variations, is much easier for plant geneticists than screening humans. A crop geneticist can plant a vast number of identical seeds in a large field, creating far more *clones* than most genetics work could dream of. The plant geneticists have easier times assessing these plants over various fields and can return each year to reevaluate. Thus, once plant geneticists have isolated one gene they suspect will improve yield through drought resistance, they can plant it thousands of times in multiple environments to evaluate the results. Human genetics obviously cannot accomplish this, but even those working with mice or cell cultures could not generate such experiments.

One of the earliest perceived successes of GWAS came from farm animals. Researchers took various chickens into a lab setting. They let the chickens breed and measured how fat they were at adulthood. Since chickens are sold by the pound, the goal was to produce heavier chickens. The researchers looked at all the genetics of their various lab chickens and sorted by weight to see genetic loci that were associated with being heavier. Using this, they were able to identify genetic signals that would point breeders towards selecting the birds that would yield larger weights and higher profits. However, this was in a lab setting—not a farm that might have more variations between which field the chickens roamed, and not even a barn that might have a draftier corner than the rest of the structure. This was a lab.

The geneticists that decided to take the chicken lab GWAS and apply it to humans were overlooking some fairly pronounced differences in human environments. Imagine you tried to do the opposite experiment and convert a chicken farm so that it has similar variations in environmental exposures as a human population in the United States. Rather than feeding all the chickens an equal allotment of seed, at least 11.6% of adult chickens and 5.6% of chicks would need to be exposed to food insecurity and intermittent starvation to match the statistics from Feeding America, a nonprofit dedicated to mitigating poverty. You would need to keep some chickens in dense environments with high exposure to car exhaust while others would need to be rural and forced to drive everywhere they go. At least a third of your chickens would need to be denied access to healthcare when sick. Some would be fed processed seeds made from refined sugar while others would have access to a local market with organic whole grains. Some of the chickens would need to be exposed to early childhood traumas, abusive homes, and whatever form of chicken discrimination one could think of. Much like Lewontin's thought experiment with the plants, the original GWAS could successfully identify what genes were responsible for differences in sizes because the conditions in a lab can be tightly controlled. But for humans, an update to Lewontin's drawing is needed.

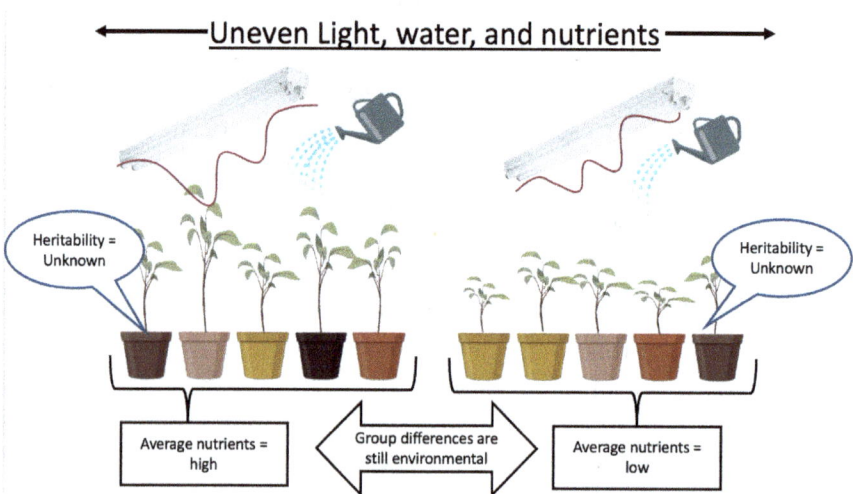

Since one cannot assume that the environments are ever identical, the only way you can ever really know that the variations you see are due to genetics is by using the kinds of crop yield assessments Dr. Bird relayed. Certainly, large differences in crop height, such as those depicted, would need large differences in environment. However, subtle differences in plant height could just as easily represent subtle differences in nutrient or toxin exposure. In short, you can't know that you are studying a drought resistance gene if you don't know your plant is experiencing a drought.

At least a few geneticists began to understand the concern that the environment may be confounding the genes presumed to be connected to human traits. One example is from the aforementioned Abdellaoui, who published in *Nature Genetics* in 2022 that educational attainment saw the greatest decline in genetic association when they accounted for where someone was born and where they currently lived. This means that taking into consideration that someone may have grown up in an under-resourced area was important enough to erase much of the presumed genetic links with education.

Eric Turkheimer replied to a tweet by one of Abdellaoui's co-authors, Michel Nivard, who was presenting how the GWAS's accuracy for predicting education differ by study design. Turkheimer read the graph the way any sensible person would, seeing that it indicated that as the environment is made *more similar*, the ability of genetics to *explain* differences plummets. This set up isn't exactly a negative control, but it does show that improving the control group by looking at people who live in more similar areas reduces the perceived contributions of genetics. Turkheimer's tweet is shown below (a reminder that h2 stands for heritability).

Eric Turkheimer @ent3c · Apr 25   ···

This is how nature-nurture ends: not with a hereditarian bang (Jensen thought h2 of IQ was .8) or some brilliant rebuttal showing it is 0. Properly deconfounded, behavioral h2 is non-zero, interesting, sometimes useful, and small enough to drown in a bathtub.

> **Michel Nivard** @michelnivard · Apr 24
>
> Replying to @SashaGusevPosts @dr_appie and @hastingscenter
>
> I don't know, I don't know the exact timing of the various papers, and whether the Howe et al. paper was read by everyone yet. @dr_appie and I have this figure in which we line up a bunch of SNP-h2 estimates, regrettably it got booted to a supplement. I would consider us "BG"

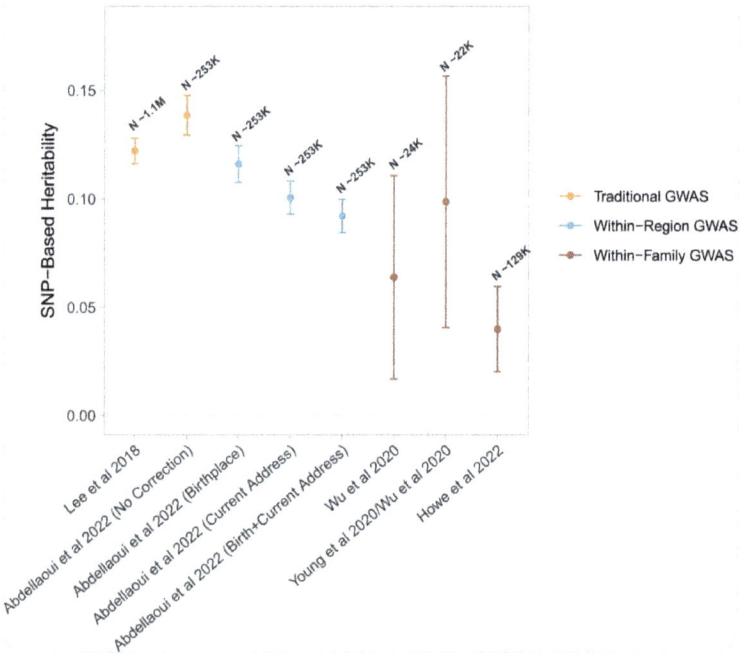

The data show that if you account for people's place of birth and their current residence, the strength of connection between genes and education declines. From an environmental perspective, you would read this as indicating that those who live in poverty or highly polluted areas would be limited in their ability to go to college. In contrast, those who move to more affluent neighborhoods or have been living in those neighborhoods all along would have an advantage. So, basically, if you take the person's surroundings into account, the ability of genes to add to the story diminishes. If you progress to compare only people in the same family (as Howe and team did in 2022 in the graph), the genetic *signal* erodes even further.

Nivard and Abdellaoui did not see it this way. They instead interpreted this to mean that people who left the under resourced neighborhoods had genes that allowed them to leave. Rather than looking at moving from a neighborhood as a possible negative control, he is basically pretending negative controls don't exist. As much as I would love to believe in a *movin' on up gene* or a *bootstrap gene*, this implies that those left behind lacked the genetic material needed for upward mobility. It implies—like Harden was hinting at in her response to the GWAS for air pollution tweet—that those living near pollution lack the genetic gifts needed to move to a better neighborhood. The authors also viewed their results as evidence for *gene-environment* interactions. They never state, nor even hint at, what might be the environmental differences between neighborhoods that might be causing differences in educational attainment, nor do they imply how these factors might operate through one set of alleles in South Londoners and a different set in North Londoners.

A similar concept was put forth by an England-based group led by Melinda Mills, who claimed to have found alleles for intergenerational occupations. They looked at the well-known pattern of people working in the same professions that their parents or grandparents worked in and concluded that there must be a gene for occupation being passed along. To believe this, you would have to wonder whether the genes for computer programmers came into being right as computers did. Would the genes for chimney sweeps all disappear from England like Mary Poppins floating into the clouds? Had they derived a negative control like one's favorite recipe or one's first initial—neither of which would be genetic—they would have compared how much more likely people were able to *inherit* their parents' profession versus inherit their interest in Royal Family gossip.

That said, the ability of geneticists to parse out the role of a more-similar environment was further improved by researchers like Dr. Alexander Young of UCLA, who modified the standard GWAS approach to look at only the genetics that differ between siblings. Rather than looking at huge populations, Dr. Young's research looked only at people living in the same home, under the hypothesis that one can fully account for the environment inside a single residence. When controlling for the environment to the level of the home, the ability for genes to *explain* differences substantially eroded. Young's paper suggested that the links between polygenic scores and either educational attainment or cognitions dropped by 25-50% when switching to a within-family analysis. Okbay and colleagues demonstrated a 70% drop using the same approach. A group led by Dr. Howe out of Bristol in the U.K. calculated a 76% drop in genetic association with education. To make matters worse, a 2023 paper by Schoeler in *Nature Human Behavior* indicated that education was the most skewed trait between the participants in the U.K. Biobank and the general British population. Altogether, this means that the standard population genetics approach relies on a non-representative sample and assumes the enrollees must have such similar environments that differences between subgroups must be due to genetics.

But here is one thing to consider: The geneticists who work on educational attainment, including many that I have derided thus far, admit that their prior results were confounded. As much as I would love to relay the results on how within-family GWAS impacts studies for allergies, rheumatology, or heart disease, I simply can't because as of this writing, no one has reported any. I do have to concede that the education attainment crowd's willingness to look into confounders allowed them to identify the impact of within-family analysis. The geneticists for mental health disorders as well as immunologic and metabolic diseases seem oblivious to the confounding in their work.

However, lest we get too supportive, there is still one major shortcoming of how researchers view the within-family results: They don't see a need for any further environmental adjustments down to the level of the individual. The genetic connections claimed by traditional GWAS erode when using the analysis that controls down to the level of the family. One interpretation of these findings is that continued adjustment for the environment will further weaken genetic signals. The second reading, and the one the geneticists like Young take, is that their within-family approach provides the *true* results and no further environmental considerations are necessary.

Young likens the within-family approach to a randomized clinical trial (or RCT for short) for a new drug. Young's claim is that the variables of interest in a within-family GWAS are the genes from a specific set of parents. With each child, the genes are *randomized* between Mom and Dad so that each sibling gets about 50% of their genes from Mom and 50% from Dad. So, for any given gene from the mother, there is a 50/50 chance that her child will inherit that gene. Young considers this akin to a placebo-controlled drug trial where for each participant, there is a 50/50 chance they will get the active drug versus the placebo. *The Genetic Lottery* also bases an enormous amount of perceived validity of GWAS on the idea that sibling studies are *randomized experiments* in the way that a drug trial would be. The notion that two siblings are otherwise identical save for the randomly inherited allele from their parents is lifted up as the justification as to why within-family studies are valid.

The first of two brief nitpicks on this is that this hypothesis of "the genes are randomly inherited" isn't even correct. Any allele that reduces the chances of an embryo thriving would be less likely to be inherited than the healthy copy of that allele. Recall the discussion about *STAT3* deficiency in which patients have one mutated copy of *STAT3* and one *normal* copy. Embryos that have the mutated copy have slightly lower chances of surviving to birth than embryos containing the healthy *STAT3*. Thus, on a population scale, the inheritance is non-random because of selective pressure against *STAT3* mutations. The second nitpick relates to the fact that every cell has an accessory DNA genome in its mitochondria. As discussed previously, the mitochondria for each cell is responsible for producing that cell's energy. The mitochondria have their own genome that is completely separate from that of the nucleus. It is a very small genome at only 16.5 thousand base pairs and 13 genes compared to the 3.3 billion base pairs and over 30,000 genes of the nuclear genome. Yet there are monogenic disorders of mutations in the mitochondrial genome that may impact up to one in every 1000 people and are often marked by progressive loss of movement and neurocognitive abilities. Importantly, your mitochondrial genome was inherited entirely from your biological mother and is thus, not randomized.

But even when we ignore the potential for influence by the mitochondrial DNA or selective pressure against mutations, does the premise that within-family GWAS accounts for the environment hold? The problem with this thinking is that just because *your* variable of interest is randomized does not mean that *every* meaningful variable is randomized. If you walk up to any researcher who is familiar with clinical trials and ask them, "What is in Table 1?" they will all provide the same answer. In clinical trials, Table 1 is nearly always dedicated to a comparison of the characteristics of people who got the active drug versus those who received placebo. The point is to show that your randomization process worked. If the treatment group had significantly more females than males, then you would need to consider that when analyzing the results. If, for example, women in the study tended to get better no matter which group they were in, then an imbalance might make it seem like your active drug worked when it was really just a reflection of poor randomization.

A lot of work goes into making sure trials are properly randomized by as many variables as are known to be relevant in the disease process. For eczema, clinical trials need to be randomized by age, sex, starting severity, and current medication at minimum. Drug trials collate known factors that influence drug responses and either show that those factors were evenly distributed between the groups in the study or show how they adjusted their analysis to account for any imbalance. I asked Young how he could be confident that his groups were balanced for important factors without Table 1. He replied:

> **Alexander Young** ✔ @AlexTISYoung · Dec 14, 2022 ···
> The point is that the assignment of treatment is random. This is true both of an RCT, where a random number generator assigns people to treatments, and of sibling genotypes within a family, where random segregation of genetic material during meiosis determines genotype.
>
> 💬 1   🔁   ♡   ᠁   ⬆

> **Ian Myles MD/MPH** @lcdriammdmph · Dec 14, 2022 ···
> YOUR variable being random does not mean your groups were analogous. Table 1 of every RCT shows subject demographics to account for confounders the researchers hoped to evenly split between groups. GWAS studies assume those (mostly unknown) confounders even out in families.
>
> 💬 1   🔁   ♡   ᠁   ⬆

> **Alexander Young** ✔ @AlexTISYoung · Dec 14, 2022 ···
> As long as your treatment is random, asymptotically the treatment groups will be balanced for confounders, and treatment effects are unbiased.
>
> 💬 1   🔁   ♡   ᠁   ⬆

The claim here is that if you just get enough people, all the differences within a family will even out to near zero. At this point, I started wondering whether population geneticists might have a severe allergy to reading anything in the field of epidemiology. The idea that everything would come out in the wash within a family would only make sense if there was data that birth order does not matter for disease.

The chances that a child will develop eczema drop with each number of older siblings they have. In the paper, "Cohort study of sibling effect, infectious diseases, and risk of atopic dermatitis during the first 18 months of life", a group led by Peter Aaby demonstrated that having one older sibling reduced the risk of eczema by just under 10%, having two reduced the risk by about 15%, and having three cut the risk in half. Even going to daycare reduced the risk by 15%. If the effects of passing on genes were truly random, then how could the older siblings seem to consistently get the eczema-inducing genes?

Researchers Bjorkegren and Svaleryd reported in *Social Science & Medicine* that younger siblings were more likely to suffer injury, abuse alcohol, and contract upper respiratory infections. Older siblings were more likely to suffer depression, ADHD, or endocrine disorders like diabetes. How could this be if the parents' genes were truly randomly assigned to each sibling? Maybe genes are only a small part of the picture, and thus not enough to account for whatever environmental factors must be causing these differences in younger siblings, but the point remains that if within-family GWAS were like an RCT, any comparisons between siblings for, say, ADHD would have to show that they adjusted for birth order in their analysis.

In the allergy world, a lot of work went into researching whether the increased rate of infections in younger children somehow primed the immune system in ways that prevented allergies. This fell under the *hygiene hypothesis* which claimed that decreased exposure to certain microbes (due to sanitation and cleaner environments) may be causing allergic disease. The hygiene hypothesis is partially correct, but getting more common colds was not the protective factor. Infections that genuinely protect against allergies are the types that have not broadly impacted industrial societies for 100 years, such as hepatitis A, tapeworms, and other parasites.

Work from numerous groups demonstrates that moving from a non-industrialized nation to either the U.S. or U.K. can dramatically increase the risk of developing eczema but only if you move before you are two to four years old. Moving after you are six years old, you will make no appreciable difference in your risk of eczema no matter where you moved from or to. Within the U.S. and U.K., there is an additional risk of living in highly industrial cities. So, say you have a family with two children, one of them six years old and the other only one year old. If this family moves from rural Uganda to urban Baltimore, the parents will have no change in their risk of eczema, the six-year-old will have no change, but the one-year-old will see their risk jump 10- to 40-fold. There is also variation with more subtle changes like the shifting seasons within the same city. Studies from both Japan (Yokomichi, et al, *BMJ Open*, 2021) and Taiwan (Kuo, et al, *Allergy*, 2016) demonstrated that children born October-December had a higher risk of developing eczema than those born in the spring. This might be due to the dryer months early in life or could be related to the timing of the various trimesters of pregnancy leading up to the late-fall, early-winter birth. I suppose the researchers could have presented their findings as a sign that Scorpios and Sagittarians had innately deficient skin.

Young does not work on allergies, so maybe he could be forgiven for a failure to realize the impact of birth order on inflammatory syndromes. But even for educational attainment and income—traits Young does research—the impact of birth order is clear. According to the *National Bureau of Economic Research* (NBER), children born earlier in the birth order will tend to have greater income, higher IQ, spend more time on homework, have lower birth weights, and make different career choices.

The reasons for this could be plentiful. Parents may be more comfortable in their parental roles with subsequent children compared to the onslaught of new experiences faced by first-time parents. Parents tend to be stricter in their discipline of first-born children. The economic impact on the parents is often difficult to sort out. It may be that parents have a second child after obtaining a level of financial security that they lacked with their first. Or the second or third child may create financial strains that were not present when only one child was in the home. All these factors lead to the observation in psychology that *no two siblings have the same parent*. This means that parenting techniques between children will, by definition, differ in the level of parental experience, financial situation, and age.

Worse still, within-family GWAS would continue to ignore the potential role of environmental toxins. Per the aforementioned NBER report, parents were more likely to have smoked during the pregnancy for the first-born compared to their subsequent children. That alone could matter for a host of traits and diseases. It would matter if a mother ate healthier during her second pregnancy or changed occupations between children. What if a family had a child in Flint, Michigan during the height of the crisis that saw millions of people poisoned by lead contamination in their drinking water? If the family had one child in Flint during the crisis but moved to get away from the situation, birth order would matter. Since lead is absorbed into the lungs better than the gut or skin, if the older child in Flint received a shower while the younger was cleaned in a bath, the older child would be at increased risk for toxicity.

The original twin study assumed the environment was the same for identical twins and fraternal twins. Those running GWAS studies in humans assumed that if one limited the evaluation to a particular area, then the environment would be mostly equivalent across their cohort whether they were looking at all of Europe or just one part of Norway. Now, the sibling-paired GWAS assumes that the environment within the home is interchangeable across families. Each wave of genetic claims is based on the assumption that even when only a few of the environmental factors relevant to their diseases or traits are known, they can still assume them to be evenly distributed across their study population. In effect, just as the PGS founders assumed that collating GWAS signals would replace the work of validating genetic mutations through functional studies, the sibling-pair GWAS assumes that obtaining a large enough sample size can replace the work needed to identify confounders and account for them in the analysis.

Some population geneticists will tell you that they account for the environment using their *shared environment* coefficients. Recall from the chapter on twin studies, that even the pre-Victorian geneticists tried to calculate how impactful the environment might be. The problem is that these approaches only look at genetic variables and then guess at the average environmental exposures. Say you hypothesized that someone's body temperature influenced their test scores and you ran a test showing temperature *explained* some portion of how well test takers performed. If a detractor pointed out that the temperature in the room where the test was taken might also matter, would it be a fair excuse to claim that you can guess your test subjects' *room temperature* based only on the *body temperatures* of their parents? Even if you think you could pull this off, it would be vital to test how well your model performed at predicting the room temperature and adjust for any shortfalls, just as meteorologists do. If your genetics-only model tells you that two people have identical environments, shouldn't you be curious enough about the model's accuracy to check it?

This lazy thinking points to the practice forewarned by Thomas Hunt Morgan, relaying that "the results are often so excellently 'explained' because the explanation was invented to explain them." As we will see in the next chapter, purposefully ignoring the environment to *explain* the disease via genetics is the true underpinning of why population genetics has failed to provide any true interventions.

CHAPTER 14
You are What You Eat, Drink, and Breathe

BOTTOM LINE UP FRONT: *Just like genetics, environmental exposures can have lifelong effects that can be passed down from generation to generation.*

I found myself standing at the nurses' station in the newly opened Ross Heart Hospital at The Ohio State University Medical Center. My white coat was still nicely creased and sparkling white from when I had received it only a few months earlier. I had only been a doctor for a couple of months, but I was asked to see a patient on the heart failure service for a chronic cough. He was a man in his early 60s who worked as a farmer in rural Ohio. Mr. Smith had the stereotypical body of a lifelong farmer. He was fairly muscular, especially for a 60-year-old. His hands were chaffed and leathered but large in a way that made it seem like he had a layer of padding just under his skin. He also had a scowl on his face for having to be in the hospital. There are two well-known jokes in medicine that applied pretty well to Mr. Smith. The first is that a farmer comes into the doctor and the doctor asks, "On a scale of 1-10, how bad is your pain?" The farmer replies, "Well, I'm here, ain't I?" The second joke is that there are only three things that bring a man into the doctor: pain, bleeding, and his wife. Mrs. Smith had insisted that he get checked out due to the week-long recurrence of severe nose bleeds, a swollen nose, and dry cough.

Mr. Smith was not happy to be in this hospital. Mind you, he was never rude to me or any of the medical staff, but it was clear that his symptoms were preventing him from tending to his farm. The patient had been transferred in from his local rural hospital for concerns that his heart might not be beating well enough to properly transit the blood through the lungs (which might explain his coughing). The hospital had run several other blood tests for other possible causes, but since the hospital was not well-equipped to treat heart failure, they sent him to a more well-resourced hospital like Ohio State. When the team at my institution tested his heart function, the results indicated that his heart was pumping just fine. So, the primary team entered what is called a *pulmonary consult*, which is a way of asking insights from other specialists for what might be wrong with Mr. Smith's lungs. In academic medicine, the lowest-level trainee usually starts the process and then relays their findings to their seniors, and, in this case, I was at that lowest-level.

I was sitting at the nurses' station combing through Mr. Smith's chart to look at the results from the various tests that had been performed. His heart screen was entirely normal, so heart failure seemed all but excluded. Flipping through the results from the prior hospital, however, I noticed that someone had sent off an *autoimmunity panel*. The results had returned just prior to when Mr. Smith was transferred to our hospital but were notable for a positive cytoplasmic antibody test, or *cANCA*. In medical school, we learned about autoimmune diseases that include positive cANCA tests, one of which was called Wegener's Granulomotosis. Wegener's is a rare disorder, about one in a half-million people will develop it, but it was more common in older white men like Mr. Smith. The classic presentation of Wegener's begins with blood vessel damage in the nasal passage but then progresses to involve damage to the lungs and kidneys. Unchecked, the disease can be fatal.

Mr. Smith had the nasal symptoms and blood work to suggest Wegener's, but his chest X-ray was clear and his kidneys seemed fine. So, given how rare the disease is, maybe I was just *hunting for zebras*, as we might say of someone in medicine looking for a rare disease to explain a more common condition. But when I looked up the diagnostic criteria for Wegener's, lung and kidney damage were not required. In fact, Mr. Smith should have been diagnosed with Wegener's per the definitions set forth by both the American College of Rheumatology and European Association for Rheumatology. I was excited to relay this to my senior resident, Brent, and he agreed. Mr. Smith likely had Wegener's and there was a real opportunity to intervene early and stave off any lung or kidney damage. Brent suggested that Mr. Smith might get a nasal biopsy just to be sure that granulomas were present, which made sense since the treatment for Wegener's was a level of immunosuppression that should be considered cautiously and because a nasal biopsy would not be dangerous.

The two of us informed the lead pulmonologist (referred to as our attending physician) about the diagnosis and plan. He sat and listened to everything Brent and I had said then, without a beat, just replied, "This does not smell like Wegener's to me." At first, I thought he was making an incredibly insensitive joke given Mr. Smith's nasal complaints. However, even after discussing the story with Mr. and Mrs. Smith, our attending decided that Wegener's was unlikely, and the team should just monitor for now. I still remember the exact computer our attending was sitting at when he said this. I remember the clothes he was wearing, and I remember the dismissive smirk on his face along with the sullen look on Brent's. I don't recall these because he was dismissive. My opinion had been dismissed before and has been dismissed since. In fact, as a trainee, you want supervising doctors to question your diagnosis and treatment plan because doing so is what allows you to become a better clinician.

I have such a vivid memory of that moment because it was on the continuum that ended a week later when I visited Mr. Smith and his family in the intensive care unit. At that point, Mr. Smith was intubated and on the ventilator because his lungs were filling with blood and he was on dialysis because his kidneys had stopped working. The ICU team, to find out what might be causing the kidney damage, had sent off a kidney biopsy for analysis. The final summary results from the pathologist contained only four words: consistent with Wegener's Granulomatosis. At that point, Mr. Smith was given the immunosuppressant Brent had proposed from the beginning, but it was too late. Mr. Smith never got to complete the tasks his illness interrupted, and his grandchildren never got to say goodbye.

Nearly three years later, near the end of my internal medicine training, I found myself back at that same computer in the heart hospital. I was waiting for my medical student, Travis, to come out of our newly admitted patient's room and relay the story he was collecting from her. Ms. Williams was only in her 30s but had confirmed congestive heart failure, or CHF for short. She was being admitted for one of the intermittent exacerbations that are common in the disease. In my mind, I had already slotted Ms. Williams' plan of treatment. CHF exacerbations are a common occurrence in a hospital and the treatments were so well established that most could be ordered with just a single click of the medical record. After over an hour, Travis finally emerged and began relaying what information he had taken from Ms. Williams. Although the admission was for the shortness of breath and lethargy that are common in CHF, Travis was particularly concerned about Ms. Williams' menstrual cycle. She had told him that all her heart symptoms began a few weeks after the birth of her third child about three years before she was admitted to our care. During those three years, her menstrual cycle had become rare and unpredictable. As her heart failure began to worsen, Ms. Williams became unable to work. She lost her job, her home, and eventually, her husband. She was trying to raise her three children on welfare and only had Medicaid coverage to provide basic support for her medications. But Ms. Williams stressed providers had never looked into her abnormal periods.

Travis was concerned about the possibility of something called Sheehan's Syndrome. In Sheehan's, blood losses during pregnancy deprive the pituitary gland of blood flow. A stroke occurs and production of the hormones made by the pituitary ceases. Some of those hormones are needed for normal heart function and can put the patient at risk for developing CHF. Sheehan's syndrome is not as rare as Wegener's, but still only occurs in about five pregnancies per 100,000 live births. But my initial hesitance to the diagnosis was not that the chances of Sheehan's was unlikely; it was that the chances that no one else had caught it in the previous three years felt unthinkable. To diagnose Sheehan's, you would need blood work (to look for the pituitary hormones) and perhaps an MRI to look for signs of a pituitary stroke. I was skeptical that Ms. Williams had Sheehan's and did not want to put her through an MRI without cause (remember, her insurance coverage was terrible). So, I asked Travis to look up how well the blood work might predict Sheehan's. After verifying that the blood tests would be good predictors of the disorder, we converged on a plan to send off all the needed blood tests and then, if those were suggestive, we would send Ms. Williams for an MRI.

The following afternoon, once more at the same computer, Travis and I logged into the records to find that Ms. Williams' blood work was exactly what you would expect for someone with Sheehan's. Her MRI later that day confirmed it. This mattered because the treatments for CHF caused by a pituitary stroke are not at all the same as those for CHF caused by something more routine like a heart attack. Travis and I called endocrine and obstetrics specialists to weigh in, and together, the teams came up with the exact medical cocktail that would be best. Ms. Williams' poor insurance coverage would continue to make affording these medications a challenge, but for once, she had been placed on the correct care.

I reflect on Mr. Smith and Ms. Williams a lot. Not only did they bookend my training, but they also played out at the same spot at the same nurses' station. Ms. Williams was given a real chance to perhaps meet grandchildren who had not even been born yet. Mr. Smith never got to see his grandchildren again. What separated these two cases was not the intellect or knowledge in their providers. Ultimately, in both cases, someone on the team was educated enough to think of the correct diagnosis. What separated these two cases was not the willingness to listen. In Mr. Smith's case, I was willing to spend time to hear about his concerns and look through his records in detail. For Ms. Williams, I will forever look back with doubts as to whether I would have been willing to listen in detail had she tried to inform me about her menstrual concerns the way, thankfully, Travis was there to listen. What really separated these cases was curiosity.

At no point in Mr. Smith's care did our attending ever offer his alternate diagnosis. He has decided, for unknown reasons, that Wegener's was unlikely but didn't seem invested in trying to figure out what else might be wrong. Brent had recommended a further test that could have solidified the diagnosis to get at what was truly going wrong. But our attending offered nothing to imply he was curious. For Ms. Williams, Travis showed the curiosity needed to figure out why a 30-year-old, previously healthy woman would have the type of CHF usually only seen in older, less healthy people. I had already determined her treatment for a CHF exacerbation and thus had biased myself against looking deeper.

Patients are often frustrated by not knowing what is really going on. For eczema, flares can happen any time for any reason. Patients and parents often express immense frustration at the inability to predict a given flare. Obviously, patients want their symptoms to go away, but often, they also want to know what is truly causing their symptoms in the first place, especially for caregiving parents. I've been in the room when doctors like Alexandra Freeman have informed parents that their child had a genetic mutation causing their symptoms. No parent wants to be told their child has a genetic disorder, but for many, there is a small sense of relief. Many of these patients had spent years bouncing between doctors who could not figure out the true cause of their illness, so even though the cause may be immutable, at least it was now knowable. Once the true cause of the disease is known, finding treatments (and more ideally preventions) becomes possible. If you are not curious enough about what is truly going on, you won't be able to find how best to fix the problem.

For the diseases where there are no clear causes, research instead focuses on abnormal chemistry. A researcher might notice that patients with asthma have an increased level of hormones in the blood known as adrenergics. They might then decide to try blocking the effect of those adrenergics to see whether they can prevent their harms and improve the patients' symptoms. Prior to testing in people, researchers test drugs in mice or cell cultures. In these studies, researchers test a bunch of genetically identical cells (known as a *cell line*) for their reaction to a drug. Or they might test their drug on a group of genetically identical mice (known as a *mouse strain*). In basic science, certainly researchers may study cells or mice with known genetic defects, and, in fact, my lab (in work led by Erik Anderson) has worked on cells and mice that were genetically deficient in *STAT3* to evaluate how the gene might influence infection control. But most of the drugs you have ever heard of were tested in *wild type mice*, which lack any known mutations.

The most common mouse wild type strain is officially called C57BL/6, but is more commonly called *black 6* due to the black fur color. Various companies breed these nearly genetically identical mice in their facilities and then sell them to researchers around the world. In 2009, researchers from New York University were busy conducting experiments on autoimmunity in black 6 mice they had purchased from a company called Jackson Labs. But the results of their experiments were entirely different when they used black 6 mice from the Taconic company. The mice were genetically identical, the mouse equivalent of identical octuplets. So, why would the mice respond differently to the same infections or drug challenge?

The answer was not in the mouse genes, but in their gut. Every living mammal has microorganisms living on and inside us. Microbes live on human skin, lungs, nasal passages, the vaginal cavity, and the gastrointestinal tract. The very first identification of microorganisms was not actually in the context of infection. Around 1677, Antonie van Leeuwenhoek created lenses of such size and quality that they could serve as the first microscope. He used this to identify and document microorganisms in water that he referred to as *animalcules*. van Leeuwenhoek faced skepticism against his claims of tiny animals living in rainwater and moved away from looking at microbes because he "did not gladly suffer contradiction or censure from others." But the criticisms from others combined with his refusal to share his methods for making the lenses sensitive enough to view the animalcules meant that his claims of microbes living among us was left hanging. Since the broader scientific community was never made aware of van Leeuwenhoek's discovery that microbes could be present on and near healthy people, science's first encounter with microbes did not come until hundreds of years later. Even then, many refused to believe that microorganisms could possibly exist, as was seen in the case of Dr. Ignaz Semmelweis mentioned in an earlier chapter.

In 1982, the Australian researchers, Drs. Barry Marshall and Robin Warren, discovered that a type of bacteria that could normally live in human stomachs called *Helicobacter pylori* (or *H. pylori* for short) was the true cause of gastric and peptic ulcers. Famously, Dr. Marshall drank a culture of *H. pylori*, demonstrated that it gave himself gastritis, then took antibiotics to reverse the harm. People had long-known that pathogenic infections could cause harm, but now it had been revealed that bacteria that might otherwise be benign could generate disease. Antibiotics became part of the standard of care for ulcer disease.

However, in 1997, Martin Blaser, author of *The Missing Microbes* and current professor at Rutgers University, discovered that the use of antibiotics had become a little too standard in their use. Dr. Blaser demonstrated that while *H. pylori* could indeed cause gastritis and ulcers, it also had beneficial properties for human health. Killing the *H. pylori* from our guts, especially in children, increased the risk of multiple metabolic, allergic, and autoimmune diseases. The dichotomy of Blaser's and Marshall's work generated a more complex and nuanced view of the microbial world. Microorganisms could cause harm but were also essential to maintaining health.

Slowly, a picture of microbiome formed that could best be compared to *Busy Town* by Richard Scarry. In Busy Town, animals are all dressed like humans walking around doing various jobs. There is a cat delivering milk, a fox running a hotdog stand, and a pig tow truck driver. The image captures numerous species living together in one part of a town, each with their own task to complete and their own role to play (I could not find a common use image of *Busy Town*, but Google it and it will be obvious). The microbiome is similar but on a microscopic scale. A wide variety of bacteria, fungi, and viruses together live on and in our bodies. Our various organs (lungs, colon, skin, etc.) are like the different towns that the microbes can inhabit. Differences in which species are living in which location can be important, but just as *Busy Town* needs a mail carrier, the human microbiome has certain tasks that microbes must perform. It might not matter exactly which species serves that role, but human health cannot exist without microbes serving in several key jobs.

When the differences between the genetically identical Jackson and Taconic mice were found, the group led by Dr. Dan Littman opted to look for a microorganism that might be the cause. Even though the mice from each company were genetically identical, they were bred and raised in completely different facilities hundreds of miles apart. In short, one specific type of bacteria that had set up residence in the guts of mice from Jackson were responsible for the differences between the two groups. If the researchers transferred the bacteria into mice from Taconic, or removed the bacteria from the Jackson mice, then the differences between the animals disappeared. While I had entered college in the GATTACA era, by the time I had started my research career, the potential of the microbiome to influence health was impossible to ignore. Or so I thought.

Recall that the first GWAS for eczema indicated an association between eczema risk and variations in the locus containing the gene *filaggrin*. If you think of your skin cells as bricks in a wall between you and the outside world, then filaggrin is one of a number of proteins that act as a sort of mortar between them. Thus, the original GWAS results induced immense excitement. Researchers went looking for a genetic cause to a skin disease and found a region of the genome that contained a protein responsible for skin health. Of course, this would not explain how the rates of the disease could increase so much faster than human genetics could change. These results could not explain why Bulgarians have low rates of eczema despite having similar *FLG* sequences as the more commonly afflicted Germans.

A slow drip of studies also undermined this one gene as a central player in the disease. Recall that mice with two genes deleted, one *FLG* and the other called *Tmem79*, developed eczema spontaneously, but *FLG* mutations alone did not create disease. Deleting *FLG* from human cells did cause abnormalities in their function, so the data was not all negative. However, there are only five reported cases of humans with deletions of the gene (and those patients don't even have eczema; instead, they have skin that blisters too easily). The specific genetic *FLG* sequences that are claimed to be a cause of disease in patients have not been shown to impact the function of the protein. Plus, the number of reported mutations in *FLG* are much larger than you would expect (as with a recipe, you would expect fewer errors with a recipe that only had three steps than if it had 50 steps). This means you would likely need two defective copies of the gene to have any problems. However, researchers typically classify people as having risk alleles even if they only have one bad copy.

In fact, people considered to have risk alleles for *FLG* express normal amounts of the protein unless their skin suffers a flare. *FLG* could not be used as a screening test for parents; it could not predict which drugs would work best, nor did replacing filaggrin protein via topical treatment provide any measurable improvement in disease. While the focus was on *FLG*, the rates of eczema jumped some three to six fold in the U.S. and U.K. from 1970 to 2020. That means that if the rates of eczema had just remained stable over the last 50 years, something on the order of four out of every five people who have been afflicted would never have suffered a day of eczema in their lives. If you could wave a magic wand and bring the rates of eczema down to where they were in pre-industrial times, something close to 29 out of every 30 people would be cured. None of these results would indicate that *FLG* is unimportant, nor do I intend to leave the reader with the impression that the gene is meaningless. However, combined with the rapidly exploding rate of disease, the conclusion should be that genetics play no more than a minor role in eczema.

This rapid rise in immune-mediated diseases in industrial societies is not limited to eczema. As previously mentioned, while no one over 40 years of age attended a peanut-free elementary school, only one generation later, over 90% send their children to a school with at least a peanut-free table, according to a 2017 *JACI* paper by Bartnikas and colleagues. In 1819, John Bostock was the first to describe hay fever (in himself), but it took him nine years to find an additional 28 cases. Today, one out of every five people in Bostock's home of England have hay fever per Siddiqui and colleagues in the *British Journal of Hospital* Medicine, 2022.

When I began my career in allergy research, I started in the gut. Several excellent books detail the impact of the gut microbiome, such as Blasers' *Missing Microbes* and *The Good Gut* by Stanford University research couple Drs. Erica and Justin Sonnenburg. From the immune system's perspective, there are a few main ways that the gut microbiome is important. Perhaps the most important is the ability of the bacteria in the gut to digest dietary fiber. The fiber that we might eat in fruits, vegetables, and whole grains is not digestible by human cells, but these bacteria digest that fiber for themselves and make what are known as short chain fatty acids or SCFA as a by-product. These SCFA have a lot of benefits for just about every organ system in the body. But for immune cells, SCFA help generate what are known as *T regulatory cells* or Tregs. Tregs are complex, but in general, they are the immune cells that act like chaperones at a high school dance. If everyone is having a good time and behaving, they can just stand on the side and watch. But if things get a little too wild, they will step in to tamp things down.

Our bodies are in a near constant onslaught of potentially inflammatory challenges from microbes, irritants, wear and tear. It is important to have a system in the body that knows when the immediate threat is cleared, and then to help return things back to normal. Without consumption of fiber, it is difficult for the body to generate the cells and processes needed to make Treg cells. As such, without fiber, the *settle down now* signal is diminished and the risk for autoimmunity increases. Much (but not all) of the harms that might be related to taking antibiotics seem to be related to killing off the types of bacteria needed to produce SCFA.

Another possible insult to the gut microbiome is refined sugar. Notice the word *refined* there, since it is different from the naturally occurring sugars in fruit, vegetables, and dairy. When you eat an apple, you consume sugar, but you consume it along with the fiber and the nutrients it is packaged in. But most processed or refined sugars have undergone the process by which the sugar has been extracted and concentrated. Indeed, sugar is still a nutrient that can nourish the body, but at such high concentrations (without the fiber to counteract the effects), it appears that refined sugar has some proinflammatory properties. Worse still, Dr. Nora Volkow, head of the National Institute of Drug Abuse, has several publications showing that refined sugars are addictive. Meaning that eating a little inflammatory sugar drives you to eat even more. No, sugar isn't as addictive as methamphetamines are, but having to walk past the candy shelf at every checkout stand makes it a hard habit to break.

Similar results are seen in refined saturated fats. Again, the word refined is meant to contrast saturated fats in their natural form (such as meat or peanuts) with saturated fats in their processed form (like candy bars). On a molecular level, there are lots of saturated fats, but processed fats typically only contain either palmitic acid or stearic acid. In a truly cruel twist of fate, palmitic and stearic acid are the most inflammatory of all of the saturated fats, but they are also the ones with the best commercial viability (flavor, texture, shelf stability). So, when you consume refined saturated fats, you are likely to be consuming only the proinflammatory versions, which have been removed from the other saturated fats that are either neutral or anti-inflammatory.

To look at the impact of saturated fats, my student researcher at the time, Natalia Fontecilla (now Dr. Natalia Biles), and I fed mouse parents a diet high in palmitic acid. Both the mothers and fathers ate the unhealthy diet but because we did not give them added sugar, the mice did not overeat, and thus, obesity was not a confounder. The mice were allowed to breed on the unhealthy diet but when the pups were done breastfeeding, the offspring were moved to a new cage where they were fed only normal, healthy mouse food. So, only the mom and dad mice ate the unhealthy diet; the pups were only exposed in the womb and while nursing. What we found was that the pups from parents that ate the high fat diet had worse outcomes in models of peanut allergy, blood infection with *E. coli*, and multiple sclerosis. Nearly all these bad outcomes for the pups were due to inheriting a proinflammatory microbiome from the mother. The bacteria in the gut failed to provide the correct SCFA training, failed to control inflammation, and failed to help keep the gut sealed so that nothing from the colon *leaked* into the blood. By swapping the *bad* bacteria inherited from the parents' poor eating with a *good* version, the mice became totally normal.

In our study, the increased disease seen in the pups came from differences in the specific species that were living in the gut. However, as stated before, it is not always the species of microbes that matter as much as the microbes' behavior. In 2018, a group led by Dr. Dan Knight published a paper in *Cell* that tracked people immigrating from Thailand to the U.S. The group compared those still living in Thailand with three other groups: the relatives who had been living in the U.S. for years, the relatives born in the U.S., and Americans who had been in the U.S. for generations. After people had moved to the U.S., the types of species in their gut slowly started to more closely resemble multi-generation Americans. However, the behavior of those bacteria changed in about nine months.

The immigrants who adopted an American diet saw a rapid reduction in the fiber processing activity compared to those who either stayed in Thailand or maintained a more traditional diet after relocating. Similar work by the Knight group has demonstrated that while the species of gut bacteria might be very different between one person and another, the functions tend to be very similar. The bacteria that digest your fiber might be a different species than the one that digests my fiber intake, but everyone has at least one microbe that is up to the task (even if they are not called upon as frequently as they should be). One interesting parallel between some microbiome work as GWAS is the focus on sequencing at the expense of this vital functional information. Knowing a microbe is *missing* only really matters if you know what helpful role that bacteria was supposed to be playing, if it was playing any role at all.

The ability for an unhealthy diet to corrupt otherwise helpful gut bacteria was suggested in early mouse studies by one of the gurus of gut microbiome, Dr. Jeff Gordon of Washington University of St. Louis. Dr. Gordon evaluated the famed mice with *the obesity gene* (called *ob/ob* mice). These mice had a single mutation in the leptin receptor and seemed to be born destined for obesity. The lay press ran wild with the idea that the *fat gene* had been discovered and soon we would all be able to eat whatever we wanted so long as we took some drug that modified this allele. However, Gordon's lab discovered that swapping the gut microbiome from *ob/ob* mice could *cure* them of their obesity. The only meaningful defect that the leptin mutation caused was to be born with an improper array of microbes in the gut. Much like in the study with Jackson and Taconic mice, if a totally normal mouse was given the gut microbiome from an unhealthy *ob/ob* mouse, then it, too, would develop obesity. So, what was billed as a *monogenic* trait was reversible with a mouse stool transplant. There was one big caveat, however; swapping out the *ob/ob* microbiome for a healthy microbiome only helped keep the mouse thin if the mouse was given a healthy diet. Planting the healthy bacteria into the new mouse only worked if you nourished them with the right foods. If the *ob/ob* mouse was given a stool swap but stayed on an unhealthy diet, then any benefit would be lost.

In humans, the bulk of the newborn microbiome tends to come from mothers during the process of vaginal delivery. This is why C-sections are thought to be associated with a number of diseases, such as food allergy.

The theory is that bypassing the vaginal canal fails to provide the needed bacteria that are supposed to be painted onto newborns. Dr. Suchitra Hourigan led a study in collaboration with my team. She discovered that babies born via C-section were not only less covered in protective bacteria, but they were also more apt to pick up potentially harmful bacteria from the environment. By swabbing both babies and the hospital environment, Dr. Hourigan identified that bacteria from the wash sink in the delivery rooms—like *Staph aureus* and *Pseudomonas*—were more likely to be transferred onto the babies born via C-section. Dr. Hourigan is also investigating if swabbing babies born via C-section with their mother's vaginal microbes can help reverse the possible unintended consequences of cesarean birth.

But there are ways for fathers to impact the newborns' health, as well. For many years, the sequence of DNA was thought to be the main—if not only—biologic mechanism for passing on protein information to the next generation of cells. We now know that while the code that determines the protein is important, there are other 'non-genetic' mechanisms in place that determine the protein expression levels in the cell. These mechanisms have also been found to be inherited in a way that is still not completely understood, but their discovery has opened a whole new field of study called epigenetics (translated directly to 'above or on top of the genes' but the term comes from the embryologist Conrad Waddington playing off the term 'epigenesis').

Today, we know that all ATCG sequences are packaged and organized three dimensionally by proteins called histones, and both the sequences and the histone proteins themselves can be decorated with biologic flags (e.g., methyl and acetyl groups). These flags have been found to be important factors in determining protein levels in the cell and, upon cell division, this epigenetic information can be passed to the daughter cells. For example, if your grandfather lived through times of famine, the genes that help extract and store calories would be flagged in such a way to assure this protein expression information would be passed on. This can get encoded into the sperm and make a baby that starts its life with cells tilted towards energy storage. Several studies have shown that having a father or even grandfather exposed to famine could increase the risk of metabolic diseases like diabetes. The theory is that the early pre-programming skewing towards energy storage is helpful during a famine, but a risk factor for obesity if you are born into a time where food is more plentiful. In our study, dad mice fed the unhealthy diet showed pre-programming towards disordered immunity. While this can be overridden by a healthy gut microbiome, I'm sorry to break the bad news to all the dads out there by telling them that their health behaviors matter for the embryo (not as much as moms, but still).

As interesting and nuanced as the gut microbiome is, diet isn't the only factor that could explain immune diseases. Certainly, plenty of people follow a healthy diet and still suffer diseases like eczema. Eating a healthy diet also isn't a cure for eczema either (regardless of what some YouTube or TikTok videos may claim). So, our group turned our attention to the microbes living on the skin. In 2012, Dr. Heidi Kong was the first to report that the microbiome of patients with eczema was distinct from healthy controls. Furthermore, the microbiome shifted during flares so that harmful bacteria like *Staph aureus* became more prominent during a flare, then subsided after treatment. Of note, the patients' genetics never changed during this time. The microbiome was dynamic and predictive of disease severity while the genetics were static and unable to distinguish having full body itch from clear skin.

When the microbiome findings were originally announced, it was unclear whether the abnormal bacteria on the skin were a cause of the flare or just a reflection of inflamed tissue. The terms "chicken or the egg" became almost as inescapable as hearing a geneticist say, "Probabilistic but not deterministic." A lot of attention had been invested in evaluating the connection between *Staph aureus* and eczema.

Without a doubt, *Staph aureus* made the disease worse but antibiotics that killed *Staph* never seemed to be helpful. Patients with eczema would be far more likely to have *Staph aureus*, but they didn't always have the bacteria. It was very possible to have severe eczema without *Staph aureus* and plenty of people could have *Staph aureus* without eczema. So, *Staph* and eczema were not so much "chicken and egg" as they might be "spaghetti and meatballs"—two things that are often found together, but not universally found together. They may enhance each other but the presence of one is not a requirement for the existence of the other.

So, the microbiome took somewhat of a backseat as researchers continued to *poke the dead fish* of the Groundwater Approach analogy. Then, several groups examined what metabolites might be missing in patients with eczema, even when their disease was under good control. Patients with eczema seemed to have defects in ceramides and sphingolipids. These are types of oils in the skin that help with regulating immune function and how well the skin stays *sealed* from the outside elements. Work that came much later eventually confirmed these lipids as the best predictors of developing eczema in newborns. Work published in 2023 in *JACI* demonstrated that while filaggrin protein levels held *zero* predictive value, the levels of sphingolipids and ceramides could predict which children would go on to develop eczema. Newborns with deficiencies in these specific oils have a 30-50 fold increased risk of developing eczema.

But when looking for why these lipid defects were present, there was no data pointing to innate human defects. When researchers took cells from the patients' skin, there was never any defect in the ability to make these lipids. Certainly, cell cultures are not always an indication of what is wrong in the person, but at least they could say that there was no innate cell defect in the production of these lipids. However, several types of bacteria make these lipids for us humans. Bacteria in the category known as Gram negative (based on the results of a specific stain use in microscopy) often make these kinds of lipids to put into the surface of their cells. When looking back at Dr. Kong's work showing the bacterial differences in eczema, it stood out to me that the spots on the body that were traditionally impacted by eczema (the folds of the arms and leg, the hands, and the face for children) were the same spots where you would be most likely to find these ceramide making bacteria. So, a hypothesis formed: If the patients have a defect in their ceramide production but the patients don't seem to have any human problem making these lipids, then maybe it is a bacteria problem. And if the bacteria that make these lipids seem to hang out on the parts of the body that are prone to eczema, perhaps loss of these bacteria and the services they provide is the cause of the disease.

This next statement is going to make me sound like a villain from a sci-fi movie, but *they laughed at me when I told them my idea*. The experts in the field noted that Gram negative bacterial DNA was found on the skin, but no one had been able to culture these organisms. The experts told me that *FLG* was going to be the key to the disease or that *Staph aureus* was the real cause. No one believed Gram negatives were going to matter at all, let alone contribute in a major way. But our group ignored them all and set out to try to culture those bacteria.

Several failed attempts at growing the bacteria later, I wondered whether the bacteria might need lipids from the skin to grow. Bacteria are grown in liquid soups we call media. Traditional bacteria media does not have any fats in it, so I wondered how to add in fats that might help a skin bacteria. I looked up which cooking oil was most similar to human skin (it is canola oil for those curious) and added that into the mix. Then, I put the culture broth into an autoclave that cooks the liquid under intense pressure to assure that everything is sterilized. That day I discovered that canola oil cannot withstand autoclaving; the media lit on fire, melted the lid of the bottle, and filled the lab with some pretty nasty smoke.

Regrettably, I failed to take a timeless adage to heart before nearly setting the entire lab ablaze: "A day in the library can spare you a month in the lab." I went to look for anyone who had successfully grown sphingolipid producing bacteria before. Strangely enough, success was found in those culturing bacteria from tropical fish tanks. Apparently, tropical fish are sensitive to bacterial contamination in water, and so, robust methods to culture any living organism from their tanks are needed to protect them. So, we copied their approach but made one final change. Rather than culturing at the standard temperatures that match peoples' internal organs, we dropped the thermostat down to the cooler temperatures of the skin. These two changes allowed us to consistently culture Gram negatives from the skin without any phone calls from the Fire Marshall.

We grew several different kinds of bacteria but could not know whether any were meaningful or just mere associations. To test this, we took the various bacteria and tested them in mice and cell cultures to see whether they impacted living systems in ways that might predict that they would be helpful in eczema. The most common form of bacteria we were able to find on the skin of healthy people was *Roseomonas mucosa*. It was from a family of bacteria that indeed made the ceramide and sphingolipids we were hoping to evaluate. We found that the isolates of *Roseomonas* from healthy people were able to improve several cell markers of eczema, including vitamin D production and protective compounds called *antimicrobial peptides*. These same bacteria were able to reverse eczema in mice, which suggested they might be useful for use in humans. As an added bonus, *Roseomonas* was able to kill *Staph aureus* using some of the same lipids it used to balance immunity in the cells and mice. In contrast, the bacteria taken from patients made the cell cultures and mice worse, even when the bacteria were also the same species of *Roseomonas*. Overall, this suggested that trading the *bad Roseomonas* for the *good* kind might provide aid to patients with eczema.

One issue that often goes overlooked in probiotic development is safety. There is a propensity to assume that because a bacteria is *natural* it will be a safe exposure. However, most probiotics use far higher concentrations of the bacteria than you would find in nature and thus present an *unnatural* exposure in some ways. So, we wanted to make sure that *Roseomonas* was safe for use. We exposed mice to massive doses, far more than we would ever consider exposing people to. We injected the bacteria directly into the vein and had the mice inhale the bacteria, as well. Nothing happened. The mice did not get sick, and they did not become infected. We even took away parts of the mouse immune system (the neutrophils) and still, the mice did not seem to be harmed by *Roseomonas*. In contrast, other bacteria from the skin, such as *Staph. hominis* and *Staph. epidermidis*, were more than capable of generating lethal infection in mice despite those bacteria also coming from healthy people. Mice are not humans, obviously, but we did feel better knowing that very high concentrations of our bacteria did not appear to do harm.

So, with that data in hand, we wrote up a protocol and asked for FDA approval to be the first people to spray children with Gram negative rod bacteria. We received all the needed approvals and went to the lab to begin making the bacteria ourselves. Mark Kieh, Eric Anderson, Danial Saleem, and I grew up large batches of *Roseomonas* and put a small amount into individual spray vials. I still remember the pain in my forearm from having to use the crimping tool that sealed the lids of each of the 1,200 vials that we made. Patients would receive a small vial of *Roseomonas* that had been freeze dried in sugar, nothing else went into the vial but that. They then got eye droppers my team filled with sterile water. For each dose, they would add the water from an eye dropper to the bacteria to dissolve the sugar and awaken the bacteria. They then sprayed it directly onto their skin.

To be as safe as possible, we began the study with adult patients. The adults in the study were assessed and provided enough *Roseomonas* to treat themselves for six weeks. One of the first people to enroll was named Jonathan. Jonathan had suffered with eczema his whole life but his symptoms became severe to the point he was covered nearly head to toe with an itchy, flaky rash. Due to the social isolation he felt from his skin, Jonathan dropped out of college before finishing his first semester. Right after he clicked "withdrawal" in the online registrar, he began searching for experimental eczema treatments and found our trial. I warned Jonathan that my guess was that treating adults may not be as effective as treating children. Since eczema was much more common in children, I worried that adulthood might be too late to intervene. Jonathan picked up a dozen of the vials my team and I had crimped and headed home. Six weeks later, I opened the door to the clinic and Jonathan almost looked like a different person. His skin was cleared of any signs of the severe rash he had when he arrived. So clear, that I insistently asked if he might have taken any other medications in addition to ours (he had not). Jonathan still has flares from time to time, but he was able to re-enroll in school and enjoy college life.

When speaking to different patient representatives, it was clear that patients did not want just another topical product that they needed to use for the rest of their lives. What they wanted was something they could use and then stop. I remember one of the first patients we enrolled came to the clinic during the summer. Despite it being a hot and muggy day in late April in Washington, D.C., the kid was wearing a long sleeve shirt and long pants. Clearly, he was trying to prevent his rash from being visible to the outside world—a pattern of shame that many patients with the disorder feel and try to address by covering up. After four months of treatment, only applying the bacteria two to three times per week, he had achieved nearly an 85% reduction in his itch and rash symptoms. His results were in line with our patients, who, on average, got about 75% better with no serious side effects. Most importantly, we brought each patient back to the clinic 4-8 months after they had stopped using the treatment in hopes that the bacteria might provide long term benefit by taking up residence on their skin. This young patient and his father came back on a freezing cold day in February, six months after he had stopped using *Roseomonas*. Even from a distance, as they came walking through the lobby of NIH's clinical center, I could clearly see two things: First, he was wearing shorts and a T-shirt, and second, he was smiling from ear to ear.

As excited as we were about the potential of a simple bacteria to treat eczema in a way that seemed to have lasting benefit, something still nagged at me. Even if eczema were caused by a failure of the skin bacteria to make the proper lipids for our skin, something in the environment must be the reason. It isn't as if human bacteria could have collectively decided to stop being so helpful in the 1970s; there had to be something that was either killing or corrupting these healthy microbes.

At first, we looked at topical products, the use of which has greatly expanded since 1970. My research assistant at the time, Carlos Castillo, tested how well bacteria could survive exposure to topical chemicals just as one might test surviving exposure to penicillin. We found that certain ingredients, such as parabens, would kill the beneficial bacteria without impacting the growth of harmful bacteria like *Staph aureus*. Even many of the products that were allowed to put "for eczema" on the label were capable of selectively killing off the good bacteria. It was not lost on me that whenever a scientist talks about the potential harms of *preservatives*, they come off as an unserious hippie flower child. But I realized that one needed to ask how preservatives work. They work by preventing contaminating microbes from growing in the product. If an organism gets in and starts to grow, eventually, the entire bottle will be rancid and ruined. So, from the perspective of the product, a preservative is a *prophylactic antibiotic*. Thus, for a topical product, preservatives are meant to protect the bottles' contents against the types of bacteria that grow on the skin (that could contaminate the product during use). So, what do you think would happen to your skin's bacterial balance when you apply a product containing chemicals that were chosen for their ability to kill off your skin bacteria?

In playing around with different ingredients, we were able to identify a way of reversing the imbalance. By adding a little bit of coal tar (termed *colophonium*) and lemon myrtle oil, we could change a lotion that previously had only killed off beneficial bacteria into one that did not impact those bacteria but instead only killed off *Staph aureus*. We presented our *recipe* to a major lotion manufacturer, but they were concerned about how their current products might look if they were on the shelf with one claiming to be "safe for the microbiome."

I kind of get their argument. When I was in college, my roommate and I used to eat those frozen pizzas that I swear are only marketed to children on a college budget—the wafer-thin ones that were somehow less expensive than the raw ingredients should be. One day, our senior year, we went into the store to buy our trusted brand and the box had a sticker on it proclaiming, "Now made with REAL cheese." Obviously, our response was, "What do you mean NOW?" Turns out REAL cheese is a brand name and not an indication that the prior ingredients were fake cheese, but the point was that having a product proclaiming "this one is safe" suggests that the other formulations were not. So, we could not find any company willing to make a "safe for the microbiome" version, since doing so would threaten their current market share. Those interested in the politics and history of the cosmetic industry should read *Clean: The new science of skin* by James Hamblin.

Again, however, while topical ingredients might play a role in eczema, it seemed unlikely that they alone could explain such a massive rise in disease since 1970. The eczema online community is strong, so I would suspect that if the problem were specific ingredients, the parents would have figured out which topicals were safe long ago. Thus, we concluded that there had to be something that could evade a typical patient's recognition that must be playing a role in causing eczema.

We took the advice of the Groundwater Approach and began to evaluate the lake in which 20% of our children were stricken with eczema. Building off what was known, we contrasted three databases: the first compared clinic visits for atopic dermatitis across the country, which allowed us to identify U.S. zip codes that were potential *eczema hot spots*. The other two databases were from the Environmental Protection Agency (EPA). These databases track about 500 chemicals that companies are required to report whenever they release them into the environment. When Jordan Zeldin, the amazing medical student working on the project, and I started planning this project, I told him, "Don't worry if the data is messy; I doubt we will have an Erin Brockovich moment." This, of course, was in reference to the famous activist and author who was played by Julia Roberts in the biopic movie. Mrs. Brockovich helped uncover that the company PG&E had contaminated the water supply of Hinkley California with Chromium 6. The townspeople began noticing various diseases including miscarriages, cognitive problems, and cancer. Mrs. Brochovich stated the following in her book, *Superman's Not Coming: Our National Water Crisis and What We the People Can Do About It*:

> *We are living in a hyper-toxic time. We aren't just dealing with the toxins themselves, but the fact that they have accumulated in the environment for long stretches of time. American children are growing up exposed to more chemicals than any other generation in history and it shows. Rates of chronic disease for children, especially those living in poverty, are on the rise. Nearly half of U.S. adults (or 117 million people) are living with one or more chronic health conditions.*

So, with all the potential chemicals out there, I assumed it would be difficult to assess which might be important for our diseases of interest. For the most part, the first analysis Jordan performed identified what we would have expected. The best predictor of how many eczema visits a zip code would have was the number of pediatricians in that area. Since eczema is a disease that most commonly impacts children, this made sense. Also important were things like allergists, followed by dermatologists, as well as the already established risk factor of population density. Yet, despite my suspicion that we wouldn't find a clear chemical suspect when we analyzed the data, two chemical classes leapt off the page – diisocyanates and xylene. The signal was so strong we were initially skeptical. We tested related diseases like peanut allergy and asthma. Both were linked to isocyanates and xylene, as well. So, to boost our confidence, we performed the most essential part of evaluating positive data which I hope the reader knows by now: We looked at negative controls.

We first evaluated psoriasis, a different inflammatory skin disease. Neither isocyanates nor xylene were anywhere near the top of the associations for psoriasis. This suggested that what we were measuring was not simply a nonspecific skin irritant (xylene had a weak link to psoriasis and thus may have some non-specific skin irritation properties). We also looked at Autism because it was a pediatric disease, it was on the rise in modern times, and it should not share any features of routine allergies. Neither xylene nor isocyanates had any associations. Another negative control was cystic fibrosis. CF is an unquestionable monogenic disease. It might be possible for toxins to worsen CF and thus might push children towards seeing a doctor, but there is zero reason to believe that CF would have any real *causal* link to pollutants. Whereas most other diseases we ran would have two to three top pollutants associated along with dozens of others that were weakly associated, CF had only one statistically significant association: anthracene. Finding only one association seemed to support CF as a negative control, but even then, a 2004 report by Ali and colleagues from the Dalton Center in Missouri identified anthracene as a direct modulator of the CFTR (the protein that is encoded by the gene mutated in patients with CF). Although the anthracene result is provocative, the paltry connection between CF and pollution supported the thought that our analysis didn't indiscriminately churn out positive results.

Next, I wanted to make sure our claims fit with epidemiology, as emphasized in Chapter 2. After all, if these chemicals caused eczema, there should be evidence for this in the population. We looked through the literature to assess what exposures were already known to either cause or worsen eczema. Living in an industrialized nation or moving to one before the age of 4-6 years old, particularly in an urban part of those nations, was the most predictive risk factor. Living near a highway increases the risk of eczema, particularly for the children of mothers living near the highway while pregnant. Fluctuations in temperature would often trigger flares but were not associated with overall risk. Exposure to cigarette smoke was another risk. Various fabrics seemed to induce flares including spandex, nylon, and polyester. Patients tended to gravitate towards cotton if they could afford to since they picked up on some of the irritating properties of these synthetic textiles.

An oddball risk factor was exposure to home remodeling prior to the age of two years; this included new furniture, paint, new hardwood floors, and wallpaper. The most recently uncovered risk factor was wildfire smoke. The fires in the early 2020s in Australia and California had measurable effects on the severity of eczema for those living in the affected areas. Dr. Maria Wei of the University of California San Francisco was even able to identify that Google searches for "itch" and "eczema" spiked in concert with the plume of ash as it blew through San Francisco. Citizen scientist and eczema sufferer, Nic Novak, independently noticed his eczema was reacting to the severe air pollution during the fires in San Francisco and won an innovation award from Global Parents for Eczema Research for his report.

Having never heard of diisocyanates, we started digging into their history to see whether exposure to these chemicals might align with the rise of eczema that began around 1960-1970. The chemical class was first discovered in 1940 and manufactured in earnest the U.S. starting around 1970. Diisocyanates are used in the manufacturing of spandex, the type of non-latex foam found in furniture, as well as paint, wallpaper glue, and wood polyurethane. The part of the molecule that has the most activity (the isocyanate side chain) has only one major natural source: wildfires. Isocyanate is also found in cigarette smoke and automobile exhaust, but only exhaust that runs through the catalytic converters, which became standard in the U.S. in 1975. For its part, xylene was also added to gasoline in the 1970s as an antiknock agent, which helps prevent the engine from rattling. Xylene is found naturally in wildfires, contained in cigarette smoke, and is made into polyester, as well as paints and sealants that may be used in home remodeling. Industrial production of xylene for these products also started to rise around 1970.

To say the least, this was starting to feel very Erin Brockovich-ish.

But at this point, we had only associations—the type of associations that would signal *mission accomplished* for a GWAS. We could claim associations that *make sense* with outside data but had yet to identify a real mechanism. So, we decided to use these bacteria like a *canary in the coal mine* to see if toxins could either kill the organisms or prevent them from producing the beneficial lipids needed for skin health. We tested *good* bacteria that make ceramides (like *Roseomonas mucosa*), bacteria that help humans make ceramides (*Staphylococcus epidermidis*), and bacteria that make natural steroids (*Staph. cohnii*). We found that whenever *good bacteria* were exposed to isocyanates or xylene, their fate was best articulated by the character known as Harvey Dent, played by Aaron

Eckhart in Christopher Nolan's *The Dark Knight*: "You either die a hero or live long enough to see yourself become the villain." Bacteria exposed to these toxins would either die or shift their metabolism away from making the lipids that protect human skin.

This was true for the fumes from these chemicals in their raw state as well as from various glues and adhesives that I purchased directly from my local hardware store. The project did make for an odd interaction with the guy at the hardware store when I insisted that I needed a polyurethane based glue, even as he informed me that a different option was cheaper (and, in his opinion, better).

Another research assistant in the lab, Grace Ratley, had the excellent idea to test how the bacteria responded to the different fabrics linked with eczema that were made from isocyanate, xylene, and a related chemical used in gasoline named benzene. She found that *Staph aureus* proliferated on nylon, spandex, and polyester but could not live for more than a few days on cotton or bamboo fabric. This suggests that patients might climb into bed during an active flare and unknowingly seed their bedsheets with *Staph.* If those sheets are synthetic fabrics, that pathogen might be proliferating even when the patient isn't in bed. Healthy bacteria could live on any of our tested fabrics but, just as with air pollution exposure, they stopped making the beneficial metabolites when grown on synthetic fabrics. Good bacteria could live on nylon, polyester, or spandex, but to do so, they had to become villains.

After identifying a potential mechanism for influencing eczema through the skin bacteria, we also looked for any direct toxic effects on human cells. Turns out, prior research on work-related asthma found that both diisocyanate and xylene were capable of activating a receptor on the skin called TRPA1. The primary function for TRPA1 is detecting fluctuations in temperature. However, like all the TRP family receptors, they can be hijacked by chemical compounds. Chili peppers taste hot because the capsaicin they contain activates a related receptor known as TRPV1. When TRPV1 gets hijacked, it sends a signal for pain to the brain; when TRPA1 gets hijacked, it sends a signal for itch. So, to review, we went searching for a chemical that might link to a disease of itchy skin and the two we found are components of every known environmental risk factor, have near perfect epidemiologic overlap with the post-1970 rise in disease, turn your heroic skin bacteria into villains, and directly activate itch signals in the skin. But it gets worse.

Like the Jackson and Taconic mice mentioned earlier, researchers often use specific exposures to induce a disease in the mouse that is similar enough to a human disease that it can be studied. The researcher will basically give a mouse a disease with one chemical, then take that disease away with a second chemical; the first chemical is therefore considered a toxin while the second chemical is defined as a drug. Just about every modern drug on the market today went through some form of this process to get permission from the FDA or similar European regulators to conduct their first human experiments. The number one chemical used to give a mouse Autism is cadmium; I'll let you guess which of the EPA database chemicals was one of the strongest associated with clinical visits for Autism when we evaluated Autism. The most common form of diisocyanates, called toluene diisocyanate or TDI, has been used for years to purposefully give mice eczema to test treatments. Xylene is also known to induce eczema-like skin disease in mice. So, it seems to me that if you can induce a disease in mice with a specific chemical, it would be worth investigating if that chemical could have a similar effect in humans.

Our clinical trial may also point to potential real-world impact of these chemicals. The placebo trial (that was licensed to the private sector after our open-label study) again found significantly sustained eczema relief compared to both the expected norms of eczema and to the placebo. But interestingly, the level of improvement experienced by the patients appeared to be influenced by the level of diisocyanates in their area. The participants in areas with higher levels of diisocyanates had greater improvement with treatment. We interpret this as meaning that *refreshing* the skin bacteria is of greater benefit for those living in the environments that are most likely to corrupt their current bacteria. Since we felt that this product needed to get into the hands of the patients in need, we opted to give away the legal patents we held on *Roseomonas* and donate the most effective isolate (known as RSM2015) to a public biobank.

A company named Skinesa, that manufactures their own oral probiotic that has independent evidence of use for eczema agreed to work with us to offer an over-the-counter version. Skinesa's formulation will also include a natural blocker of TRPA1 (ground cardamom seeds) to hopefully improve the itch relief more than what *Roseomonas* alone can do. We still need to study the bacterial treatment further, but in about six years I went from almost burning down the lab looking for any bacteria that might make ceramides to seeing our clinically vetted bacteria being prepped for over the counter sales (even though I would not profit from those sales).

So, with all this environmental data in hand, let's revisit the idea that twin environments would be so similar that only genetics could differ between them. Take a hypothetical scenario for parents bringing home twins from the hospital for the first time. I want to present a scenario in which one could create the largest difference in isocyanate or xylene exposure between the two twin children. When you read this scenario, I want you to pay attention to whether any of the life choices I present seem outlandish. Ask yourself if it feels like I'm painting a hypothetical that does not feel like it would be a common issue for new parents, or if these issues might happen every day.

Prior to the birth, you decide to purchase two new car seats. Maybe, since you know you are having twins, you decide to get two different car seats— or maybe two colors of the same model. Since you want the child to be comfortable and safe, you make sure to get car seats with foam padding all around. You finally decide on a make, model, and color but even though you bought them at the exact same time from the exact same store, one of the seats you purchased was manufactured six months before the other. Thus, one of the seats had been off gassing for an extra half-year. We don't know the batch-to-batch or brand-to-brand variation in how much diisocyanate you would expect to come off the foam padding, nor do we know how long it might take to off gas (assuming it ever does).

Next, you secure one seat behind the driver seat and one in the middle of the back. So, as you are driving, when your window is down, one twin will get a little more ventilation than the other—nothing dramatic, but different nonetheless. Thus, the exposure to the car fumes on the road as you drive will be different to a very small, but measurable degree. You make sure to assign one seat to one twin, and one to the other so they don't get mixed up. Thus, you have assured that one twin will most often get the window seat and the other will most often get the middle.

You take the children inside and put them in their new polyester onesies. Maybe you made sure to get distinct ones for each, thus having different brands. Or maybe you just got one set and decided they can share. Either way, we also do not know how much variability there is in the xylene release from polyester clothing. The same would be true for the nylon sheets you got, as well as the foam mattress and newly lacquered cribs; each of these almost certainly varies in its chemical release in ways we don't yet know. And, unlike perhaps the shared sheets, the crib and mattress are likely *assigned* to one twin or the other.

You were so excited for your first children that you made sure to remodel their room with new wallpaper and paint. These, too, filled the home with chemicals at levels and durations we do not yet know. But at least for the remodeling chemicals, you would expect them to disperse throughout the room evenly, so perhaps the decision to remodel wouldn't have a different impact on the twins. Oh, except that the way the room is oriented, one twin's crib is closer to the window than the other's, so that twin gets slightly better ventilation when the window is open. But then again, with the window open, the gasses from the wildfire 500 miles away, as well as the toxin output from the factory 30 miles away, can waft into your children's room; for those days, the kid closest to the window would get the larger exposure.

Do any of these choices parents make seem ludicrous? Doesn't each feel like an extremely *normal* decision that a parent might make that influences the exposures of their child? I purposefully did not specify whether the twins were fraternal or identical because, for the most part, it would not matter. On some level, it might matter in terms of whether parents decided to share clothing (identical twins are famously more likely to be dressed into identical outfits). However, overall, the differences could come up for either type of twin. Most importantly, the scenario I described might only represent the first 48 hours of life; each of these subtle exposure differences could compound over time to become meaningfully large by one or two years of age.

If you don't think twin environments would be that dissimilar, let's consider the potential differences between siblings. Remember, the foundational premise for within-family GWAS presented by Young and Harden is that two siblings—even if born years apart—are effectively a randomized controlled trial for the parents' genetics. But to assert that the environment would not be meaningfully different for eczema risk factors within the same family, you would need to ask:

- Were both children born in the same country, city, and home? If they were born into the same home, did the parents renovate the nursery for the first kid and then leave it as it was for the second?

- Were the parents' diets and/or exercise habits less healthy with one kid versus another? The demands of childcare may impact the time devoted to preparing healthy, home cooked meals.
- Did the parents use different cleaning products? Maybe the parents could only afford certain products with one kid versus the other. Or maybe a specific product had not even come to market when their first kid was born.
- Similar to cleansers, did the parents use different moisturizers, soaps, shampoos, or detergents?
- Did either parent smoke during one pregnancy but not others?
- Did the mother get antibiotics with one kid and not her others?
- Was there a difference in being born by vaginal or C-section delivery?
- Did a factory move in during the time in between the birth of the siblings?
- Was a highway built nearby during the time in between the birth of the siblings?
- Did they have a dog living in the home at the time of some of the children's' births but not others?
- Did the mother's travels differ between her pregnancies?
- What kinds of furniture were the siblings exposed to? Did the first kid get a new foam mattress and freshly lacquered crib, whereas the later children just got to use what their older siblings didn't destroy?
- What about different textile fabrics? Were the clothes new for the first kid and hand me downs for the subsequent children?

After all our work on the environmental causes of eczema, a friend texted me a tweet by the comedian Roy Wood Jr, who both predicted our findings and encapsulated the results perfectly:

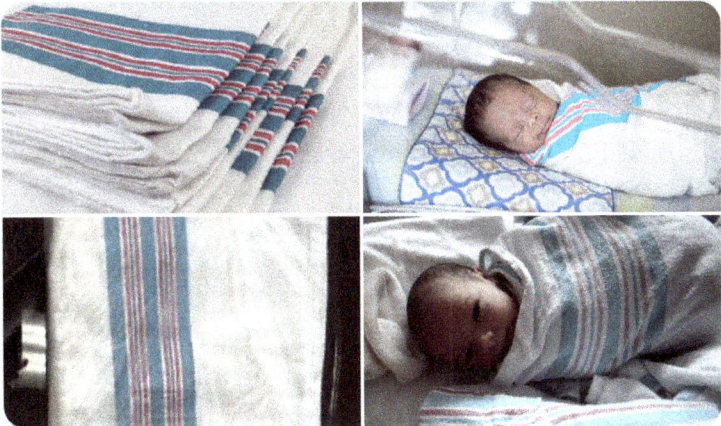

The point here is clear: neither twin nor sibling-GWAS studies can assume that environments were interchangeable for the participants. These examples are only for the main chemical exposures linked to eczema. Other diseases would have their own unique questions and various chemicals could cross react. If the types of polygene-*ish* scores supported by some geneticists were a tally of all the potential alleles that might contribute to a disease, what might happen if we combined the effects of multiple pollutants? The published PGS for eczema struggles to capture 31% of variation if the researchers wipe away the effects of race, poverty, and age. However, when we made a *poly-pollutant score* from adding all the different chemicals linked to eczema, we were able to capture 91% of the variation across the entire U.S.

This doesn't even address the elephant in the room of any discussion around environmental exposures: climate change. In the cases of allergy and infectious diseases alone, the rising CO_2 levels influence temperature, humidity, pollen production by plants, mosquito exposure, and wildfires. These rising CO_2 levels would also, therefore, be different between siblings born years apart. Yet one day we will see a GWAS for having lost your home to a wildfire or for living under water shortages. When families are killed in severe storms the geneticists will tell us that since the dead were genetically related, we cannot fully blame the flood waters for their demise.

The connection between pollution and allergic disease should force us to revisit the claims that Black Americans might have innate defects leading to eczema that were presented by Burchard. Dr. Elizabeth Matsui of the University of Texas at Austin and many other researchers have demonstrated racial disparities in exposure to pollutants like diisocyanate, xylene, particulate matter, and many others. Therefore, it is concerning that the explanations for the increased risk of eczema seen in Black communities consistently ignore environmental injustice in favor of genetic determinism.

Some might read these scenarios and point out that we cannot know that the exposure to these chemicals is different between the children described. But that is the point. We can't possibly assume the environment is insignificantly different until we actually measure it, and especially until we know what to measure. No one would run a study on COPD or lung cancer without accounting for cigarettes smoked and air pollution exposure. But if you don't know which exposures influence eczema, or Schizophrenia, or irritable bowel disease, then how can you just assume that the environments were the same between those with the diseases and those who are healthy?

There is an old saying that I initially heard from Dr. Uday Nori of Ohio State, "The eyes cannot see what the mind does not know." If you don't know to look for something, it will go unnoticed. If you didn't know to be worried about putting your newborn onto a foam mattress with polyester sheets, why would your eyes see doing so as a problem? If you don't know there is a factory nearby with an output that can cause your kid's eczema to flare, how would you recognize it when it happened? Furthermore, the fact that some of the products we buy might cause harm highlights the shortsightedness of the common practice of only adjusting for socioeconomic status when claiming to account for the environment. If poor people live near the factories that make the toxic products that rich people put into their homes, adjusting for income alone is doomed to failure.

A GWAS on eczema might (but not always) be adjusted for poverty and, to a less common extent, whether the patient was from a rural or urban area. However, do you think they adjust for the possibility that the controls might include more dog-owning farmers with all cotton hand-me-downs living far removed from highways or factories while the patient groups might have more city dwellers living near factories who wrapped their child in "a rough ass Motel 6 blanket"? The flaw in the GWAS approach—as it was with the twin studies before them—is to assume all the relevant environmental factors are either known, or so small at scale they can be ignored. So, when they run their analysis, they come away thinking they found a gene linked to eczema, when they may only have found a gene variant associated with the population stratification.

I would like to make one final comment to highlight how extensive our need to consider the environment is if we are truly curious about finding the causes of a patient's disorder. When a baby girl is born, she is born with every egg she will ever release in her life. Those cells have been developed to a set point and then lay frozen until puberty begins releasing them for possible fertilization. This means that every time a mother is pregnant with a daughter, she is responsible for three generations at once. Every exposure that the pregnant mother faces carries the potential to harm her, her future baby daughter, **and** the eggs that will become her biologic grandchildren. Harm can also come if the unborn boy suffers damage to his developing testicles and/or sperm producing cells. One infamous example of this is Diethylstilbestrol (or DES). DES was thought to prevent miscarriages but instead caused multigenerational harms. The mothers who took it had an increased risk of uterine, breast, and ovarian cancers. The unborn fetuses exposed suffered similar sex organ cancers, as well as increased rates of infertility (in both boys and girls). And the grandchildren born of the eggs exposed during their grandmother's pregnancy were found to have an increased risk of neurologic disorders, premature birth, physical disabilities, and infant death.

Realizing that every pregnant mother bears the potential burdens of three generations at once should ignite not only a sense of responsibility but curiosity—a curiosity about how our environment can create ripples that travel through time and space. A curiosity that goes beyond our innate genetic programming to wonder how our actions might improve the lives of children that have not even been conceived let alone born. Wegener's was able to take the life of Mr. Smith because his care team never asked what might be truly going on. In 2013, the journal *Arthritis and Rheumatology* published a GWAS for Wegener's that suggested HLA type and the gene *SEMA6A* might *contribute* to the disease. Yet I'm left wondering how knowing Mr. Smith's genotype could have improved his care. What benefit would knowing they may have inherited a miniscule genetic risk for Wegener's provide for Mr. Smith's family?

Meanwhile, work by Lee and colleagues in the *Annals of Rheumatic Diseases* identified a risk factor for Wegener's that was vastly more impactful than any *genetic predispositions*: exposure to the pesticides used in farming.

CHAPTER 15

The Hidden Price of Our Genetic Obsession

BOTTOM LINE UP FRONT: *Gene-centric thinking prevents advancements in medical science and causes economic, political, and moral harms.*

Neither of my parents grew up rich. My father grew up in the poor parts of Cleveland, Ohio that, at the time were termed "colored". My mother grew up in the slightly-less-poor—but still not great—Polish, Italian, and Irish parts of Cleveland. Their worlds did not offer a lot of culinary options back then. The local diversity in my Dad's neighborhood afforded access to Indian, Native American, Mexican, and Jewish foods, but my mother's only food options were limited to Italian, American, and kielbasas. People often make fun of the poor for being *uncultured*, especially when it comes to food. But the major factor that limits people's ability to try unfamiliar dishes isn't culture or intellect; it's money. When you don't have a lot of disposable income, gambling money on a type of food or even a dish that you are not certain will be palatable is a major risk.

By the time I was born, my parents were more firmly middle class and living in Colorado. The increased exposure to people of Latin American backgrounds facilitated my mother trying a burrito for the first time in her life. However, they continued to mostly keep to the foods that they were comfortable with. That meant I grew up with a lot of pasta. One dish in my mom's frequent rotation was a chicken alfredo, which is a pretty simple recipe of chicken breast, pasta, and cream sauce. But she would insist on mixing in a can of peas. Oh, how I hated the addition of those peas. I would plead for her to put them *on the side,* but she saw right through my conniving plot to eat the chicken and pasta then claim to be *too full* to eat the peas. Googling "chicken alfredo with peas" yields plenty of results, so it is not as if my mother were the only one preparing the dish in this way. But I tell you, my dream was for a pea-less chicken alfredo.

Moving to the college town of Fort Collins, Colorado greatly expanded my exposure to various cuisines. While Fort Collins is far from cosmopolitan, there were Indian restaurants, as well as Ethiopian and Vietnamese places. The first time I tasted a mango lassi at the Indian diner a few blocks from campus, I was left wondering where this had been my whole life. It may be quaint, but my favorite dish was chicken tikka masala. One summer, when living back home, I told my parents they needed to try this amazing meal from the other side of the world. Usually, my mom needs to be coaxed into trying any food more exotic than oatmeal but, to my parents' credit, they both gave it a try. It was a smashing success and to this day, they still rotate Indian food into their newly expanded menu.

There are dozens of genetic studies claiming to have identified genes that influence dietary preferences. One published in *Nature Communications* claimed to have even mapped out the genes for *acquired tastes* for 139 total foods. The authors considered their results valid, in part, because: twin studies suggested dietary preference was 50% heritable; the authors found correlations such as 'the genes for eating junk food were associated with the genes for body weight and inversely related to the genes for strenuous exercise'; and that many of the genes for beer drinking are expressed in the brain. If you paid attention during Chapter 9, you know why all these arguments are trash. Yet overall, they predicted that their results could *explain* a few percentage points of the variation in how yummy different people found different foods. Seeing these claims, I'm left to wonder how many bites of chicken tikka masala consumed by my Italian-Irish mother do these researchers think they can account for using genetics?

At some point, you have to translate the mathematical into the real. If you claim that genetics can *explain* 5% of an outcome, then if your outcome is that my mom ate 1,000 forkfuls of tikka masala in her life, you must think 50 of those forkfuls may not have happened if she had different genetics. But my mom would have never been exposed to Indian food had I not brought it home. And I would not have brought it home if the food I tried had tasted bad. Also, if the restaurant I tried was not priced for college students to afford I would have never bothered. So, how could one make an analysis for genetic contributions in a way that accounted for "she had a son that made her try this"? The authors of the *Nature Communications* paper might claim that they are only talking about foods people continue to like, claiming that they are measuring how many bites after the first one someone might try. This too falls short on logic, since I've since been to Indian restaurants where asking for *mild* must be seen as a joke because the food is so spicy it burns your palate. If that level of spice had been my first experience with Indian food, it might have also been my last. How can you possibly account for the chance that someone's negative impression of a food might be due to a bad first experience?

If researchers want to claim that genetics *explains* real world impacts, then it would need to be measured not by correlation coefficients but in forkfuls of food, days of education, and children dying of asthma attacks. When those working in public health report how many people *die from cigarette smoke* per year, they are translating the mathematical associations into the real world impact of reduced smoking. While the deaths caused by eugenics may dominate the headlines, the damages of the gene-centric thinking reach into many more aspects of economics, politics, and morality that are far less humorous than the idea of tikka masala genes.

The economic cost

Resources and money are not infinite in research. Money spent on researching the genetic *contributors* to people dying of lead toxicity is money that could be spent on replacing contaminated pipes. Every day spent trying to determine the inborn reasons why Black Americans who have been historically redlined into the most polluted parts of America's most polluted cities have more pulmonary disease is a day that every American continues to huff inflammatory particulate matter.

Even performing a full investigation of automobile crashes would require assessment of air pollution. Have you ever had one of those days driving where you could swear everyone else on the road had completely forgotten how to operate their automobile? If not, then have you ever felt like some days, everyone was honking at you as if you had forgotten how to drive? Well, in 2022, two papers (one in the *International Journal of Environmental Research* and the other in the journal *Public Health*) demonstrated that a quartile change in particulate matter that was 2.5 microns or smaller (called PM2.5 for short) can be linked to a 28% increase in fatal car crashes. This was more impactful than bad weather in their analysis.

Increased exposure to PM2.5 also drives up the rate of occupational accidents, accounting for 5.7% of non-fatal and 19.9% of fatal accidents at construction sites, according to Matthew Chambers of Vanderbilt University. A decrease of one microgram per meter cubed would be expected to reduce workplace accidents by 7% and reduce fatal car accidents by 2.4%. Translating these reports into real impacts, indicates that if Los Angeles were only as polluted as Seattle, about two people every month would still be alive who otherwise died in a traffic accident.

In 2021, the journal *Scientific Reports* found that between the years 2003 and 2016, there were 50,223 total excess deaths in America due to PM2.5 road pollution. This number includes the traffic accidents as well as deaths from inflammatory disorders like asthma, heart disease, or stroke. These results mean the deaths of 10 people per day in America can be directly tied to the accumulated harms of breathing in *just one* of the many types of air pollution we experience.

Eczema directly afflicts 30 million people in the U.S., and that number would not count the impact it has on their caregivers. When you tally the price of medications, doctor's visits, transportation, missed school, missed work, and reduced productivity, the cost of eczema alone was estimated by Dr. Adewole Adamson in 2017 to be *over five billion dollars per year* in just the U.S. *The Journal of the European Academy of Dermatology and Venereology* similarly estimated in 2022 that eczema directly impacts over 40 million Europeans and costs over 30 billion euros (about 32 billion U.S. dollars). Numbers for Canadians are less robust, but older estimates from the *International Journal of Dermatology* combined with reports from the Canadian Skin Patient Alliance suggest about two to three million Canadians are affected with an annual cost of over 2.2 billion dollars. So, if you were trying to summate the burden of eczema on just some of the North American Trade Organization (NATO) nations, you would arrive at nearly 40 billion dollars paid by the impacts on 73 million patients. This calculation would not include all the costs from eczema-related diseases like asthma and food allergies.

Lowering the rates of eczema down to levels from ancient Greece is probably not going to happen but recall that if the rates of eczema in the world were just where they were in 1970, then about four of every five patients would not experience the disease today. That would mean that if the rise of eczema could be reversed, NATO nations would be expected to spare about 58 million people from the disease and save 32 billion dollars every year. 32 billion dollars is more than the gross domestic product of Iceland. That is more money than the profit for all of football—and I mean *all* of "football." The annual profits of the National Football League in America *and* the Fédération Internationale de Football Association (FIFA) globally come to only 24 billion per year (at least the over-the-table profits for FIFA). So, the entire global "football" economy could not afford to take on the economic cost that just eczema places on humanity.

So, from a purely economic perspective, which makes more sense? Should we continue to spend more than the annual budgets for the world's most popular professional sports to find expensive treatments for a disease that is at least 80% preventable? Or might it make sense to spend money on preventing the disease? Reports from Dr. Bill Foege of the CDC indicate that smallpox eradication cost the globe about 300 million dollars but had since saved 39 billion dollars (a 130-fold return on investment) and saved the lives of 150-200 million people since 1980. Maybe one of those people is someone you care about.

It will take time, of course, to turn the tide of pollution that has been driving eczema for the past four decades, but as they say, "Rome wasn't built in a day." Then again, if we really could save all the money preventing eczema, humanity could build a brand-new Roman colosseum every week and still come out with hundreds of millions of dollars to spare. Some of the newer medications for eczema are life changing. People who are suffering from the disorder are suffering a lot less because of these new treatments. No one would argue that having new medication options is a bad thing. However, perhaps we should all take a step back and reflect on how our society reached a point where slathering our children in repurposed chemotherapy to treat a disease that could have been avoided is seen as a success.

The political cost

U.S. and British politics have been streaked with the stains of eugenics for nearly a century. Besides the harms of slavery, forced sterilization, and bigoted immigration policies aimed at marginalized groups, *natural selection* has been invoked as an argument against the social safety net. In the U.S., *The Bell Curve* was directly cited as reasons to cut off minorities and impoverished white Americans from any social or welfare programs. The crux of the argument was that some people will always naturally be *at the bottom of the curve*, and thus, intervention programs aimed at helping would always be doomed to failure. However, the political pitfalls I'd like to mention here are those that occur not in the Oval Office, but in the physician's office.

"Patients in these chat groups are worried their doctors don't know anything that the pharmaceutical companies don't tell them," I'm told by Amanda, a patient's mom. Amanda's son had suffered severe eczema that was worsened by the use of medication given to her by her healthcare provider (topical steroids). Amanda is highly educated and leery of believing anything that the internet tells her. However, she is also hesitant to expose her child to new medications with only partial information. Amanda had tried numerous medications for her son's eczema and was willing to try others, but something stopped her: "Deep down, I knew that there must be something more going on with my son's environment that I hoped to correct so that he didn't need drugs."

I had found myself listening to a lecture on the role of the environment in the development of allergic disease in the gut by Dr. Cezmi Akdis. Dr. Akdis is the director of the Swiss Institute of Allergy and Asthma Research and the editor-in-chief of the scientific journal, *Allergy*. Among many toxins he covered that day, Dr. Akdis spoke about the potential harms of a detergent that can go by many names but is most commonly referred to as sodium lauryl sulfate (or SLS). SLS erodes the ability of the gut cells to form a tight seal. This allows bacteria from the gut along with any other ingested chemical to transit from the gut to the blood through unnatural channels. Dr. Akdis presented data on the exact impacts of SLS in cell cultures and mice to demonstrate which specific proteins and metabolites were damaged or altered by SLS exposure. Then, he outlined the many sources of SLS in our industrialized environment.

Dishwashing detergent, soaps, shampoos, and even toothpaste might contain SLS. Mind you, SLS is not an essential ingredient in these products, so there are plenty of versions out there that lack SLS. Thus, as he was speaking, I was planning to head to my medicine cabinet when I got home to check on whether my family's toothpaste contained any SLS. But I was also in the room with other researchers and NIH administrators. One of them was a higher up in the department that would judge NIH grants for allergic disease. This meant that universities that file for NIH funding to research allergic disorders would be scored by this guy and the teams of judges he might put together. When Akdis was finished with his lecture detailing how toothpaste contained chemical gut rot, the guy who would hold the purse strings for NIH grants had only one question: "Can we combine this information with GWAS to identify genetic predispositions to these effects?"

I was floored. This man sat through the same 45 minutes as I did. Forty-five minutes of listening to Akdis detailing the epidemiologic and biochemical evidence that select modern exposures are harmful to the health of children around the world and the guy the NIH trusts to vet research projects wants to know about innate *predispositions?* Unfortunately for him, I was not in a mood to let it slide. I asked, "Why would we possibly care? Whether my child had a predisposition to a bad reaction to toxic toothpaste or if my child was genetically impervious to toxic toothpaste, the solution in both scenarios is to avoid the toxic toothpaste."

"But we need to know the pathways impacted," he countered.

"We already do. The *RNAseq* and proteomics presented give us that information. How does knowing whether some allele modifies the risk further help anyone?" I asked.

"Some people will always fall at the extremes of the bell curve, so knowing the genetics means we might be able to make a targeted therapy for the genetic pathway," he offered as his final attempt to justify his proposal.

"You want people to give their children a drug to counteract the presence of a completely optional and fully avoidable toxin?" I asked in a tone that had turned from inquisitive to condescending. At this point, a senior researcher stepped in and wisely changed the subject, but the thinking on display is the opposite of the aforementioned Groundwater Approach.

In a *New York Times* profile, Dr. Thomas Insel reflected on his career as the head of the National Institute of Mental Health. The author of the profile, Ellen Barry, outlined that Insel had "bet big on genomics" 20 years prior, but now admits that little-to-nothing had come from it. Barry relayed a story about when Dr. Insel realized the failings of his molecular focus.

> In the book, he describes an "epiphany" during his last year at N.I.M.H., after he had delivered a PowerPoint presentation to a group of advocates, touting researchers' progress on genetic markers.
>
> A man in a flannel shirt got to his feet and reeled off the story of his 23-year-old son, who has Schizophrenia – a cycle of hospitalizations, suicide attempts and homelessness. "Our house is on fire," the man said, "and you are talking about the chemistry of the paint. What are you doing to put out this fire?"

"In that moment, I knew he was right," Dr. Insel writes. "Nothing my colleagues and I were doing addressed the ever-increasing urgency or magnitude of the suffering millions of Americans were living through – and dying from."

These paragraphs feel like they are building towards a wonderful comment about how Dr. Insel realized his errors, commanded his research institute to better evaluate the best firefighting practices, and began saving patient's lives, right? Instead, all the author provides us with is an admission, "I should have been able to help us bend the curves of death and disability, but I didn't," and then informs us that Dr. Insel still believes we should "double down on brain research"—even as we are told that the critics who *correctly predicted these failures* are still being overlooked today as the care provided to people with serious mental health disorders continues to worsen.

Healthcare providers only have so many *touch points* with their patients' disease. Obviously, each office visit is a point for interacting with the patient, but any prescription written will also generate a kind of touch point whenever the patient takes or applies that treatment. However, for patients and their caregivers, there are vastly more disease touch points. Patients and caregivers need to remember to take their medicine, need to worry about affording or refilling that medicine, might need to consider special dietary limits for their disease, and may need to consider different exposures that might trigger their symptoms. Patients with allergies, for example, might need to keep their asthma inhaler and EpiPen on them, check any food product for peanuts, and avoid soaps or lotions with ingredients they know they are allergic to. So for patients, their disease is far more than a drug target.

When your paradigm is only capable of viewing the patients' physiology as the root of the disease, then the only solutions you will look for are those that can *correct* the patients' physiology. By definition, a substance aimed at beneficially altering someone's physiology is a drug, and thus, a form of pharmacologic tunnel vision has overtaken medicine. While the average clinician—especially those in primary care—would love nothing more than to prevent their patients from getting sick, often biomedical research fails to outfit them with the data needed to do so. For most diseases, various professional associations will put out what are called *treatment guidelines*. While not legally binding, guidelines aim to give practitioners an up-to-date understanding of the most important aspects about diagnosis and treatment of the disease in question. The guidelines are not meant to be exhaustive—that is, they may not cover every single aspect of a disease. However, they should cover the core factors that someone managing the disorders might need to know to best care for their patients. Using eczema as an example again, here are the guidelines for atopic dermatitis as put out by my duty station (National Institute of Allergy and Infectious Diseases, or NIAID), the Association for Family Medicine, the American Academy of Pediatrics (AAP), the American Academy of Allergy Asthma and Immunology (AAAAI), and the American Academy of Dermatology (AAD). The print is too small to read but I have color coded the results by topic.

Even from a low magnification, it is clear that the overwhelming majority of the text is dedicated to the details of prescription and nonprescription drugs. These sections might inform your provider about what treatment options are approved, what doses are advised, and perhaps what complications to be on the lookout for. So, I don't mean to imply that this information lacks value. But concerns arise when comparing how much ink is dedicated to medication instead of topics that are within the patient's control.

The guidelines form NIAID, Family Medicine, and AAP do not contain a single word of dietary advice. The guidelines from allergists (AAAAI) and dermatologists (AAD) have only a few paragraphs that can be summarized as: Don't randomly test for food allergies unless the patient is having anaphylaxis; don't randomly avoid foods, but if the patient is certain the food worsens their eczema, then consider avoiding it. Summarized further could be a sentence stating, "Only avoid foods you know are a problem," which is helpful, but I think more than a little incomplete.

Only three of the five guidelines offer any insights into environmental triggers for eczema, and the content there can be summated as: We don't have the evidence to make recommendations on what fabrics or cleansers patients should use, and if you are proven to be allergic to dust mites, then avoiding dust mites may be helpful. Like for food, this is valuable but staggeringly incomplete.

Meanwhile, looking at the guidelines, you might notice that more ink is dedicated to discussing mutations in *FLG* than environment and diet combined—a gene that offers providers no insights on diagnosis, no value in selecting the best treatment, and has not proven to be a therapeutic target. The guidelines suggest that telling your doctor about how your kid might have an innate *probabilistic* defect promoting eczema is more valuable than providing insights as to how parents can reduce their kid's suffering in ways other than medication.

For patients with eczema and their providers, eczema intrudes into all kinds of decisions for which providers have nothing concrete to offer.

Go to any U.S. department or grocery store and look down the topical lotion aisles. Can you tell me which of the 100-some options are best? How about something simple like whether you would be better with a lotion, cream, or ointment? Would some patients respond better to products formulated as ointments while others better to lotions? Are there ingredients that you should avoid in topical products? Are there ingredients you should opt for? Your provider might be able to offer opinions or give insights from their years of talking with other patients, but they will not be able to offer any solid evidence to answer basic questions patients face daily.

Many bottles for topical products will say "for eczema" right on the label, but do you know why they claim that? One specific regulation from the FDA, known as Title 21, section 347.10, dedicated the ingredients one could put into topical products and make the claim that it was "for eczema." So long as the company could show that their topical had the correct percentage of any of the compounds on that list, they could label the topical to directly market to the eczema community. For 40 years, that list included known allergens like lanolin on it, meaning that companies were incentivized to add a potential inducer of contact dermatitis into products they would then sell to families with children prone to eczema and allergic disease. Thankfully, this was amended by the CARES Act in March of 2020 to no longer include anything harmful. Now, only colloidal oatmeal with or without mineral oil can earn the "for eczema" label.

It is important to note that having a better understanding of day-to-day exposures for patients is not mutually exclusive with recommending medications. In fact, many patients are more than happy to take medications but are still curious as to how to get the most benefit they can while avoiding harms. As a basic example, most people know not to drink alcohol while taking Tylenol or to take Advil with food as ways to avoid harms from the medications. How might you best maximize your outcomes on any of the numerous medications listed for eczema in the guidelines?

If you were to start 100 people on dupilumab (the first injectable antibody for eczema), you could expect about 10 or so would have a side effect or other general opposition to the medication that would prevent them from continuing the drug. Thankfully, these side effects are not long lived. So, while it is unfortunate that those 10 patients are not benefited, at least the side effects wear off. Next, you would expect about 15 people not to notice any difference at all. They would not have any issues but would not notice any improvement either. This leaves about 75 people who would see a good response to the medication—"good response" here being defined as at least a 50% reduction in symptoms. While you always root for 100% of people to get 100% better, seeing 75% of patients halve their disease is something to celebrate. The drug, at this point in time, is very expensive—upwards of 40,000 U.S. dollars per year. So, there are people still unhappy with the new medication's limited access. Yet only the most anti-pharmaceutical people in the world would deny the success of seeing so many people who were previously suffering from eczema now suffer less due to access to dupilumab (albeit hemmed in by cost). However, there is a range of improvement seen in those 75% of people who did well on the drug. Some people improved by 50% while others improved by 100%. On average, people improved by around 80%. So, one might ask, "Did the people who got 80% better use different topicals than those who only got 50% better?"

Certainly, if you were open to investing the time and money to use a new injected medication, you would like to know whether switching soaps might maximize your benefit, right? Or maybe we might wonder whether the people who saw a 90% improvement ate different foods, slept on different bed sheets, or were exposed to less air pollution. Unless the drug is 100% effective with only one dose, every patient or parent will wonder how they can assure the prescriptions they take will be safe and effective for as long as they are using them. Yet when patients like Amanda go to her provider, the provider has no evidence-based opinion beyond "take this medication." Patients and parents are then left without much guidance for how to handle all the ways in which their disease manifests but is not controlled by medication. This, I believe, fractures the relationship between providers and patients.

Best case scenario after a new medication is started is that the patient is happy with the improvements seen but still needs to search for answers to any concerns that remain. Worst case is the medication creates a side effect, but the patient is told either "take this medication to counter the harms of the prior drug I gave you" or "take this drug, instead." The provider may care deeply about the patient and their needs but when they cannot provide insights beyond their prescription pad, the patient will misperceive the provider as being "in the pocket of pharma." Patients will turn to online chat rooms to get opinions on the questions their providers are simply not trained to answer. Some of the insights on these chat rooms are good, but in an era of digital grifters and professional online dipshittery, many of the comments are nonsense. Patients who are not perceiving improvement from their medications will be more likely to entertain ideas from outside the mainstream. Again, some of these will be potentially insightful even if unstudied, while others will be hatched from people looking to take advantage of patients' untreated suffering. Here, the risk of conflict peaks. The patient may bring the alternate idea, gleaned from internet comments, to their provider hoping to get an evidenced opinion. The provider, however, might only offer what is viewed as *the party line* by recommending nothing beyond prescription drugs.

The perception of *doctors as drug dealers* will be made worse when the profession parrots the overpromises made by pharmaceutical companies. Drug companies are pouring money into finding the next Gleevec, looking for a genetic marker that signals a drug has a unique subpopulation of patients it works best for. Conceptually, this is born of a real desire to maximize benefits for their patients. In practice, Gleevec presented a truly rare find given that it involved a chromosome-level change that created a new gene that existed *only* in the cancer cells. In contrast, drugs targeting gene markers with small impacts will be only slightly better than other medication options. Yet, these drugs will be hailed as precision medicine wonder pills. As outlined in *The Tyranny of the Gene*, many patients are going broke discovering that the "a drug that is just right for you" slogans their drug was sold under was more about marketing than miracles. Additionally, providers will risk appearing as the drug company's spokesperson when they are inevitably forced to admit to their patients that the drug they might have mortgaged their house to afford isn't working.

In his beat poem, *Storm*, the stand-up comic Tim Minchin relays a story about his humorous rant at a dinner party aimed at the titular guest who spewed numerous accounts of anti-intellectual and anti-science nonsense. One line from that poem is frequently passed around medical circles: "You know what they call 'alternative medicine' that's been proved to work? Medicine."

I applauded the first time I saw the clip of Minchin's performance of *Storm* and still highly recommend it. However, this line, and the mindset it represents, gets thrown around far too frequently. I can personally attest to the frustrations of having to deal with the level of ignorance about medicine and science portrayed by the character, Storm, in the poem. But most patients are not looking to alternative approaches because they are dumb; they are doing so because they are desperate (Minchin never implies otherwise in his poem, however those who intend to imply that patients are dumb use his poem frequently). Most patients are unsure of what to do but are not being dismissive *of* modern medicine, and therefore, do not deserve to be dismissed *by* modern medicine.

According to the National Center for Complementary and Integrative Health (a different NIH group that once contained the term *alternative medicine* in its name), Americans spent 30 billion dollars out of pocket on *complementary health approaches* in 2016. That number was only eight billion in 2002, per a report in the journal, *Health Affairs*, by Davis and colleagues. Patients like Amanda traffic the chat boards with the same critical eye that they bring to their provider. However, patients and parents are being stigmatized by guidelines that cement the paradigm that the patient's innate biology is the root cause of their disorder with little regard for the environment. The patients and caregivers are showing the type of curiosity inherent to wondering, "What is really going on here?" but are left feeling *unseen* by their providers who don't seem to share that curiosity.

The feeling of disconnect with medical providers is the key opening exploited by the charlatans in the supplement industry. That is not to imply that every company in the over-the-counter market is exploitative—far from it. I've personally met many who truly understand the science of their product and take it seriously. As Minchin points out in his poem, most alternative medicine "has either not been proved to work or been proved not to work." Studies proving that the effectiveness and safety of supplements are often lacking or of poor quality. Thus, sales are often driven by word of mouth. But why are so many patients open to suggestions from online strangers?

Even when data supporting safety and efficacy do exist, the major limit for the non-prescription world is the failure to assure dosing. Drugs and supplements, like genes, work through molecules not magic. So, when a product is made by generating an extract, it is important to know how much is in there. Maybe variation doesn't matter, like for coffee, where a little more caffeine from one bean versus the other won't make that big of a difference. The range of caffeine output per bean isn't wide enough to create a problem. But I've heard plenty of patients report that a supplement was helpful but then *stopped working* and I'm compelled to wonder whether they just ended up with a dud bottle devoid of whatever ingredient had been helping. So, when you buy a supplement, even one that has evidence of being helpful, the major piece of data you lack is an assurance that the correct amount of the correct molecule is in the bottle.

One legal hurdle for over-the-counter products, at least in the U.S., is the limited ability to make any claims about helping a disease irrespective of whether they have evidence to support their claims. Supplements with genuine data showing they can benefit eczema can't say that outright. Instead, they have to say *for skin health* or some other marketing euphemism. On paper, this policy is meant to prevent snake oil salesmen from flat out lying to people and saying, "This cures eczema," when it does not. And indeed, the policy has been effective in blocking false claims. However, it has generated the unintended reality in which both the snake oil and the Balm of Gilead are sold under identical terminology that make it impossible for the patients to sort good from bad.

Thus, when the type of data exists that might be able to redefine something from *alternative medicine* to *medicine* (as *Storm* implies), it can't filter down to the providers. The expert panels who write the guidelines must focus on the treatments with the highest level of evidence, which are and will always be prescription drugs that can afford to invest in big trials that test their drug on large numbers of people. But this means that your average provider can only offer evidence-based opinions about drugs in ways that cause them to be misperceived as Big Pharma shills by some patients. The patients, frustrated by the system, turn to a supplement industry ripe with con men who also fail to offer a transparent and balanced presentation about their product's strengths versus limitations. Often, these con men structure their sales pitches to highlight that *modern medicine* seems more aimed at profit than patients. I don't write this as an excuse for the Storms of the world, but patients who have drifted towards medical distrust due to biomedicine's lack of curiosity surrounding non-prescription prevention and treatment options are not going to float back into the fold upon waves of stigmatization.

Yet the *poking the dead fish approach* can also be stigmatizing to entire groups of people. The AAD eczema guidelines shown before contain the line, "Being of Black race does appear to increase risk. A higher level of parental education is a risk factor for disease, but the effect of socioeconomic status is unclear." As stated before, the claim here that "being of Black race does appear to increase risk" is only true if one ignores the entire African continent and its over one billion current inhabitants. Furthermore, using this phrasing implies that "being of Black race" is the problem, especially when juxtaposed with the claim that poverty did not explain the differences. A correct phrasing would have been to state, "Rates of eczema are higher in Black Americans than in white Americans and are also far greater compared to Black Africans. Therefore, differences related to environmental justice likely play a significant role in eczema." This phrasing would be non-stigmatizing and would correctly identify that the time has come to investigate the lake instead of continuing to poke the dead fish.

A paper from the group led by King Jordan from the motte-and-bailey fallacy example previously discussed was published in 2022 in the journal, *Gene*. The authors found that an increased percentage of genes indicating African ancestry was associated with an increase in serum creatinine (a marker of kidney function in the blood). Since this was not true for the percentage of European or Asian ancestry, the authors concluded that "genetic ancestry contributes to creatinine disparity, but the social environment does not." However, their analysis was only adjusted for age, BMI, deprivation index, biologic sex, and height. Are those factors sufficient to *exclude social environment* as a contributor to their findings?

Might we counter that increased African ancestry predicts increased discrimination and thus an increase in stress from racism, increased environmental injustice of heavy metal and pollution exposure, and reduced quality foods leading to increased salt intake? Results suggesting that African ancestry may be a marker of hardship and poor care for asthma were reported in the *JACI* in 2022 led by Dr. Robert Kahn of the University of Cincinnati. To believe that adjusting for the zip code deprivation index was sufficient, it would presume that being poor and white is interchangeable with being poor and Black. Certainly, poverty will have many shared features for every group of people in a given country. Yet even considering access to universal healthcare in the U.K., no one would espouse that poverty is color blind. Furthermore, adjusting for only deprivation assumes that being rich and white is interchangeable with being rich and Black.

The starkest example of how wrong the assertion that wealth is an equalizer for racial disparities is the increased mortality after childbirth in Black women. Even after adjusting for income, Black women were still overall 2.6 times more likely to die during childbirth than white women in New York, per Elizabeth Howell of the Ichan School of Medicine at Mount Sinai. But here, too, there is an assumption that *adjusting for income* indicates that a rich white woman and a rich Black woman would have interchangeable experiences in a healthcare setting. Tennis star and very wealthy Black woman, Serena Williams, was able to win a tennis championship while pregnant but almost died from not being able to get her providers to check on her after giving birth.

Considering the history of *racial diseases* and the frank teaching in some medical schools that Black people experience less pain, there is strong reason to have anticipated that adjusting only for income would not be adequate. Of additional interest from the same Dr. Howell report is the finding that the disparities in death varied greatly across the state. In the worst areas, Black women were 6-fold more likely to die after childbirth while in other areas, the risk was the same for everyone. If we are to assume some major contribution to innate biology or genetics in the racial disparities in maternal mortality, is it reasonable to assume that Black women in Queens have different genetics than those in Manhattan? How many subway stops apart would Black women need to be before we would guess their genetics became fatally different? Why is the default assumption that seeing a 6-fold difference between cities assumed to be environmental, but a 2.6-fold difference between races is assumed to have *genetic contributions*?

Before the geneticists pull their *probabilistic and not deterministic* card, I will note that there is one healthcare system in the U.S. in which Black women do **not** have an increased risk of death. That would be the system run by the U.S. military. Publications in the *American Journal of Obstetrics and Gynecology* by Dr. Erin Keyser, the *Journal of Obstetrics, Gynecologic and Neonatal Nursing* by Dr. Candy Wilson, and the *American Journal of Perinatology* by Dr. Anju Ranjot all reported that while disparities existed for cesarean section delivery, ICU admission, and need for blood transfusion, there were no racial differences in either neonatal mortality or maternal mortality in military medicine. Military medicine often gets derided in academic circles and suffers from front page news stories of genuinely poor care at outdated systems at Veterans Affairs (VA) hospitals. The system does indeed have its flaws. But when the National Academy of Sciences and Institute of Medicine (NAS-IOM) released a report on disparities in the healthcare system, the VA took it seriously.

Initially, the report received political pushback and was partially redacted by the then head of the Department of Health and Human Services. The politicians were concerned about the report's conclusion that a notable portion of health disparities may be due to implicit bias in healthcare providers. Physicians' organizations were not pleased with the implication that people dedicated to providing care may carry such harmful bias, even if it only manifested unintentionally. You might translate the American Medical Association's response to the report as, "Physicians are gentlemen, and gentlemen's *actions* are clean."

In a speech at the Association of Military Surgeons of the United States (or AMSUS), then executive-in-charge for the Veterans Health Administration (or VHA), Dr. Richard Stone, openly discussed how the VA system was trying to address its shortcomings. He noted that, in addition to maternal mortality, the military system was the only U.S. system *without* racial disparity in surviving prostate cancer. Disparities still existed in the system, but Veteran's Health had set up task forces to stomp them out and improve care wherever possible.

In his lecture, Dr. Stone emphasized that the VA realized that "all of their soldiers and their families deserve the best care." While other hospitals had assumed a combative attitude towards the NAS-IOM report, the Veterans Health Administration had taken a combative attitude towards implicit bias. The VHA felt that they should show the same dedication to protecting their soldiers from preventable healthcare deaths as they would from enemy combatants. Dr. Stone was quick to point out the ways in which the VHA still had work to do, but they did not shy away from the need for change. So, how can we explain that women giving birth in a military system lived through childbirths that might have killed them in other hospitals? The VHA isn't just for soldiers, so the explanation that maybe the women were all in better shape due to the demands of military service doesn't hold up. So, are we to believe that genetics differ between service members and civilians and also differ by New York blocks?

Although the military did not have disparities by race, there was a difference in maternal mortality between the Air Force and the Navy. Should we assume that the *love of sailing* genes claimed by twin studies researchers a century prior also correlate with genes for death during labor? The argument for the genes only being *probabilistic* does not work here because the racial gap is non-existent in some places while enormous in others. The claim from population genetics would have to be that Black women across these healthcare systems all share a similar rate of genes that *can kill them*, but those genes are unable to have an effect in the military system. To this, I would simply reply, "Exactly." *The one hospital system that put in the effort to address racial disparities in death just so happens to be the one environment where all these supposed genetic predispositions melt away.* The success in the VHA is not simply because every woman has assured access to care, either. In the U.K. system, Black women are still three times more likely to die than white women. A 2021 report from the U.K. government also noted that Black women were more likely to receive substandard care, even in their universal coverage system. So, if addressing social determinants can render genetics irrelevant, then why are we funding genetics research? Why should the U.S. government spend money on deriving a PGS for *dying in labor while Black* when they could spend that money implementing the changes that the VHA has shown to save lives?

At the core of the population geneticists' fallacy is biologizing. Disparities in health outcomes along ancestry lines exist. Genetic differences between ancestries exist. To assume that some portion of the disparity will be biologically mediated is to *biologize* health disparities. Imagine a researcher claimed that genetics play a large enough role in the racial disparities in *incarceration* that he decided to dedicate his career to uncovering those genetic components and would like government funding to do so. I suspect most *respectable* geneticists would distance themselves. However, there seems to be no hesitation to collaborate with those claiming that the racial disparities in maternal mortality found in Black American women will be *genetic enough* to deserve time and treasure. "I hear this from women's advocates," Angela Saini tells me when asked about her interactions with self-reported liberals who assume genetic causation for health disparities in ways that are statistically indistinguishable from comments made by eugenicists. "They believe they are helping, and it is very difficult to convince them otherwise," she continues. Whether the advocate intends to suggest it or not, believing that genetics plays a meaningful enough role in Black maternal mortality definitionally implies that Black women must carry some innate inferiority when it comes to surviving child labor.

Yet, even if you were okay with biologizing disparities, does a biologic explanation for Black women's increased mortality make scientific sense? Biologically speaking, Black women are dying in childbirth due to bleeding and/or infection. So, if you want to make an argument that Black women are predisposed to dying in labor, you need to show *some* difference in their ability to fight infection or regulate clotting. Yet, there is not one single biologic assay for infection or coagulation in existence that can sort people by race. You cannot guess someone's ethnicity by looking at coagulation protein levels or clotting function, nor can any immune cell give you insights on ancestry. So, how, *exactly*, can you claim that genetics would be able to cause a three-fold increase in death primarily through bleeding and infection without creating any difference in the coagulation or immune function? Even if we were to take the persistent racial disparities in blood transfusions within the military system as a sign that Black women may have an increased risk of bleeding, the military was able to erase the gaps in death by realizing that preventing bleeding from becoming fatal requires that providers catch the problem before it's too late.

This lack of engaging with the Black community in an inclusive way does not go unnoticed. Most attention is paid to situations in which a disease is framed as Black despite the Black community recognizing that the issue is environmental. The first wave of the COVID pandemic predominantly hit non-white Americans working front-line jobs; while some noted the racial disparity in who had to report to work and who could stay home, plenty of supposedly well-meaning officials implied that

COVID was hitting the Black community because of innately entrenched issues in Black communities. A year later, COVID had become a disease of rural white, unvaccinated conservatives, but no one shrugged and implied those Americans were born predisposed.

However, it is also possible for racialized disease claims to cause Black communities to misperceive a disease as *white*. Minority women with family histories may be less likely to seek screening for breast cancer risk alleles like *BRCA1* or *BRCA2*, in part because their communities may not appreciate that genetic screenings for cancer is *for them* and instead incorrectly perceive such screenings only apply to white women.

Yet while misplaced faith in genetics may cause people to ignore genetic screens with proven benefit, it may also inflate genetic screenings that lack value. In her book, *A Crack in Creation*, Nobel Prize winner and co-discoverer of CRISPR gene editing technology, Dr. Jennifer Doudna, outlines numerous sensible uses for her technology. Using CRISPR to treat rare genetic disorders like cystic fibrosis may be the most obvious, but potential applications abound for the types of plant yield improvements noted by Dr. Bird earlier. But Dr. Doudna also proposes using CRISPR to help treat HIV by altering the protein CCR5 in the embryos of women in hard hit regions. A rare mutation in CCR5 can make a person naturally immune to HIV because the virus needs CCR5 to enter human cells. People with these CCR5 mutations can still be infected and test positive for HIV, but they seem to repress and perhaps even clear the virus. So, in theory, if humanity could CRISPR everyone so that they possessed this mutation, we could make the world HIV resistant.

Far be it for me to disagree with a Nobel Prize winner but I can see a couple of problems here. The first is that while we may know that CCR5 mutations are an advantage in HIV, we have no idea what tradeoffs might occur in these patients. Could CCR5 mutations worsen outcomes in other ways that perhaps the law of unintended consequences might want to have a comment on? Second, Smallpox provides both an example and a blueprint for ridding humanity of a deadly pathogen. As of this writing, HIV still lacks a viable vaccine; however, several promising candidates are undergoing clinical trials. But even if a vaccine is never announced, eradication is possible with antiretroviral treatments.

If the world could muster the will power *and* earn the good will to get everyone who has HIV on antiretroviral treatments, then HIV could be eradicated in a generation. The cost of these treatments can vary depending on the program and medication, but some estimates from the Bill and Melinda Gates Foundation report that treatments can be provided for as low as $75 per year (other estimates can be as high as $2000 per year). HIV persists because of failed access to medications, funding shortfalls, and a lack of trust in outsiders present in some nations.

The conservative estimate for standard in vitro fertilization (more commonly known as IVF) is around $3000 to $6000. The cost for CRISPR based gene editing would be an addition to that $3000-6000 price tag. So, we live in a world that has not been able to get treatments that might only cost $75 per person into the hands of people in need, but we expect that a bunch of European and American teams are going to fly their doctors into foreign countries and convince the citizens to allow them to edit their embryo's genome at an up-front cost that may be anywhere from two to 100 times greater? To be fair, there are CCR5 targeting therapies under evaluation, including the possibility of using CRISPR to edit CCR5 into bone marrow for patients with HIV as a form of a treatment. Of course, the knowledge of CCR5 as a potential treatment target was not discovered by GWAS, but by traditional methods. So, maybe enthusiasm for gene editing could be tempered a bit under an agreement to render unto CRISPR the things that are monogenic, and unto public health the things that are preventable.

The idea that the innate biology of patients or populations will always be the main cause of their disease breeds a political mindset that the solution is always just some Superdrug away. This leaves patients and populations increasingly concerned that the medical system has become more of a pharmaceutical industrial complex than a source for healing. As that concern mounts into distrust, people become increasingly willing to ignore genuinely expert medical advice to follow the recommendations coming from something akin to the *Storm Wellness Podcast*.

The moral cost

Morals run deep in cultures but are informed by knowledge and critical thinking. That is not to imply that morality is always derived rationally, but people don't often debate if something is "good or bad" before first deciding if something is "true or false". Population genetics has managed to avoid both debates. The claim that behaviors and common diseases have meaningful genetic components is taken as a truism within population genetics without any real evidence to support it. Yet, when pressed on the implications of their claims, geneticists dodge the moral implications by hiding behind their uncertainty.

If you ever come upon a behavioral geneticist, the most illuminating question to ask them is: "Only in terms of education, what is the degree of innate genetic advantage that your calculations predict exist for the children of professors like yourself compared to the children of average blue-collar workers?"

Their first reaction will be to pretend to be an *everyman* and feign disgust at your amoral phrasing of implied superiority. But one can ask about an advantage without implying innate inferiority or superiority. So, after their faux histrionics about value judgements, their next excuse will be to note that their predictions are terrible on an individual level. In this case, they are 100% correct. But the question was not "How accurate are your predictions?" the question was "What do your predictions say?" Per the online magazine, *Insider*, people from the Netherlands have the tallest average height at 5-foot and 7.69 inches, while the nation of Timor-Leste comes in the shortest at 5-foot and 1.28 inches, on average. So, people from the Netherlands might be predicted to be just under six and a half inches taller than someone from Timor-Leste. If, as the population geneticists claim, height were 40% genetic, then you could predict that in terms of height, someone from the Netherlands would have an innate genetic advantage of just over two and a half inches compared to someone from Timor-Leste. These calculations may stink on an individual level. It is completely possible that some child born in the Netherlands today will end up being shorter than the average person in Timor-Leste. Maybe a child who is destined to be seven feet tall was born last week in Timor-Leste. However, that would not change the fact that your population-level predictions would guess that the children from the Netherlands have a genetic advantage for height.

Similarly, if the children of professors are expected to have four to six more years of education than the children of blue-collar workers, then if education were 50% genetic, you'd be claiming that (on average) a professor's kid would be born with an extra two to three years of education. If you think education is 15% genetic—as Harden claimed—then they are asserting that professors' children have genetics that would translate into about one to two innately encoded semesters of schooling. It does not mean that every professor's kid will go on to graduate law school or even high school. And the success of the American experiment has shown time and again that children from parents of little means can go on to thrive in higher education. The question for these professors of population genetics remains: "How much genetic *advantage* do you predict the children of people like you have over others?"

They will then certainly pull out the favorite *get out of jail free* card for population geneticists and say that genes are *probabilistic and not deterministic*. But this dodge also falls flat because the question is "How probabilistic is it?" Is it probabilistic as in it might still rain even on a sunny day with a weather report saying 1% chance? Or is it probabilistic as in getting hit by a meteor while riding a unicycle? Both a single chest X-ray and a lifetime of smoking technically *increase the probability* of lung cancer without being *deterministic*. Would anyone conflate the two? When discussing probabilities, what matters is whether a result is actionable. Finding actionable insights is the entire purpose of science and this often gets lost in the process of science.

The scientific process is one of forming questions as hypotheses, testing those hypotheses with experiments, and then assessing the results to derive answers. For every question answered, more questions arise, and the process goes on, never ending. The stand-up comic Dara O'Briain, states this brilliantly: "Science knows it doesn't know everything, otherwise it would stop." But recall that the scientific purpose is to solve problems and improve quality of life. Answer just one question: "With all that we know right now, how should we best address this problem?" The answer to this can be humbling because the answer today may not be the same as next year or next week, since new information is constantly coming in and changing the "what we know right now" part of the equation. But the purpose, at least in biomedical science, is always to ultimately either reduce the suffering for those impacted by a disease or to prevent anyone from suffering from that disease altogether. Slightly improving the correlation coefficient in your linear model to go from *basically zero* to *just above basically zero* may improve the probabilistic predictions, but it isn't actionable. As Albert Einstein is credited with saying, "Not everything that counts can be counted and not everything that can be counted counts."

Furthermore, when probabilities are translated from populations to patients, the distinction between probability and determinism disappears. Take a glass of water and fill it to the top. Now, start adding more drops, but just one drop at a time. Eventually, the molecular attraction of the water will cause it to swell above the edges of the glass. At some point, a final drop will prove to be too much and will cause the water to run down over the edges. Is that final drop probabilistic or deterministic? Physicists could calculate the exact balance of intermolecular attractions and demonstrate why the final drop tipped the balance of forces towards the spill. But the spill only happened because of all the drops that had come before it. Once things run over, then everything involved both contributed to the probability and determined the outcome.

The chances of getting eczema in the U.S. is 20% for the population, but the risk in your child is either 100% or 0%. If 1000 factors combine to cause your child to develop eczema and if genes play some role in that development to cause the cup to run over, then all 1000 of those factors are both deterministic and probabilistic. A few drops of *genetic predisposition* in an otherwise empty vessel will never be deterministic of anything and do not change the probability of a spill. Similarly, a few drops of genetic predisposition in a vessel already overflowing would not change the odds of an already determined outcome. But if your child's cup spilled over and genetics were even a small part of the array of factors involved, then they were part of what *determined* the disease. Every individual raindrop is responsible for the flood.

Eventually, the population geneticists you have cornered will Gish gallop all the way to saying that the genetic advantage they might calculate would be context dependent. They will simultaneously claim that they cannot give you an answer for how impactful genetics are without knowing the environment, but also will tell you that they calculated the impact of genetics by *controlling for* the environment. But is there a moral hazard in believing that the entirety of someone's experiences, both good and bad, can be wiped away with a few lines of analysis code? If the goal of your research is to help people, should you ever wipe away the impact of culture, life experience, and environmental exposures?

Certainly, it is fair to control for such factors to look at what else might impact a trait or disease, but it seems immoral to then never add those factors back into your model and speak as if they never existed. This is particularly problematic when *adjusting for* culture and environments that the researcher is entirely unfamiliar with. Controlling for *poverty* based on a year's worth of income does not come close to capturing the spectrum of experience by the impoverished. People of similar incomes may differ in many ways, such as access to healthcare, family support, language barriers, cultural acceptance, access to public transportation, social program eligibility, and toxin exposure.

So, ask the genetics professors again—with all the caveats about how terribly unreliable or contextual their calculations are—what do those calculations predict the level of advantage in education is for the children of people like them? What do their calculations tell them the genetic advantage is for their children's lungs, pancreas, GI tract, and skin? By erasing the impact of environment, culture, and social interactions, these approaches create a tunnel vision focusing scientific research on what might be described as the anti-serenity prayer. The serenity prayer asks, "God, grant me the serenity to accept the things I cannot change, the courage to change the things I can, and the wisdom to know the difference." The modern population genetics version might read, "GWAS, grant me the serenity to dismiss the things I can change, the ability to inflate things I cannot change, and the wisdom of the journal editor to not care about the difference."

THE HIDDEN PRICE OF OUR GENETIC OBSESSION

When the population geneticists are asked to defend the validity of their claims, they act as if genetic elitism of certain populations is an established fact and that their approaches adequately account for environmental differences. When the population geneticists are asked to face the moral implications of their proposed genetic elitism, they act as if their boisterous claims are nothing more than hypotheses waiting to be tested.

Any claim that the genes of my son's Italian-Irish grandmother will influence how much tikka masala he eats can't be separated from the multiverse worth of potential *what if's*. But you better believe that my son will get plenty of servings of chicken alfredo with canned peas. Cooking for loved ones isn't about winning a Michelin star. Yes, you want to provide love and enjoyment for a meal, but my mom tossed in the peas under the hope of making the meal just a little more nourishing. "You need to eat something green every day," she would say as I pushed the peas around with my fork. Her north star for that meal was doing what she could within her means to help maximize my chances of a healthy life.

Doing what we can, within our means, to help maximize people's chances for a healthy life is supposed to be medicine's north star, as well. But the insistence on looking inward for the causes of diseases has caused us to veer off course. Indeed, discovering new medications is a means of maximizing health. But is it less of a victory to clean our drinking water so that people don't need medication in the first place? Indeed, creating molecular tools to diagnose diseases before they do irreversible harm is a means of maximizing health. But when someone's zip code can better predict disease than any blood test, is it less prestigious to work towards mitigating the environmental causes? We spend pounds on cures when an ounce of prevention might suffice, and in so doing, have distanced the healthcare system from a public hoping that their healthcare providers will be a partner in health rather than a conduit for prescriptions. Furthermore, we suffer a moral rot in our belief that researchers can *explain* cultural nuances with complex mathematics that they cannot even so much as *describe* with simple words. As we will discuss in the next section of the book, it has also misaligned our moral compass to guide us into a less equal society.

PART III

Harvesting Hatred

The prior section of the book used some of the more absurd examples in behavioral genetics as a contrast with medical diseases. The goal was to demonstrate that while it is easy to laugh at behavioral genetics for claiming there are genes for liking green tea, the foundational assumptions of such work are no different than those made by researchers believing they will find a gene for Black American asthma. Furthermore, when a racist online or in real life acts on behavioral genetics, people are quick to attack the behavioral genetics while the geneticists using the same flawed assumptions fly under the radar. So, it may feel a bit unfair that the behavioral geneticists are the only ones forced to defend their assumptions and approaches. This section of the book aims to assure that no such perceptions of unfairness manifest in the reader. While those wasting time and treasure looking for "Black asthma" genes have, relative to their behavioral colleagues, thus far dodged criticisms, the behavioral geneticists still have not received the level of criticism they deserve. In the glass onion analogy, the population geneticists looking at diseases may provide top cover for the behavioral geneticists, but the behavioral genetics provide direct cover for the bigots hoping to mainstream their hatred more fully into both the public discourse and public policy.

CHAPTER 16
Guilt by Genome-Wide Association

> **BOTTOM LINE UP FRONT:** *The ideas presented in The Genetic Lottery, and behavioral genetics in general, fuel bigotry through misunderstandings of what racism and eugenics are truly about.*

The term *Chinese water torture* was first described in either the 15th or 16th century in Italy. The term described a form of torture in which a person is tied up and a bucket of ice water is put over their heads. The bucket slowly drips water down onto the face but in a random and unpredictable manner. The person being tortured knows that the next drip is coming but does not know when. The confusion and unpredictability are said to drive the person insane. It is not clear that *Chinese water torture* was ever Chinese, and it may not even be torture, but it is an excellent analogy for reading *The Genetic Lottery*. *The Genetic Lottery* inherits the propensity for poorly thought-out analogies that were woven into population genetics by *The Selfish Gene*. It opens with a false analogy in the very title and then slowly drips logical fallacies onto the reader's forehead like a maddening blast of ice water. As you read, you know that another drop of flawed logic is coming; you just don't know when. *The Genetic Lottery* attempted to convince both racists and anti-racists that the field of behavioral genetics does not provide endorsement of eugenics and discrimination. The book was released prior to the murders in Buffalo, but I read the book afterwards, so I read it with full knowledge of how spectacular the author's eventual failure would be.

One passage of the book that is central to the message relayed the struggles that Harden's son has with reading. She commented that her older daughter is an avid and successful reader while her son has struggled to the point of needing special assistance. Since Dr. Harden believes twin studies hold validity, she notes that the heritability score derived from twin studies indicates reading skills are 90% heritable (and by extension, therefore, 90% genetic). She also believes sibling studies are akin to a randomized controlled clinical trial and that even though they were born seven years apart, her children share the same environment. These two beliefs combine to lead to the conclusion that since the environment could not explain her son's reading troubles, then he must have simply *lost the genetic lottery* when it came to whatever reading alleles she and her husband passed down. In the passages about her son, you can feel the love she has for him and the admiration for his hard work on overcoming his struggles. But just when these sections begin to endear the author to the reader—*drip*—you get smacked in the face with the false dichotomy. Dr. Harden implies that the only two conclusions are either her son lost the genetic lottery or that her pre-kindergarten child is somehow to blame for his reading struggles.

For those interested in a dissection of the book's statistical flaws, the October 2022 edition of the journal, *Evolution*, presented a technical critique by the population geneticists Graham Coop and Molly Przeworski. But the work of Harden and her fellow behavioral geneticists did not lead to a mass murder based on whether the online racists knew how to adjust a p-value.

Broadly speaking, there are five overarching flaws. Lifting all of these examples from *The Genetic Lottery* may give the appearance that I am focusing only on this one book. However, the book functions as a good encapsulation of the flaws in behavioral genetics that led to the murders in Buffalo and, more specifically, how these ideas get whitewashed into the mainstream. Thus, the book is an exemplar of the flaws in the field and therefore earns a level of scrutiny that could just as easily be aimed at other texts.

1) Words have meaning.

The first major fail point in the book was the definition of the word *causal*. To Harden's credit, she notes that she does not approve of the use of words like *explain* or *predict* when describing the relationship between genetics and outcomes. She prefers the term *capture*, which I admit is a better term for describing a mathematical relationship without implying cause-and-effect. At least, it would have been a better term if Harden did not constantly use the terms *cause* and *matter* throughout the book (and even in the subtitle). Her definition of cause is "what would happen if X, and only X, were changed." In respect to gene sequences, this is not an unknowable question. All the gene validation studies, like the one described for proving *STAT3* mutations cause immune defects, involve changing genetics to see what happens. The advent of CRISPR technology made these kinds of studies easier and cheaper, but the ability to manipulate only one gene has been around for decades. The entire candidate gene era, which Harden herself said "did not work out," attempted to answer the question: "What would happen if this gene, and only this gene, were changed?" The answer came back with a resounding 'nothing'.

She mutates (pun intended) the definition of "cause" to include mere statistical associations. If the allele in one child is X, but in the other child is not-X, then she determines that only X has changed. But there are lots of things that might have also been changed in such scenarios. If a researcher in toxicology said, "Smoking behavior captured 90% of lung cancer risk," what she would be implying is that if everyone stopped smoking, rates of lung cancer development would drop by 90%. Cause, meet effect.

Harden presents a study in what is called the *word gap*. This is the finding that children raised in poverty know fewer words by elementary school than children raised in wealthier homes. Harden then asks us to consider the possibility that the rich parents may have genes for speaking that get passed on to the child. The speaking genes in the parents cause them to talk to their children more. Then, the speech genes inherited in the child cause them to, as she puts it, "babble back." She bolsters this with an experiment of her own in which her team observed a parent-child interaction when the child was presented with a new, educational toy. For 10 minutes, they watched and then rated the parent on how much they tried to teach the child the new toy. Her team found that the children with more involved parents at age two had better reading skills at age four. Here, again, you feel a pull of sympathy reading between the lines that she feels she was interactive with her son, and thus, his reading troubles must be from something beyond her control. But can we look past this drip of false pretense to think of any other reasons a 4-year-old with more involved parents might end up speaking more words beyond the idea that the parents' genes are the ones having the conversation?

Take the experiment to the extreme and never talk to the child. How many words do you think they would learn? And if I told you that one kid from an English-speaking family had a nanny who spoke to him in Spanish and another kid from an English-speaking family had an English-speaking nanny, which one do you think would know more Spanish words? Or perhaps the parents *won the genetic lottery* and inherited genes that *cause* them to have a Spanish-speaking nanny?

She insists that researchers prove cause and effect, but limits those demands to those researching literacy programs. She likens the idea of spending money researching literacy programs without accounting for genetics to "robbing a bank" but then denounces the idea that she is advocating for genetic determinism. She only offers one study design proposal: to test if moms reading to their adopted children also causes a rise in the kids' language skills. What if her proposed experiment suggests that the impact on reading to an adopted child is less potent than reading to a biologic child? Then what? Can you think of any other confounders for why an adopted child might have early life exposures that could influence their reading other than not sharing their adopted mother's genes? The idea that a literacy program should be scrapped if a highly confounded test of genetic contribution suggests as much is the *definition of genetic determinism*. Throughout the book, Harden rightly bemoans *The Bell Curve* for its overtly racist claims and propensity to argue against social welfare programs. However, canceling programs because the genes might be a contributor is exactly what *The Bell Curve* suggests. Genetic determinism isn't just about genes determining health, it is also about genes determining public policy.

Dr. Harden relays a study that demonstrated that children with high PGS for education who attended low-income schools performed worse than children with low PGS at rich schools. Turns out, at the rich schools, the PGS has no statistical relationship with completing calculus before graduating high school, but at the poor schools, the lower PGS is associated with fewer years of math completed. Harden presents this as a way to argue that any study into why children in rich schools do better needs to take their genetics into account rather than simply assessing the schools as a whole. Yet she never says what, exactly, she thinks this might uncover, even in a hypothetical sense. What environmental factor does she propose might be so impactful as to influence high school graduation but weak enough to be masked by the *failure* to adjust for PGS?

Another thought experiment: let us make a new metric called the Genetic Researcher Imagined Environmental Factor test, or GRIEF, for short. Imagine in the rich school, every kid has a low GRIEF score. But in the poor school, the children with the lower PGS for math class attendance have high exposure to GRIEF while the other children have low to average amounts of GRIEF. If you analyzed the schools with no adjustment for genes, you might incorrectly conclude that GRIEF did not play a role in math class completion. But if we were to listen to Harden's siren call and adjust for PGS, we would uncover the mystery of GRIEF. We would see that the poorest children in the poorest schools have a worse GRIEF score than nearly every kid in the rich schools. And by separating those children out for analysis, we would reveal the groundbreaking findings that GRIEF plays a role that had been previously overlooked. We would be able to see exactly how GRIEF is able to knock an otherwise gifted kid off their course for innate success in life and how wealth may help overcome GRIEF. So, what do you think an example of GRIEF might be? Harden is insistent that PGS be considered when assessing environmental factors but offers no examples, neither real nor theoretical, in which an environmental factor could be overlooked. She only seems to imply that failure to account for PGS might lead one to overvalue the role of GRIEF.

Devoid of a negative control, the PGS for education could be nothing more than a marker of population stratification or differences between sibling exposures. Harden seems intent on demanding that genetics be a covariant for the environment under the conclusion that educational reform has thus far failed to rescue struggling schools. I'm sure many of her detractors in social sciences would agree that poor schools continue to struggle in America, but they would likely point out that poor schools are often in struggling neighborhoods that have many other social and political determinants working against them.

But if resources and environmental mitigations can render genetics moot when it comes to math aptitude, claiming genes are still causal harkens back to the beginning of reform eugenics. The very title of her book is a euphemism that was common in the times when eugenics was trying to distance itself from the blood stains of the Third Reich. The argument from reform eugenics was that if the environment brings out a genetic defect, then the problem is still genetics. Even if environmental mitigations completely override genetic predispositions, the reform eugenicists argued that the genetics were the reason the mitigations were needed in the first place. This was used to justify sterilization of unaffected children of people with Schizophrenia. The argument was that if society could not assure the environmental trigger was eradicated, then it would be *safer* to sterilize any kid with a risk for carrying perceived Schizophrenia genes. Like Harden, they assumed the environment was set and then concluded that genetics sorted the winners from losers.

2) Less than great expectations

Professor Harden presents behavioral geneticists in a way that implies she sees her field as one filled with social scientists who use genetics rather than geneticists who work on social sciences. Harden herself is a clinical psychologist who has entered the field of genetics, not a geneticist. Anytime a shortcoming of behavioral genetics is mentioned, she makes a comparison to a social science metric. Whenever a potential strength of the field is brought up, Harden makes comparisons to biomedical science. For example, Harden admits that heritability scores vary over time, noting that her ability to graduate college as a woman is far different today than it was 100 years ago. So, the heritability score for education has shifted over time, in part due to increased access to college education. To defend her claim that education is simultaneously woven into our genetics and so variable it could be overcome by a scholarship program, she invokes the GINI index.

The GINI index is a measure of the income inequality within a single nation. The index ranges from *total equality* to *get the guillotines*. Yet, no one using the GINI index claims that income inequality is an immutable characteristic stitched into the very fabric of every society. The fact that the GINI index can change is exactly why social scientists measure it: hoping to find a way to reduce the gap and make society more equal. If you wanted to make a comparison to a measure claimed to be a reflection of innate biology, perhaps select body temperature. No one knows what a thermometer might read for people living in ancient times, but we could probably guess their average temperatures were in the 96-99 degrees Fahrenheit range we have today. Certainly, people 100 years ago were not walking around with temperatures in the 80s. Harden consistently chooses to compare a measurement designed to reflect a rapidly changing world to one claiming innate biology.

Social sciences have been termed *soft science* to contrast them to fields like biology or physics as the *hard sciences*. I agree with Harden's assertion that these names misrepresent just how challenging social sciences are. I would prefer the term *biomarker science* over *hard science* because it does not imply softness of other fields and focuses on how certain diseases have one or more chemical or physical features that help define the disease.

Wegener's granulomatosis has cANCA, diabetes has glucose, and COPD has pulmonary function tests. Those working in biomarker science have an expectation that any claim made about the biology of their disease will alter some, if not all, of those biomarkers. If you claim your drug improves diabetes, then the patient's glucose levels should be altered. If you claim that a single gene causes COPD, then the patient should have abnormalities on the pulmonary function tests. But there is no blood test for Schizophrenia and certainly no objective measure for things such as *happiness* or *grit*. So, when Harden makes claims that genetics impact *grit*, those in the biomarker sciences are unimpressed.

A similar flaw that lowers expectations for GWAS and PGS comes from the fact that the field of population genetics is still mostly staffed by statisticians who are expert in study analysis but not always in study design. *The Genetic Lottery,* and behavioral geneticists in general, often present the fact that their math is reliable as if that indicates that their assumptions must be correct. In the times of tobacco science, early studies exposed mice to very short bursts of cigarette smoke and then concluded that smoking did not cause cancer. It does not matter how well the statistical assessments were run; such studies were never going to be able to answer questions about the potential harms of chronic cigarette smoke. Study designs that assume that sorting people into groups, measuring SNP frequencies between those groupings, then assuming that *any* SNP with a p-value below an arbitrarily determined cut off is *causal* for any trait is a bad study design. Biomarker science, particularly cancer, is protected against such flawed logic given the demands of additional work to evaluate proteins and metabolites, along with cell or animal modeling.

3) Worst. Ally. Ever.

One line of unfair criticism leveled at behavioral geneticists is that they all secretly harbor racist views that they suppress in public. Seemingly, the racists and anti-racists alike both imply that researchers like Harden agree with arguments of white supremacy but don't want to admit it. Harden and many of those in her field, to their credit, argue against racism in general and against attempts to use genetics as justification for racism specifically. They are all just very, very bad at it.

Harden's first argument against racism is one that is now standard for those in her field. *The Genetic Lottery* outlines that the social construct of race is not a true reflection of genetic ancestry. Harden notes that while 98% of Black Americans have African ancestry genes, 90% have some degree of European ancestry. She also mentions that ancestry is *granular* and that even European people can be divided into smaller and smaller groups. However, Harden never justifies the common practice by population genetics of using ancestry groups that are a near perfect overlap with those used by the founders of race science. Her claims of denouncing a genetic basis for the *black, white, yellow, brown, and red* used by racists would go a lot further if her work did not consistently use *African, European, Asian, and admixture* as categories. Saying genetics is an imperfect metric for race isn't the same thing as saying genetics does not support the concept of race.

Harden details an experiment in which two primates were rewarded with food for finishing tasks. When both primates were given cucumbers, both were happy. When both primates were given grapes, both were happy. But if the researchers gave one monkey a cucumber and the other a grape, then the one getting the cucumber would get angry. Harden presents this as evidence of an innate sense of fairness and desire for egalitarianism. But one must ask: *what about the other monkey, Paige?* The monkey that gets the grape while the other one receives a cucumber must also recognize the lack of fairness, but seems happy to enjoy it. Is there a PGS for an *I got mine* mentality?

The book relies heavily on John Rawls and his thought experiment, the Veil of Ignorance. This thought experiment proposed that if you did not know how resources would be shared, you would opt for equality. Harden offers a personal story about asking her children to split a cookie, then putting the two pieces behind her back and having the children pick a hand to earn that half of the cookie. Her children, in good Rawlsian fashion, always opt to split the cookie as evenly as possible. Opting for a fair split when you don't know which half you will get could be a sign of egalitarian bona fides, but it could just as easily be a sign of self-interest. Opting to split the cookie fairly assures that the kid will not risk getting the lesser half. An adorable experiment no doubt, but if you wanted to prove the children were innately interested in fairness, let them pick the subsequent cookie halves without the veil and see how they behave. Underneath every veil of ignorance, you might find the face of hypocrisy.

Dr. Ijeoma Ononuju (Dr. IJ for those that know him) is an assistant professor of education at Northern Arizona University. Dr. IJ offers a better analogy when he likens white supremacy to a family business. The family's great granddad built it, passed it to his son, who then passed it to his son, and so on. It may very well be that the great granddad was a monster in the way he built the business, but it is in the hands of the descendants now. Every parent would defend the benefits of this family legacy so that they, too, could pass it along to the next generation. So, inheriting a system that gives you an advantage, even if you realize it is an unfair advantage, will be protected with a mindset more akin to "it's just business" than to overt bigotry. If genes can *make you rich*, then couldn't they also make you selfish? How does Harden propose we get over that hurdle? Should Dr. Harden's home state of Texas decide that all children in each district should be placed into a literal lottery and then assigned to a school by random ping pong balls? Surely, bussing some of the children from the rich schools to the poor ones will make things more egalitarian, but you will forgive me if I'm doubtful that Harden or any other card-carrying liberal in her school district would approve of such a program. Elie Mystal, legal correspondent and author of *Allow Me to Retort: A Black Guy's Guide to the Constitution* put it best:

> If you want to see a white liberal drop the pretense that they care about systemic racism and injustice, just tell them that their privately tutored kid didn't get into whatever "elite" school they were hoping for. If you want to make an immigrant family adopt a Klansman's view of the intelligence, culture, and work ethic of Black folks, tell them that their kid's standardized test scores are not enough to guarantee entry into ivy-draped halls of power.

Since U.S. schools are primarily funded by property taxes, and since minorities were prohibited from even owning high value property until the Civil Rights Era, the good schools are built on generations hoarding their grapes while others got cucumbers. Harden asks us to dream about a more equal society without addressing that a more egalitarian world is worse for one specific group: the primates who are content hoarding their ill-gotten spoils.

Since Harden so badly misunderstands the nature of eugenics, she fails to grasp the times when her writing is unintentionally encouraging it. Eugenics has never been about morality. Eugenics is about modeling society to match the visions of perfection by those who inherited their place atop the social hierarchy. Any group or disorder that stands in the way of the vision of a perfected society is to be removed.

As another thought experiment, pretend that science found a 100% predictive allele for being a child molester. Pretend that everyone with this allele would be assured to at least attempt to sexually abuse a child. But no one who lacked this allele would ever even consider such a thing. If the world put you in charge of writing policy around this finding, what would *you* write? Would you require teachers and priests to be screened? Should dating sites flag anyone with this gene? Should adoption agencies warn prospective adoptive parents if the child carries this allele? Would you implement a tracking system for anyone afflicted? Might you consider full-scale eradication of anyone with the allele in hopes of sparing future generations the pain their actions might cause?

So long as the idea of a *dysgenic pressure* presented by a group of people is entertained, the eugenicists will argue that, in the long run, society, overall, would be better off if those pressures were removed. The desire to sterilize everyone with or related to someone with Schizophrenia was not done in a moral vacuum. Those performing the procedures knew the people being sterilized were being harmed. They justified their actions as the lesser of two evils, not as an unequivocal moral good.

Harden attempts to alleviate the fears of racists who think that Black people present a dysgenic force in American society by offering a version of the ecologic fallacy so tortured that those specific passages could go on trial at the Hague. The upshot of her point is to say that it might be possible that people of African backgrounds have more of the supposed IQ-genes but their effect has been suppressed by the environment. I'm not sure how anyone could be surprised that white supremacists were not convinced to change their views by presenting the possibility of Black supremacy. But even if she were able to convince the bigots that the structures of America are suppressing the innate superiority of Black people, is she really so naïve as to think that this would inspire the bigots to deconstruct those suppressive structures? Yet, whatever allyship a reader might have believed Harden's book presented was undone with the line:

> *Let us not flinch, from considering what seems like the worst-case scenario: What if, next year, there suddenly emerged scientific evidence showing that European-ancestry populations evolved in ways that made them genetically more prone, on average, to develop cognitive abilities of the sort that earn highest scores in school?*

This comment comically echoes a now running gag from *The Bell Curve's* co-author Charles Murray, who has been promising that genetic proof of white supremacy was five years away for nearly three decades. In a 1994 interview with Robert Siegle, Murray claimed, "I would ask for you to have me back in three or four years, and let's see who is right on this issue" regarding IQ genes and racial differences. In 1995, he said, "I have confidence that in five years from now, and thereafter, this book will be seen as a major accomplishment." In 2014, he said that it would "probably take less than a decade." In 2016, Joseph Graves relayed that Murray told him, "The five years are in the hope of the appearance of better genomic studies to buttress his claims." In 2019, he Tweeted the issue would "definitively settle within a few years." So, in 2021, Harden continues the trend, postulating that it could be within a year.

This quote is another example of how Harden fails to engage with the version of white supremacy that exists in reality. Why would the evidence she presented be the worst-case scenario in a book that claims over and over again to be advocating for egalitarianism? What if, instead of the foundations of white supremacy, Harden's ecological fallacy claims that Black people were the master race were proven next year? Does she think the racists will all start reading Ta-Nehisi Coates while getting a tattoo of James Baldwin? What if the racists' favorite fig leaf of the Asian *model minority* were proven? What would the response be from the same swath of society that curses out random fourth-generation Vietnamese-Americans in their grocery store because they were mad about how the Chinese government handled the Coronavirus outbreak? Given that Jewish people are constantly the target of the Elders of Zion type conspiracies that claim Jews are super villains, I could foresee some pretty dark policies coming from data claiming Jews were biology's chosen people. In the earliest days of racism, Northern European *scholars* believed that Northern Europeans were the first humans on Earth and all others thus represented a degradation of God's original plan. Once anthropology completely obliterated that belief, their conclusions flipped to claiming that Africans were God's rough draft and Northern Europeans came last and were thus a perfection of God's plan. The notion that any data would sway these people off their conclusions that people like them are innately superior isn't one worth dabbling in.

Kevin Mitchell, author of *Innate*, is a frequent defender of genetic determinist ideas surrounding mental disorders but is an equally fierce denouncer of those using genetics to make racist claims. In 2018, Mitchell wrote in the *Guardian* that any genetic differences in IQ between groups were *unlikely*, and he debunked the idea that cold European winters could generate more selective pressure on IQ than any other austere environment. When it comes to pushing back against eugenic trolls, Mitchell is a *value added*, so my criticism here is tempered, but I'm left to question what he means by *unlikely*. Unlike the term *significant*, unlikely has no universally accepted statistical definition. Dictionary definitions range from 1-49%, so do we mean unlikely as in one in 10 or unlikely as in one in a billion? These are people who want to murder all Jewish people out of concern that their genetics are a threat to gentiles; it's not helpful to tell them that they are *probably* wrong, but we might not know for sure until we run more tests. Harden isn't saying that the white supremacists are wrong; she is only saying that the neo-Nazis don't have the data they need to support their claims, then suggests that they might get that data any day now.

To be fair, scientists are trained to speak in a way that gives deference to any possibility. Researchers will proudly boast of their findings at an academic conference but then turn into Doubting Thomas the second they need to interface with the public. It is obviously good for scientists to accurately relay the uncertainty in their findings. However, when engaging with the public, there can come a point at which nuance serves only to muddy the issue. Saying that it is *unlikely* that Hitler was correct can mean different things to different people who might act on those words in different ways. If someone is not comfortable implying impossibility, one might say instead that the odds that Hitler was correct are lower than the odds of having a winning lottery ticket snatched out of your hands by a reincarnated dodo bird owned by Tom Hanks. I am not trying to argue that scientists should tell people what to do. In fact, science should avoid paternalism when speaking to the public. But people do appreciate learning how new information might guide their actions.

I once had a terminally ill patient who was being set up to go home with Hospice. The transport was scheduled to arrive that evening to take her home, but the patient's status had been worsening all morning. At one point, her daughter asked me whether she should wait for her mom to arrive home or not. I didn't think telling her, "The future is unknowable" was going to be helpful. So, I said, "I think that if you want to be at her side when she dies, it would be best if you came here now." The daughter arrived and was there to say goodbye as her mother passed well before her mom would have transitioned home.

While her book does deal with racism, Harden is remarkably silent on discrimination against women or people with disabilities. She does point out that women, like herself, face discrimination. But she never pontificates on the possibility that we might discover that men have "evolved in ways that made them genetically more prone, on average, to develop cognitive abilities." While gender is far more complex than a binary, it is true that most women carry two X chromosomes, and most men have an X and a Y. Most heads of major companies are men, most of the presidents of major universities are men, every U.S. President has been a man, and 48 of the 49 U.S. Vice Presidents have been men. If we ran a GWAS for educational attainment and then didn't adjust for gender, what would it show? Harden seems open to discussing the possibility that the foundational premise of Hitler's Germany might be proven true *next year* but does not wonder whether the genes that *cause* leadership might be more common in men. To be clear, I don't say this as a means of endorsing misogyny. Instead, I would like to contrast Harden's openness to biologizing racial differences in scholastic performance while remaining blind to how misogynists use those same arguments to insult the ability of women.

I asked Angela Saini, who has written books on sexism (*Inferior*), misogyny (*The Patriarchs*), and racism (*Superior*), about why racists claim their findings are *genetic* while misogynists opt to invoke *biology*. She pointed out that the early eugenicists modeled their belief on breeding programs for animals. The male/female dichotomy is a shared feature of animals and thus may not have seemed as *genetic* as the white/black or scales/feathers differences seen in animals. Harden seems to fall into this same behavior by writing a book about genetic discrimination and then paying little attention to sexism, the lone version of genetic discrimination that has an actual genetic difference between groups.

The Genetic Lottery botches the arguments against racism, ignores sexism, and then frequently (albeit unintentionally) emboldens the ableist claim that dysgenic pressures are a true concern. Harden cites Alex Young, introduced in an earlier chapter for his belief that two siblings are interchangeable. Young wrote a *PNAS* paper noting that the link between the PGS for IQ and the direct measurements of IQ has been weakening over time in Iceland. One possible interpretation is that the aristocracy is eroding. Those from family trees that used to signal low access to education are now gaining access, thus *reducing* the correlation between education and pseudo-nobility. Young instead presents this as evidence that the opening of the movie *Idiocracy* has befallen the Icelandic population through dysgenic pressure of smart people having fewer children. The researchers claimed that the drop in PGS for education could lead to a 0.04-point drop in IQ per decade. Yes, 0.04, per *decade*.

Mind you, they did not *measure* a drop in IQ or claim that one was happening. Instead, they extrapolated between the frequency of those with high PGS for IQ as measured by generations gone by and determined that in 4,000 years, the average person in Iceland will meet the medical definition of severe intellectual disability. Iceland presents an interesting case for Young's IQ study given its rapid rise in abortions for fetuses with chromosomal abnormalities. A CBS News report indicated that 85% of Icelandic women opted for prenatal screening and nearly 100% of those with a positive screen for a fetus with Down's syndrome elected to terminate the pregnancy. Of note, there are no government requirements for screening or abortion and this social pattern pre-dates Young's paper. Thus, I do not mean to imply that Iceland started aborting babies in response to his work. However, a society that believes it is under dysgenic threat typically responds.

This makes it all the more frustrating to see population genetics continue to try to rehabilitate the term *eugenics*. Imagine you discovered a new drug that was able to work its way into the bone marrow of a patient that might otherwise die of an immune deficiency. This drug would change out the faulty gene for a functional one in all the afflicted cells and restore the patient's immune function. What might you name this drug? I'm guessing you would not call it *Final-solution* to invoke Nazi ideology? Eugenics was defined as the "study of how to arrange reproduction within a human population to increase occurrence of heritable characteristics regarded as desirable." Eugenics was never about aiding a patient in need. Dr. Harden is correct that Eugenicists would recoil at the idea of providing increased resources for people with genetic disorders, but the decision of her colleagues to attempt to place gene therapy under the terminology that was the central motivating factor for the Nazis is an egregious error.

Similarly, imagine you designed an IVF program that allows parents to know the chances that their children will have a serious genetic illness and use that information to decide if they want to carry the baby to term. The program might also help select an embryo for IVF that would be free of any major genetic diseases. If you made this program, would you name your clinic *The Institute for Imbecile Genocide* to invoke the practice of killing people with cognitive impairment? Why affiliate parental choice with a legacy of state-sponsored breeding laws? People who pay to have an embryo selected for IQ are just proving money can't buy intellect, but it is their choice to waste their money. As the biotechnologist Keira Havens pointed out, eugenics is the opposite of reproductive choice:

> Eugenics is a top-down organized effort at "optimization", a team sport, one that requires both a population to shift and individuals with a shared goal of shifting it. Reproductive freedom is the exact opposite, the ability to make your most intimate, personal, and

*private decisions in accordance with your own values and not those
of myopic hereditarians.*

Giving people the freedom to decide what is best for their family, rather
than being forced to decide based on what was perceived best for society
is not in line with eugenics. As Havens stated, the goal of using the term
eugenics to describe non-eugenical approaches is "to convince you that
social problems have a genetic basis and that eugenic policies are the
right tool to fix them." So even if *The Genetic Lottery* aimed to be an ally
to women and minorities, its failures were predictable.

4) Allergic to History

If most of the book is like Chinese water torture, where it slowly drips
logical fallacies, the section in which *The Genetic Lottery* offers *anti-eugenics*
solutions is analogous to being water boarded with platitudes by a
strawman.

Harden offers the conclusion that we all must accept the "inescapable
conclusion that genetic differences between people cause social
inequalities." Her response to genetic inequalities is a noble one. She
states that we should strive for a world in which "treating each person as
an individual" allows us to "arrange society to the betterment of all
people." Yet her only policy proposal is for universal healthcare. Don't
get me wrong; I completely agree with her argument that the only way
for society to avoid mistreating people due to their disease is to make
sure that everyone is assured high quality care. However, the continued
racial disparity in maternal mortality in the universal healthcare in the
U.K. attests that equal access to care does not assure equal quality of care.
The issue with this claim is that, once again, she is using biomarker
science as a comparator when it suits her in a book about social science.

Outlining that people with genetic disorders should be provided equal
care is admirable, but what about the genetic contributions to social
issues? Harden mentions that twin studies peg criminal behavior as 80%

heritable, and thus claim crime to be 80% genetic. Even if she caveats it by claiming she is only talking about the degree of genetic contribution of modern white people committing crimes, her claim is exactly one that the eugenicists of old would be familiar with. Again, if the question is about efficacy for eugenics, ask yourself, "Would society be mostly free of all crime if every person with the criminal genetics was rounded up and launched to the moon?" Presumably, crime rates would drop for some short period—if nothing else—due to people's fear that jaywalking might get them sent off. But do you really think that crime would be meaningfully diminished? According to the National Museum of Australia, between 1788 and 1868, Britain sent 162,000 convicts to Australia. Part of their reasoning was that doing so would rid England of their criminal stock. Today, compared to England, Australia has less violent crime, fewer prisoners, lower murder rates, and a higher percentage of people reporting that they feel safe walking at night. Poms got that one wrong, didn't they mate?

Harden discusses the previously mentioned MAOA "warrior gene" but only in the context that it made her "uncomfortable" when a defense attorney claimed their client should get a lighter sentence due to having this variant. Her discomfort aside, acting upon the MAOA defense requires policy decisions. If a criminal with a *high PGS for committing crime* was sentenced the same as everyone else, would that be *ignoring* genetics? If the criminal justice system is to account for genetics, how? If one of her loved ones was harmed by someone who invoked the MAOA defense, would she testify in favor of a lighter sentence? To be fair, her overall point is that if people have mishaps that are beyond their control, society should work to help them rather than condemn. On this point, Harden is correct. But here is where the decisions must be made about how deterministic genes are. Harden mentioned studies that show people are more charitable towards others if they believe their shortcomings were bad luck instead of bad choices. But just how genetic is poverty or crime? If genetics really account for 80% of crime, then the public will be more likely to view them as beyond rehabilitation. If genetics play only a small role, then the public will continue right along judging people for whatever portion of their fate was caused by bad decisions.

The lack of understanding of eugenics by behavioral geneticists causes them to completely misunderstand both racism and anti-racism. No one is opposed to proven genetic findings being placed into society. Most would not even be opposed to potentially anti-egalitarian genetic information with proven biology if we already lived in a truly egalitarian society. What people are opposed to, however, is putting anti-egalitarian genetic information supported only by spurious correlations into the society we have today.

Furthermore, basing policy off of PGS for behavioral traits that have never put into pathways, risks an opportunity cost of selecting the wrong treatment. What happens if you label children as *genetic lottery losers*, provide them with tutors, and they still struggle? Adding only a tutor could not be expected to overcome the potential challenges for children in unstable homes or who have medical ailments. So if deciding which student gets a tutor based on their PGS is only a viable idea if the intervention the student needs is getting a tutor. If the gene-based tutoring fails, it will be quickly eliminated by those who would prefer to take resources away from those deemed biologically inferior.

For pure sake of argument, let us pretend that one actually did bundle the genetic signals into pathways and revealed that genes associated with educational attainment were mostly found in the processing of vitamin B. This finding would provide a clear path for action: first children could be screened for genetic signals linked to abnormal vitamin B metabolism; next children with low PGS could then be given vitamin B supplements. Controversy and discrimination would both be less likely since such programs would be more likely to be seen as medical than political. A clear screen, with a clear read out, leading to a clear and related intervention, would almost certainly be more tolerated than a nebulous program to provide college-prep tutors to children you've labeled as biologically destined for non-college tracks. Yet if high school dropouts just needed a little more leafy green vegetables, what exactly would be the value of providing them with dedicated tutors at 10-fold the price?

Harden's suggestion that society should be structured for equality of opportunity so that genetics can provide the means to sort people might initially sound great. But it also sounded like good phrasing by the eugenicists who needed to rebrand and reform after the Holocaust was forever tied to their work. It was then, as it is now, an illusion. Harden may be earnestly presenting reform eugenics whereas the eugenicists were being disingenuous. However, society does not need PGS to justify universal healthcare or to improve marginalized neighborhoods. Society does not need PGS to justify removing lead or coal ash from the water. And society certainly does not need a PGS to justify providing proper nutrition to our children. To claim that the ever possible genetic nuances must be disentangled from the environment prior to taking policy actions is to assure that these egalitarian efforts never happen.

5) Genetic Tunnel Vision

One year, I was fortunate to present some of my findings at the Aspen Allergy Conference in Aspen, Colorado. After the conference was over, we took a river kayak ride. Nothing with rapids, just an easy float down a calm part of the river. I overheard two of the boat guides chatting about where they had been working prior to coming to Aspen and what they might do after the summer season was over. One mentioned that prior to being a kayak instructor in Aspen, he was a surf instructor off the coast of Hawaii. After the summer, he was debating either being a guide in the Canadian Rockies or just "going wherever the wind takes me". It is amazing that when people say they are just "going where the wind takes them," it always ends up on a beautiful beach or picturesque mountain town. I lived in Ohio for several years and it is a wonderful place. However, while I can assure you that the wind blows *in* Ohio, no one ever claims the wind blew them *to* Ohio.

I carry a similar doubt when I hear people who have wrapped their life's work and livelihoods into population genetics tell you they are just "going where the data takes them," only to see that the data always seems to take them to a place that is better for people who have wrapped their life's work and livelihoods into population genetics.

The chemicals causing eczema outlined in the previous chapters, xylene and the most common form of diisocyanate known as toluene diisocyanate, are part of what are known as BTX compounds (for benzene, toluene, and xylene). The BTX compounds were all anti-knock agents for gasoline additives. The anti-knock agent they replaced was lead. There may be no greater indicator of how ingenuine claims of "agnostically searching for truth" are from the educational attainment wing of population genetics than their indifference towards lead.

An excellent overview of the history of lead was provided by the online video company Veritasium called *The Man Who Accidentally Killed the Most People in History*, which can be found on YouTube. Clair Patterson and George Tilton set out to discover the age of the Earth by using the phenomenon that uranium predictably decays into lead. But their measurements detected an amount of lead so high that it was impossible to have been natural. In brief, Patterson would go on to discover that the lead in his samples was coming from the lead that had been added to gasoline as a means of improving engine performance at a high profit for the gasoline companies.

Lead mimics calcium in the body and can deposit into bones and teeth. This means that early life exposure can end up poisoning people for decades as the lead slowly leaches from the bones into their blood. Patterson used bone and teeth samples from ancient humans to compare with those recently deceased and found that modern humans, including children, had 1,000 times the amount of lead as our ancestors. The neurotoxicity of lead was known even in the days of Benjamin Franklin, and thus, researchers quickly set out to look for potential harms of the modern lead exposures.

Among the many harms from lead that researchers discovered include: children with higher lead levels in their baby teeth were more likely to drop out of high school; at least half of the U.S. population (especially those born between 1951 and 1980) were exposed to unsafe levels of lead in childhood; preschool lead levels predict violent crime rates in the U.S., U.K., Canada, and Australia; children arrested were four times more likely to have high blood lead levels than controls who had not been arrested; and lead probably killed at least 25 million people from cardiovascular damage alone. Blood lead levels over 30 micrograms per milliliter are associated with a 10-point drop in IQ and reduction in the volume of the corpus callosum in the brain. As little as 5 micrograms per milliliter may reduce IQ by two IQ points; that is more impactful than the most generous adjusted PGS for IQ. For context, 5 micrograms per milliliter is what you would get if you put *two pinches (1/8th of a teaspoon)* of salt into your bathtub. Mice exposed to lead struggle to complete mazes, have abnormal brain structure, and show increased violence against other mice in their cage.

Iodine supplementation has the opposite effect of lead. An NBER working paper from 2018 summarized the work of Dr. David Murray Cowie. Dr. Cowie convinced Midwest states to add iodine to salt as a means of preventing goiters. These states had water that was deficient in iodine because the iodine had been filtered out by the journey from the ocean to the inland. The areas that supplemented their water saw an 11% increase in income due to increased work participation. Related work by

James Feyrer and others indicated the iodine supplementation may have shifted the average IQ a full standard deviation (e.g., 15 points) in the one-quarter of Americans that were living in low IQ areas in 1924. Before you go running out to take iodine, I will note that they also noted an increased rate of thyroid related cancer in older populations, so supplementation was not without unintended consequences.

The realities of the impact of mineral exposure on cognition is what makes Harden's term "the environmental lottery" the most offensive comment in her book. A *Consumer Reports* investigation in 2021 found that 25 million Americans drink from water systems that "fail to meet federal health standards, including by violating limits for dangerous contaminants." These Americans were disproportionately of Latino background and in rural areas, but the most predictive demographic of people drinking poor water was poverty. As was made evident (and perhaps criminally evident) in the Flint water crisis, tainted water pours onto poor communities due to bad people making bad choices, not *bad luck*.

Research into air and noise pollution have also shown direct toxic effects on cognition. Steffen Kunn, Juan Palacios, and Nico Pestel from the Netherlands found that an increase in PM2.5 of $10ug/m^3$ caused a 26.3% increase in errors among pro chess players. This wasn't some mathematical association. The researcher brought chess players into a controlled environment and measured their error rate. They then intentionally pumped pollution into the room (at levels high enough to be impactful but not enough to cause cough or be otherwise perceived by the players). Similar studies on noise pollution showed that for every 10 decibels increase in noise, productivity declines by 5%. Here, again, was an interventional study by Joshua Dean in 2020 from the University of Chicago School of Business. He measured how fast factory workers could sew pockets onto shirts, then put a giant noise maker outside the factory and intermittently turned it on. He could thus directly measure the pockets sewn per hour with and without the excess noise.

These findings undermine both the behavioral geneticists and racist pundits who claim they are *just following the data*. One would think that the combination of lead, low iodine, noise, and air pollution could provide a perfect means of claiming superiority over marginalized communities. Poor white communities, and especially poor minority communities, have far higher burdens of all these exposures. So, the "just asking questions" style race scientists could—if they wanted to—argue that marginalized communities having worse outcomes in education and violent behavior could be explained by the biology of the various neurotoxins in those communities. I don't pretend that if they did make such claims, they would do so without push back and condemnation from the political left. However, the obvious reason that none of these pundits ever invokes the actual data on group differences in cognition and behavior is that they only care about advancing the idea of *innate or immutable* differences. If marginalized communities could be uplifted by cleaner air, water, and better city planning, then political will and attention would turn against drug development and discrimination. Lead and similar neurotoxins provide a tailor-made biologic explanation of how marginalized communities struggle, but the bigots ignore them because the biology comes from fixable actions instead of God.

Harden only mentions lead once in her book and that is to claim that IQ tests can be helpful to marginalized communities because they can be used to prove lead poisoning impacted the neurologic abilities of those poisoned. Pointing out that a test *might* help poisoned marginalized communities in court without mentioning that the same test is far more often used to justify the initial marginalization is a little tone deaf. And failure to account for lead, iodine, or other environmental challenges simply because she assumes differences in siblings' exposures will vanish with enough study participants borders on absurd.

This is made all the more absurd by the intentional miscommunication between how much genetics can actually *capture* educational attainment. As pointed out by Dr. Matias Kaplan, a Stanford-based RNA researcher, the correlation between polygenic score for education and years of education completed is poor, as presented.

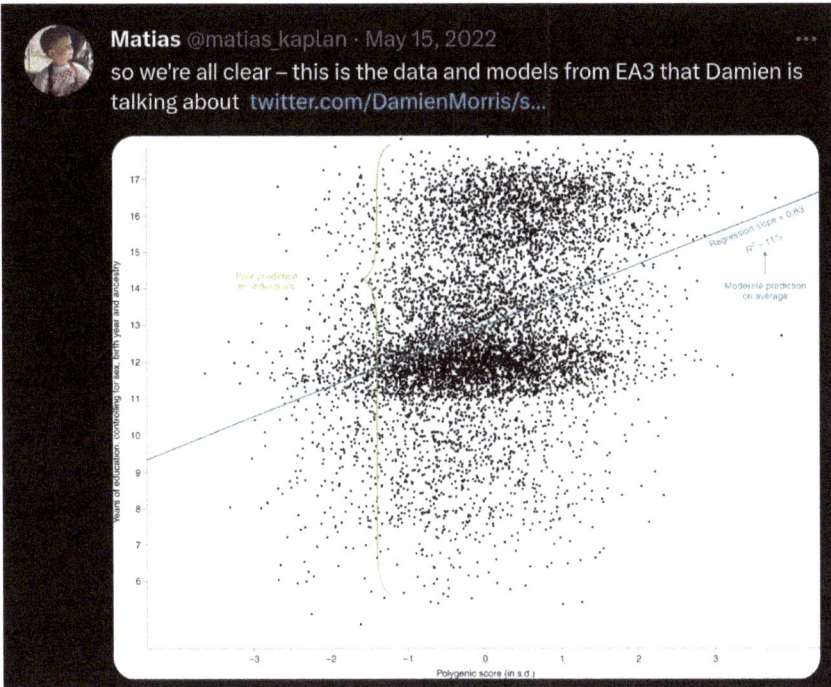

Matias @matias_kaplan · May 15, 2022
so we're all clear – this is the data and models from EA3 that Damien is talking about twitter.com/DamienMorris/s...

The dots appear nearly random except for the clear grouping for college graduates and two- and four-year grad-school programs. Even though linear regression can draw a line between the dots, there is no single person on the graph for which you would be able to reliably predict education based on genetics. This data is what the education attainment crowd claims indicates that they can *capture* 11% of educational attainment. Yet take note of the label on the Y-axis that states "controlling for sex, birth year and ancestry."

That means that the analysis can—at best—explain the variation that *remains* after accounting for those factors. The same *PNAS* paper professing concern over a theoretical 0.04-point drop in Icelandic IQ reported that women born in 1920 with 20 years of education were so uncommon they might as well have been unicorns. But by 1960, women with 20 years of education were the most common group. Access to higher education was vastly restricted among minoritized groups prior to the 1960s as well. So, it is not that genetics can "explain 11% of education"; it is that "if we lived in a world where sexism, racism, and generational change did not exist, we could explain 11% of the *remaining* difference in educational attainment." If you saw an advertisement for a candy bar claiming that it had "the same calories as a piece of broccoli" but then had incredibly small print stating "*after adjusting for added sugars and fats," would you consider that fair advertising? The education attainment crowd doesn't even provide the small print.

This is an important point to harp on because Harden puts forth that she was initially *skeptical* of GWAS as a means of identifying genetic causes for behavioral traits. She claims she was converted after seeing the correlation for the PGS on education attainment was 10% compared to the correlation with parental income being 11%. Her conclusion was that genetics are just about as important as your parents' income in relation to going to college, and therefore, her initial skepticism was wrong. She points out that she wanted to know why college education is associated with lots of good outcomes. So, when she claims, "We don't know why children of people with more money are more likely to go to college," the finding that genetics may play an equal role was enough for her to stake her career and reputation.

However, the comparison here is between an *adjusted* correlation for PGS and an *unadjusted* correlation for parental income. The parental income correlation posits that if you knew *nothing else* about a group of people other than their parents' income, you could capture 11% of the differences on who went to college and who did not. To get a correlation of 10% for PGS, however, you have to mathematically erase the impacts of a century of civil rights movements. A real comparison would require that the PGS crowd put forward their unadjusted values, which their papers never do. Environmental research is expected to show both values—the unadjusted risk and the adjusted. This is to transparently communicate how important factors were on their own versus when you begin to construct a model.

But are ancestry, age, and birth year the only factors that one would need to adjust for? What about income? Distance to the nearest college? Quality of high school education? How about having nearby family members? My child did not come into my life until I was well past medical school, but I still appreciate the advantage of having caring grandparents, aunts, and uncles nearby who can help provide childcare in a pinch. Maybe you only lean on the in-laws to facilitate a date night, but those trying to get through college with a child are greatly benefited by having a caring loved one who might help watch their kid while the parent studies or attends class. Did this study account for the strength of one's social circle (and how on Earth could you)?

Here is a final analogy to demonstrate the difference in methodology between biomarker science and association science: what is the most important factor determining whether someone gets emphysema? Hopefully you said smoking. If you wanted to look at genes that might influence emphysema, would it make any sense to assess a non-smoker the same way you would someone who smoked a pack per day for 40 years? Any assessment for genetics involved in emphysema would have to account for how much and for how long someone smoked. Thus, the analysis would be *controlled for smoking*. Let's say you also found literally every single environmental factor that played a role in emphysema and mathematically accounted for it in your analysis.

Even if the PGS could explain 100% of the *remaining variation* in emphysema after accounting for environment, does that mean genetics explain 100% of emphysema? Of course not. And yet, population geneticists consistently use language claiming to be able to *explain* large amounts of the variation in their data while skipping over the fact that they intentionally removed numerous important factors prior to their analysis. They operate without acknowledging that attributing residual variation to genetics assumes no further environmental adjustments are needed. Acknowledging additional environmental factors would potentially weaken the genetic signal they claim to find. Recall from Dr. Murthy's work on using other social factors to predict BMI that the more meaningful environmental variables you add to your model, the less variation remains for genetics to even attempt to address.

The entire foundation of the work of the within-sibling geneticists is that children are just gene randomization trials. While *The Genetic Lottery* genuinely conveys the author's love of her son irrespective of his reading struggles, her assumption that his struggles could not have an environmental source is, and seems to forever be, the failed understanding of population geneticists. Harden mentions her Austin, Texas home throughout the book. Interestingly, Texas was the worst offender on that *Consumer Reports* water investigation measuring how many people are given unsafe water. The EPA maintains a website called "Where You Live," which allows people to look up what the top pollutants released by factories in their zip code or city are. So, I looked up Austin County, Texas and provided the output below.

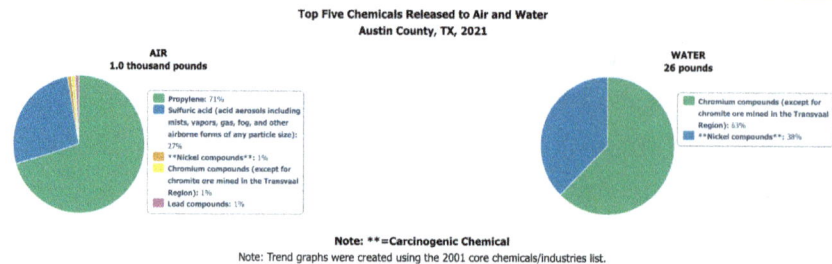

Top Five Chemicals Released to Air and Water
Austin County, TX, 2021

AIR
1.0 thousand pounds

Propylene: 71%
Sulfuric acid (acid aerosols including mists, vapors, gas, fog, and other airborne forms of any particle size): 27%
Nickel compounds: 1%
Chromium compounds (except for chromite ore mined in the Transvaal Region): 1%
Lead compounds: 1%

WATER
26 pounds

Chromium compounds (except for chromite ore mined in the Transvaal Region): 63%
Nickel compounds: 38%

Note: ** = Carcinogenic Chemical
Note: Trend graphs were created using the 2001 core chemicals/industries list.

Each year, about 1,000 pounds of the air pollutants monitored by the EPA are released into the air in Austin County. This number would not account for toxins from car exhausts, any chemical pollution that isn't covered by EPA reporting rules, or any pollutant released in a different county that blew into Austin. The top air pollutant was propylene, which a 2012 study in *Pediatric Research* found induced neuronal apoptosis in mice. A similar mouse study in 2020 on the propylene exposure in electronic cigarettes also found neuronal death and brain damage.

The human data for water contamination, however, is even more revealing. The top water pollutant in Austin County is chromium (a different version than the one Erin Brockovich investigated). A 2020 study published in *Environmental Health Perspectives* found chromium was among the top three best biomarkers (along with lead and manganese) for lower IQ in people with high exposure. Chromium, lead, and manganese blood levels captured 2.9 points worth of decline in IQ. In 2019, *Environmental Pollution* published a study by Rafael Caparros-Gonzales and colleagues concluded as follows:

> *Exposure to chromium was associated with neuropsychological impairment in children living in polluted areas of southern Spain. The results of this study reveal a greater neurotoxic effect of chromium on neurodevelopment among boys. Our findings provide evidence of the detrimental effects of postnatal chromium exposure on children's neurodevelopment and support gender-related differences in the neurotoxicity of metals among children... since boys showed more pronounced neurotoxic effects.*

Is it possible that a factory was releasing poisonous amounts of chromium into the Harden family's water, either when Dr. Harden was pregnant or during her son's critical windows for neurodevelopment? A chemical shown to hold the potential to cause worse developmental damages in boys should probably be evaluated prior to assuming that genetics are the only reason her son has reading issues that are not present in her daughter, right? If it was found that a factory was the source of the toxic levels of chromium and her politicians did nothing to prevent it, would she be comfortable saying that her son "lost the environmental lottery"?

The Genetic Lottery aimed to convince racists and anti-racists alike that genetics does not support bigotry. However, it misrepresented the nature of racism, endorsed the possibility of Aryan supremacy, presented reform eugenics under the banner of *anti-eugenics*, and failed to acknowledge the environmental injustices that leave marginalized communities with the kinds of toxin exposures that might have found their way into the author's own home. It is little wonder the book, and all similar articles of its kind, have failed to provide the allyship the authors hoped for.

CHAPTER 17
No going back from Buffalo

BOTTOM LINE UP FRONT: *The racist murders in Buffalo changed the conversation around social genetics forever.*

In 2016, a team led by Daniel Benjamin published their results looking for specific genes that predicted success in school. In this study, the researchers sorted 300,000 people by how many years they had attended school, documented the loci signals for each person, and then looked to see if there were any genetic variants that were significantly more likely to show up in those with higher years of school attendance. The researchers claimed to have accounted for the *relevant environmental factors* like family income, local school quality, two parent households, and whether the participants' parents had attended college or grad school. The results they reported were that 74 loci *contributed to* how long people spent in school, so long as the population they evaluated was exclusively white European.

In the next chapter, we will dissect this report in more detail, but for now, I will briefly state that the effects claimed by Benjamin's team are so weak that they are likely within the background noise that would have been detected had they used a negative control. While they assume they have adequately accounted for environmental factors, their adjustments for environmental influences were superficial at best and could not possibly have accounted for all the known, let alone unknown, factors that influence intellectual development. The work did not account for the influence of assortative mating. Furthermore, their work failed to connect their findings to an understanding of biology either with models for mechanism or an explanation as to why their findings would apply to white Europeans but not to non-white Europeans attending the same schools.

In partial defense of Benjamin's work, the Harvard geneticist, David Reich, wrote an op-ed that was regrettable even by the standards of *The New York Times* op-ed page. In May 2018, Reich wrote:

> *Groundbreaking advances in DNA sequencing technology have been made over the last two decades. These advances enable us to measure with exquisite accuracy what fraction of an individual's genetic ancestry traces back to, say, West Africa 500 years ago – before the mixing in the Americas of the West African and European gene pools that were almost completely isolated for the last 70,000 years. With the help of these tools, we are learning that while race may be a social construct, differences in genetic ancestry that happen to correlate to many of today's racial constructs are real.*
>
> *I am worried that well-meaning people who deny the possibility of substantial biological differences among human populations are digging themselves into an indefensible position, one that will not survive the onslaught of science. I am also worried that whatever discoveries are made – and we truly have no idea yet what they will be – will be cited as "scientific proof" that racist prejudices and agendas have been correct all along, and that those well-meaning people will not understand the science well enough to push back against these claims.*

The savvy reader should notice similar failings as *The Genetic Lottery* that would come out years later. Reich says race is a social construct, and thus, has no place in genetics. Yet, he then immediately turns around and implies that genetics are a correlate of race. You will note the conflation of the terms *genetic* and *biologic*. Reich notes that genetics are different between groups, then shifts to the "possibility of substantial biological differences." He then offers an olive branch dipped in manure by claiming he is *worried* that proof of *substantial biologic differences* would be misused by the types of people who have built their entire castle of hatred on the idea of *substantial biologic differences*.

Reich defends his position by discussing that prostate cancer is more common in men of West African descent than European men. He postulates that the genetics of West Africans might explain the difference. Reich's group found a region of the genome that contained gene variants more common in those of West African descent prostate cancer. Next, he explains:

> When we looked in more detail, we found that this region contained at least seven independent risk factors for prostate cancer, all more common in West Africans. Our findings could fully account for the higher rate of prostate cancer in African-Americans than in European-Americans. We could conclude this because African-Americans who happen to have entirely European ancestry in this small section of their genomes had about the same risk for prostate cancer as random Europeans.

Reich's analysis didn't adjust for any other factors that might influence disease risk, such as access to care, access to prostate screens, and environmental exposures that might impact prostate cancer risk. His conclusions are Black Americans with more European DNA had been liberated from their ancestral prostate cancer risk. Perhaps, had Reich spoken to social scientists outside of his academic silo, he would have asked if Black Americans with more European DNA might have other differences that would impact cancer risks?

Or, as Reich believes, is it that these gene variants really do cause prostate cancer and Black people just lost the lottery and have more of them? If this were true, certainly an institution as prestigious as Harvard would afford Reich an opportunity to splice these gene variants into prostate cells to see if they become cancerous or silence these variants in prostate cancer cells to see if they are viable treatment targets. But even then, let's assume Professor Reich is correct that these are causal gene variants dictating risk on a population scale. Once again, how is this meaningful? If Reich already knew seven gene variants associated with prostate cancer *before* beginning his West African search, how did uncovering this help anyone? Reich didn't claim to have found a new gene target, not even one specific to West Africans. If the implication is that spending money showing that West Africans have genetically inferior prostates may lead to targeted gene therapies, wouldn't it have been more sensible to research therapies for the gene targets already known?

If a gene treatment were made for prostate cancer based on any of the seven targets indicated in Reich's study, what is the clinical utility of knowing that West Africans have a higher frequency of these gene variants? Any patient who was diagnosed with prostate cancer would be screened for genetic targets; if those targets were present, then regardless of ancestry, the treatment could be tailored to his cancer using genetic medicine.

There may be some benefit to population level analysis. If we look at populations with higher rates of prostate cancer compared to populations with lower rates, and if we look at their genetics, perhaps we can find genes that are in higher abundance in the high-risk population, assume those to be potential causes of prostate cancer, and conclude we found gene variants worth researching for cures. To some degree, this makes sense until you realize that you would have to account for several other factors that might make the high- and low-risk groups different. For example, you could argue a similar logic for looking at the dietary differences between West Africans and Europeans, or the environmental exposures. There, too, you would expect to find differences between the groups, some of which would also be more common in the high-risk group than the low risk-group. But it would be an obviously absurd leap to go from "West Africans eat more cassava leaf and have more prostate cancer" to "cassava leaf contributes to prostate cancer"; it would be even more absurd to then pen an op-ed stating that denying dietary differences between ancestry groups could undermine efforts to find a treatment for prostate cancer.

Had Reich limited himself to just talking about prostate cancer, I'm not sure he would have become one of the favorite citations for racists across the web. However, Reich decided to venture into genetics and intelligence by bringing up Benjamin's 74 white people college SNP paper. Unlike Benjamin's group, Reich failed to outline that the findings could be from reverse causation or population stratification caused by people with graduate degrees preferring to mate with other people with graduate degrees. Reich may have meant well with his op-ed, but he effectively juxtaposed imagined dangers of denying *biologic differences* between races with an uncritical discussion about *genetic differences* in IQ. He would have to be criminally naïve not to anticipate that he would be seen as endorsing IQ differences between races. Reich, like many in his field, intended to communicate that his work does not endorse race science or eugenics. However, as soon as his words hit print every race scientist and eugenicist stood up to shout their disagreement. When your only associates are people you swear you don't want to be associated with, it might be time for a moment of introspection.

As the neo-Nazis praised his writing, Reich got substantial pushback from those in the fields of genetics and social science. Never one to miss an opportunity to defend discrimination against any group other than the ones he identifies with, online blogger and caliper fancier, Andrew Sullivan, echoed the position that failing to discriminate on genetics would worsen racism. The podcaster, Sam Harris, chimed in, citing his training as a *neuroscientist* based on a single publication which over extrapolated results from the brain scans of religious people asked to think about giraffes. The policy-wonk Ezra Klein got involved to present the opposing view but made the decision to put spectacle ahead of science when he agreed to debate Harris on his podcast despite Harris' refusal to include Klein's requested genetics experts in the conversation. This performative punditry served to assure the mainstreaming of the claim that the European intellect alleles had been found without presenting any of the limitations that the paper's authors presented.

There are only two possible alternative conclusions that would give merit to Benjamin's work and Reich's defense of it. The first was put forward by the researchers themselves. They claimed that while white Europeans had these 74 intelligence alleles, it did not indicate superior intellect as a group. Instead, they put forward that Africans and other non-white groups would have a different set of intelligence alleles more reflective of their environment and ancestry. However, as stated in the prior chapters, the claim that ethnicities have different genes for intellect would imply that they have different proteins for intellect; this would imply that white people and non-white people have a different brain biology. Such a claim is the essence of what is euphemistically referred to as *race realism*. This should be more correctly termed *biologic ignorance* but when we live in a world where neo-Nazis get to call themselves *alt-right*, I suppose we can't be shocked by white-washed phrasing. The *race realist* online put forward the false claim that ethnicities have different biology, and thus, any breeding between ethnicities or races will definitionally change white people's biology. Their fear is the same as those of the founders of eugenics—the fear that white peoples' *stock* will be diluted by a multicultural society.

The second possible reading of Benjamin's findings came after an anonymous online account took the 74 SNPs and contrasted their frequency between Europeans and Africans. This account claimed that only 54 of the 74 were of high frequency in both European and African populations, and thus concluded that Africans lacked 20 of these *intelligence loci*. Their extrapolation was predictable. They claimed that Europeans must be more intelligent because they, as a group, possess these 20 special intellect alleles that other groups rarely possess. Again, the concern for the online racists was that mixing with other races would dilute these 20 loci and cause the average intellect of white people to decline.

The racist online account took loci that associated with an outcome without any biologic validation, tallied the frequency of those loci between populations, and then claimed that *some* of the disparities between groups must be attributable to the differences in allele frequency. This process should sound familiar. This is exactly what the genetics of racial disparities researchers do for diabetes or asthma instead of for education or cognition (or three-point shooting). Aaron Panofsky, associate professor from UCLA and author of *Misbehaving Science,* has numerous publications detailing that the biggest audience for research on genetic contributions to racial differences (including differences in common diseases) are the online forums that radicalized the Buffalo shooter.

However, the limitation of the allele frequency and gene-environment excuses become evident if you imagine that heavy metal exposure impacts allele 1, 2, and 3 to limit educational attainment. Let's also say that allele 1 is predominantly in Africans, allele 2 predominantly in Europeans, and allele 3 predominantly in Asians. Such distribution could allow for a response to the environmental challenge (heavy metals) to operate through completely different gene signatures. However, you will still be left with three problems. Firstly, if these alleles have real effect, you expect that protein 1, 2, and 3 that are encoded by those alleles would be different, along with several metabolites in the signaling pathways containing allele 1, 2, and 3. If you don't have any biomarker signal, the genetics are likely spurious. Secondly, while the distribution of alleles might be different between populations, the frequency of an allele in a single person is either 0, 1, or 2 copies. So, again, if the alleles are real, then there is no need to judge people by racialized genetic assessments. If they have the allele, the impact should be expected. Thirdly is that the differences could be nothing more than spurious reflections of population stratification. But if you refuse to admit any of these limitations and insist that Africans might have alleles that are a disadvantage within the environments of Europe or the U.S., then you are going to present the liberal eugenics position that groups *would be better off* "going back to where they came from" and that white Americans and Europeans might dilute their stock with interracial mating.

On May 14th, 2022, a man felt inspired by the work of the likes of Benjamin, as well as Harden, Vissher, Abdellaoui, Young, Murray, Sullivan, the linguist Steven Pinker, along with many non-geneticists. Having been radicalized by those framing the education attainment GWAS results as a call to protect the white gene pools, he walked into a Top's grocery store in Buffalo, New York just after 2 p.m. He killed ten people:

- Ruth Whitfield, age 86, who was a great-grandmother on her way home from visiting her husband in the nearby nursing home.
- Pearl Young, age 77, who was a substitute teacher who took on learning Zoom during the pandemic so she could teach a special ed class for those with severe learning challenges. She was shopping, in part, to pick up food for a weekly food pantry she ran.
- Katherine Massey, age 72, who was a tutor, activist, and school board member.

- Celestine Chaney, age 65, who was shopping before heading home to where her family was planning to surprise her with tickets for the type of cruise she had always wanted to go on.
- Geraldine Chapman Talley, age 62, who was a mother and an aunt, and who had worked as an executive assistant.
- Andre Mackneil, age 53, on his way to pick up the birthday cake for the family reunion he had traveled into town for.
- Margus Morrison, age 52, a stay-at-home father of seven children.
- Roberta Drury, age 32, who was helping her brother recover from a bone-marrow transplant.
- Heyward Patterson, age 67, who was a deacon at the Church giving rides to community members from their homes to the store.
- Aaron Salter, age 55, who was a retired Buffalo police officer and aspiring inventor. Officer Salter would have stopped the gunman if not for the body armor that the murderer wore.

The shooter also injured Zaire Goodman, Jennifer Warrington, and Christopher Braden. At one point, the killer aimed his gun at a white cashier, only to apologize and move on.

Sadly, America is no stranger to mass shootings, including those motivated by racists. In fact, we are so familiar with such killings that the Buffalo shooting passed through the news cycle in under a week. However, while the Buffalo shooting faded just as quickly from the American mainstream media as the dozens prior and since, the massacre had a special staying power within the academic, biomedical research community.

The racist murderers that came before who tried to justify their actions pointed to supposed *evidence* that had been filtered through books, movies, or blog posts. However, the Buffalo shooter cited the academic papers on the genetics of intellect and education *directly*. He explicitly outlined that the argument put forth that Africans have only 54 of the 74 *intelligence loci* was what first started him down the path leading to mass murder. No longer could academic researchers pretend that their work was being misrepresented to the racist readers online. This killer read their actual papers firsthand and concluded that white *stock* needed protecting.

Professor James Tabery, author of *Tyranny of the Gene*, attended a meeting of geneticists and ethicists that had been assembled to discuss ways genetics has been used to support racism after the murder of George Floyd. He told me that he had expected that the invited genetics supporters like Harden and Benjamin would be DNA hardliners. Instead, he found them to be genuinely interested in being *the good guys*. Thus, after it was known that the Buffalo Killer cited many of the geneticists who were on the report working group, Tabery said that there was a sense of "befuddlement" and that the geneticists were "heartbroken": "They truly could not understand how this could have happened," he told me. Tabery expressed some frustration that the non-geneticists on the working group as critics all very much could see "how this could have happened." However, Tabery also stressed that the population geneticists who were most aligned with bigotry or who claim DNA is the only important factor in human behavior would have never participated in such a working group. For example, Harden, to her credit, fiercely reviewed Robert Plomin's *Blueprint* for a level of genetic determinism so absurd the author claimed DNA was a "100% reliable... fortune teller". Overall, Tabery felt like the geneticists like Benjamin and Harden take the brunt of the push back in their field because they are most open to criticism in their profession.

I will admit that Tabery's stance softened my frustration with the work of both population and behavioral geneticists alike...for about an hour. The legacy of the massacre in Buffalo was to show academia that the Internet had torn down the walls of the Ivory Tower. No longer could scientists pretend that their methods, results, and interpretations could stay siloed within scientific journals or conferences. Those who were well-meaning but who were unquestionably providing bigotry with a façade of legitimacy do not earn forgiveness just because they did not intend for their work to be seen as an endorsement of eugenics. Sensible people recoil from Plomin's unfettered genetic determinism because they recognize the absurdity of claiming parenting styles play no role in the type of adult a child may grow up to be. But watered-down genetic determinism may be more insidious because writing "DNA matters", without ever detailing how much it matters, can be easily interpreted in different ways depending on the reader and context.

When I was newly out of medical school as an intern, my resident and I had a miscommunication about a medication order for a patient. I heard my resident, Dr. Moneal Shah, tell me to order 50mg of the blood pressure lowering medication, metoprolol, intravenously. 50mg is the dose you would provide as a pill whereas 5mg is the IV dose. The pharmacist for the team, Dr. Tracy Macaulay, called me just after I put the order in the computer. "What are you guys doing?" she asked. Tracy, like all pharmacists in hospitals, oversaw vetting any medication order before sending it to the nursing staff. 50mg IV would have absolutely killed anyone who it was injected into. I do not care what your PGS for heart rate showed; 50mg IV would be lethal. Tracy flagged it and made sure it didn't go any further. To this day, Moneal and I still disagree as to whether he told me the wrong dose and I correctly (but ignorantly) transcribed it, or if he said the correct dose and I misheard him. But it didn't matter which of us got it wrong; our system failed the patient.

It was great that Tracey did not fail the patient. Had Tracy failed, it would have been up to the nurse to catch the error. So, one could conclude that the system worked—a system designed in anticipation of human error and built to catch it before anyone is harmed. But for Moneal and me, our system had failed. So, from then on, even after we moved onto different teams, he and I kept a system of what is called *closed-loop communication*. If he said, "Order a chest X-ray on Mrs. Johnson," I'd repeat, "Chest X-ray on Mrs. Johnson." It didn't add much time, but was an admission that the way we had been going about things was flawed enough to risk harm, could be foreseen to create the same risk again, and thus, needed to be changed.

So, with almost a year to reflect on discovering that the way behavioral geneticists had gone about their work could directly inspire a mass murder, you would hope that there would be a change in approach. Instead, Harden was given space for an opinion piece in *Nature Reviews Genetics*, which spent nearly all its ink outlining distinctions without differences between genetic determinism, essentialism, and reductionism.

Rather than address the fact that she, and her colleagues in her field, were directly cited as inspiration for mass murder, she offers only a single sentence reading, "More recently, belief in genetic determinism was evidence in the writings of a mass shooter who targeted Black victims." Notice how a story about a *racist who murdered* 10 and injured three Black Americans while claiming to be inspired by the author's work along with a specific *Nature* publication was sanitized for the reader into a "*shooter who targeted*" due to genetic determinism in general. This opinion piece defending genetic contributions of intelligence against accusations that it fosters bigotry was peer-reviewed by Abdellaoui and Vissher, both of whom would be more akin to co-conspirators in the face of such accusations than objective judges. Her prior *befuddlement* must have resolved because she went on to distance her field not from the need to change their approach, but distance it from the accusations that their approach needed to change. She wrote:

> If genetic essentialism is directly and causally related to prejudicial
> attitudes towards social groups, particularly towards racial

minorities, then this suggests that researchers linking genetics with phenotypes that are involved in group-based stereotypes (as well as editors, funders, journalists and others involved in scientific knowledge production and dissemination) have a particular responsibility to justify the benefits of their research as outweighing its social risks, and to describe results in ways that do not support beliefs about the naturalness or entitativity of social groups. However, if genetic essentialism does not actually cause social bias but is merely correlated with it by virtue of the historical events and cultural narratives or is invoked as a post hoc justification for existing prejudices, explanations of why research does not support essentialist beliefs, or curtailing the research altogether, might have limited or no benefits to reducing stereotyping and prejudice. This is, of course, an area of active enquiry and debate among scholars and other stakeholders.

To translate her comments from the language of privilege to the language of English, Harden is arguing that only if the frequent association between the genetics of education researchers and murderous bigots is *real-world causal* should researchers be required to justify why their work is worth more than the lives of those lost to hate crimes. But if the bigots are not really inspired by GWAS-IQ studies but merely use them as justification, then it would harm society more if Harden and her colleagues were asked to stop their work.

The first glaring issue here was best captured by Professor Abbika Kamath, an evolutionary biologist at Colorado University who quipped, "Seems a little bit ironic, no, for someone who's made a career on GWAS to be like, 'Oh, correlations don't matter if we don't know the mechanism'?" The second issue is the failure to understand the role of behavioral genetics in mainstreaming eugenics, even when serving as *just* an excuse for racism. Dr. Jedidiah Carlson, a post-doctoral researcher in genomics from the University of Washington, demonstrated in *PLOS*

Biology that genetics-focused papers, particularly those around neurologic disorders or education/IQ, are by far the most shared by Nazi sympathizing circles online.

What if we ran a GWAS for the victims of the Buffalo shooting; what would it show? For that matter, what if we ran a GWAS contrasting people killed by police while unarmed versus those who harmed police and lived to stand trial (like the Buffalo shooter)? Using *causal* in the way statisticians do, devoid of biologic validation, seems viscerally out of place here. What if we ran a GWAS on mass shooters? Would we conclude that white men have causal genes for domestic terrorism? Can behavioral geneticists really pretend that it would be unforeseeable for, one day, some state to use PGS for education as a screen to exclude children from college or specific professions? As much as Harden may advocate for a world where children with a low PGS for education are being offered a scholarship instead, can we really be that naive? To her credit, in her review article, Harden admits that PGS can be confounded by historical discrimination, especially when spread over generations. However, we are left wondering if they are still willing to continue to write dangerous prescriptions in hopes that the system will catch it before it becomes deadly?

Even if studies on the genetics of educational attainment were mere top-cover for racist and eugenics beliefs, it would not absolve such researchers from responsibility. One could easily argue that it is, in fact, more damaging to provide an avenue to mainstream such beliefs under a pseudoscientific façade than it would be to force those advancing those beliefs to state them outright. Providing an outlet for bigots to claim they are *just following the science* was what Reich did with his op-ed, which was taken up by Sullivan with his blog and Harris with his podcast. Allowing bigotry to hide behind your science allows it to fester, grow, and spread beyond the borders it might otherwise be confined to. Perhaps the American politicians who decided to reject thousands of Jewish refugees fleeing Hitler's Germany did so because they had been convinced by the eugenics science of the time, or perhaps they did so because they were just antisemites using eugenics as an excuse. I do not imagine the difference carries a distinction to the thousands who were returned to Nazi control, as was outlined Danial Okrent's book *The Guarded Gate*. Nor do I imagine nuance between direct versus indirect causation would sway the families of those killed in Buffalo, or the Tree of Life Synagogue in Pittsburgh, or the Emanuel African Methodist Episcopal Church in Charleston. Can behavioral genetics offer a counterfactual example? What might happen if this type of work is *curtailed*? Who (other than those making their living dredging the U.K. Biobank) would be harmed if Benjamin's 74 intellect allele paper had never been printed?

If the goal of their research is to generate publications, professorships, job advancement, and speaking invitations to conferences allowing university-sponsored travel, then the field has sprung forth with endless success. If, however, the goal is to take genetic information from populations and translate it into improved outcomes for individuals, the field has been a monumental failure.

The population geneticists, with behavioral geneticists in particular, seem to go beyond the stance that Dr. Frankenstein cannot be held responsible for the actions of his monster, to imply that any resultant destruction should be met with a change in words instead of a change in actions. They have become fanatics as reflected in the quote from George Santayana who stated, "A fanatic is one who redoubles his effort when he has forgotten his aim."

CHAPTER 18

Not Every Geneticist is a Racist, but Every Racist is a Geneticist.

> **BOTTOM LINE UP FRONT:** *Population geneticists continue to use the term "causal" when their data only show "association".*

Researchers in the field of social and behavioral genetics are very fond of telling their fellow researchers that they need to be mindful of communicating both their findings and limitations to the public. Often, the scientific publications dealing with social behavioral genetics will come with extensive frequently asked questions (FAQ) sections that attempt to educate the lay public about the correct interpretations of their work. However, for all the focus on communication, the field continues to rely on euphemisms and terms misappropriated from other disciplines. The murders in Buffalo put the researchers hunting for the gene for intellect in a spotlight that they would have otherwise opted to avoid. The conversation had shifted to people asking whether the theoretical benefits their field had been promising for decades could offset the tangible harms of today.

Michelle Meyers co-lead a team of researchers and philosophers on a report for The Hastings Center that was meant to tackle the pros and cons of social behavioral genetics (SBG) research. The report came out after the Buffalo shooter cited several of the working group participants as his inspiration, and they had months to go over exactly what they wanted to say. And yet, the report provided no meaningful definitions one could use to determine the pros versus cons for social genomics by any metric. The report contained a figure that indicated the key question for any behavioral genetics study was: "Does the study design permit sufficiently unconfounded results?" Everyone in their group agreed that if the answer to that question was "no," then the research should not be conducted. Thus, you would think that the report would contain a clear definition for what "sufficiently unconfounded" meant. You would think there would be a metric by which someone could define what was, and was not, "sufficiently unconfounded." You might also guess that there would be clear examples of work that was unconfounded contrasted against work that was too confounded to be justified. There were no such definitions in the report.

The group laid out other standards for how genetics research should be assessed such as asking whether the work was "valid", "carefully conducted" or that it was "well communicated" while holding "a greater chance for benefit than harm". However, they also did not provide guidance on how researchers would objectively define those standards or determine who gets to judge if researchers are meeting them. The report mentions that a "compelling justification" would be needed to run a study on SBG that separated people by race. But given that race is not a biologic concept, most genetics insist that studies *never* sort people by race so what exactly constitutes a *compelling justification*?

Like *The Genetic Lottery*, the report defends a mathematics-only definition of causation when it stated:

> *When we say that a genetic variant has a causal effect, we are using the term in a way that is standard across the sciences. By definition, in a given environment, a genetic variant has a causal effect if the phenotype would have been different had the genetic variant been*

different. Since researchers cannot run experiments that compare two human individuals who are the same with respect to every genetic variant except one, this counterfactual cannot be observed for a given individual, and it is thus impossible to know the causal effect of a genetic variant on a given individual.

You should notice that when they stated that it is "impossible to know the causal effect of a genetic variant on a given individual," they are saying you can't test this and thus it lacks falsifiability. Furthermore, the term *causal* would never be used "across the sciences" to mean a mathematical association. If someone said to you, "Frosted Flakes are causally related to colon cancer," how would you interpret that statement? How do you think the defamation lawyers at Kellogg's would interpret that statement?

The report went on to offer its most memorable statement when it put forth the following analogy in relationship to the geneticists use of the term "causal":

Notice that the definition of a causal effect is completely silent on the mechanism through which changing the genetic variant would have changed the phenotype. In a famous example of a causal effect that operates through a mechanism that involves genes but is not "genetic" or "biological" in the intuitive sense of those terms, consider a genetic variant that causes a person's hair to be red. In a society in which people with red hair are discriminated against, that genetic variant could have a (negative) causal effect on many phenotypes, including how much education a person gets. In the standard framework used by behavior geneticists, this would be counted as a causal effect, even though it operates largely through a social or environmental, rather than biological, mechanism (the redness of the hair is biological, while the response is social or environmental). Some causal effects of genetic variants likely operate through biological mechanisms in the intuitive sense of "biological," but many others, perhaps especially but not only for SBG phenotypes, likely operate through environmental mechanisms, much as a causal effect is imagined to operate in the red hair example. In other words, environmental mechanisms can be baked into the causal effects of genetic variants (and into the causal effects of PGIs that we discuss later).

There are at least three major problems with this analogy.

1) Victim blaming

Would anyone say that having two X chromosomes is *causally related* to an increased risk of being sexually assaulted? How about if a victim of domestic violence doesn't have her abusive husband's meal ready on time? Does the abuser get to say that her lack of timely cooking was *causally related* to his violent outburst even though it was only indirectly related via the environment? I get that the genetic statisticians have decided that correlations can be referred to as *causal*, but this report was supposed to be about how to *better* communicate genetics information to the public.

A true cause of something should be—at least theoretically—the target for fixing the outcome. If a mutation in *STAT3* in the bone marrow cells is the cause of someone's illness, then fixing the *STAT3* in those cells should be a means of treating the disease. But in the redhead analogy, what would need to be fixed? Should the redheads undergo CRISPR? Should parents start screening embryos for redheaded-ness and then aborting any baby with the gene? Or should the society that arbitrarily and unfairly decided to discriminate against redheads stop being anti-ginger bigots?

The legal system worked out *causal* claims for harm befallen to *predisposed* victims over a century ago. These are collectively referred to as *eggshell skull doctrine* and based on several different cases over the course of U.S.

and British history. One basic case was 1891's Vosburg v Putney, where 14-year-old Andrew Vosburg was kicked in the shin by 11-year-old George Putney while in their classroom. Vosburg, however, had a previous knee infection that had required surgery, and thus, that particular leg was not as healthy as might be expected for a kid his age. So, the kick from Putney might not have been expected to cause harm normally, but it actually caused lifelong problems for Vosburg given his underlying condition. The ruling was that Putney was still responsible for the harms he caused, even if they were unexpectedly more severe due to Vosburg's prior history.

How this would apply to genetics is best demonstrated by a case in which a person shot someone with the monogenic disorder of clotting, hemophilia. The victim ended up bleeding to death because of his wounds. The perpetrator's lawyer admitted his client was guilty of attempted murder, but felt that the court could not charge murder because the victim's blood disorder was *causally related* to their death. The defendant who made that claim died in jail because the courts, unlike the population geneticists, are not foolish enough to believe that finding genetic predispositions after the fact has any bearing whatsoever on how one should judge intentional harms.

2) Again, are these people allergic to history books?

There is simply no need to come up with a fake analogy for discriminating based on physical characteristics. History has plenty, but of course, none as genetically encoded as red hair. Hiding behind a fake analogy that uses a true genetic link obscures the reality that most discrimination is not actually linked to any true genetically encoded phenotype. By using the analogy of redheads instead of any of the numerous other historical examples of discrimination based on arbitrary physical characteristics, the report puts forth a fake example in an attempt to address minorities suffering real harms of discrimination.

3) Genes travel via chromosomes, not chance.

The redhead discrimination analogy would be true for *any* gene more common in redheads—not just the one that genuinely influenced hair color. If redheads were being discriminated against, then all the genes that were more common in redheads would be linked to discrimination as well. So, genes that signal being of Irish or Scottish background would also associate. I suspect this is why they did not use the analogy of *being Irish* as being linked to discrimination, since it would then be obvious that lots of genes would be linked. This is the most insidious failure in their analogy and the key to how they will attempt to spin their work going forward; that is by implying that their work on genetics will make it easier to unlock the environmental causes.

The premise the redhead analogy is trying to put forward—one echoed by Harden in her book and by the authors of the report in their subsequent interview with *Science Magazine*—is that genetics can be used to better understand the environment. Let's take an example of how genetic information might actually be helpful. According to a pair of reports from the European Commission in 2019 and 2020, Ireland had the highest rate of breast cancer in the European Union while Cyprus had the lowest rate. However, Ireland also has a higher rate of *BRCA1* and *BRCA2* mutations, as reported in the *British Journal of Cancer* in 2003 by the Scottish/Northern Irish Consortium. So, if you completely ignored genetics and saw that Ireland has a higher rate of breast cancer than Cyprus, you might be misled into thinking that there must be some sort of environmental factor in Ireland that was causing the higher breast cancer rate. Or you might overestimate how protective *the Mediterranean diet* was if you failed to account for Cyprus' lower rate of *BRCA1/BRCA2* mutations. But if you were able to account for the increased rate of *BRCA1/BRCA2* mutation, you could better evaluate the environmental contributions to the differences in breast cancer rates. Your adjusted calculations would not be perfect because even *BRCA1* and *BRCA2* have nuance in their risk presentation. Yet on the broadest scale, the geneticists claim that genetic information could be useful in uncovering environmental contributions could be correct. The problem is that this only works for situations where you know for a fact the gene is *directly* linked to the outcome.

Take a second example of the rates of Irish dancing between the two nations. Given that the rate of Irish dancing is much higher in Ireland than Cyprus, one might conclude that an environmental factor contributes to the differences. Living in Ireland makes one more likely to be exposed to Irish dancing, to have friends who participate in Irish dancing, and to have opportunities to study Irish dance. So, the environment (in this case, the overall Irish culture) is clearly the cause for the increased rates of Irish dancing in Ireland. But what would happen if one ran a GWAS for Irish dancing?

The genes that are simply markers of being Irish—the types of genes all the ancestry groups might use to label someone as partly of Irish heritage—would also be associated with Irish dancing (including, by the way, the gene for red headedness). The fact that genes could be associated with family trees rather than the diseases being studied is the reason why the negative controls are so important and assortative mating is so confounding. Since participating in Irish dancing would be linked with the genes for being of Irish background, it could give one the impression that the redheaded gene was *causally related* to Irish dancing. Just as the report admits, if a discrimination against redheads would link the redhead gene to the negative outcomes of discrimination, then positive outcomes could also be linked if redheads were more likely to Riverdance. So, what might be the harms of evaluating the environment in this example?

In the BRCA1/BRCA2 example, it made sense to adjust evaluations for national breast cancer rates to account for differences in a major genetic risk. But if you were to do the same for Irish dancing and adjust the analysis for the redhead gene (or the poly-Irish-gene score), what would happen? If you did this, you would incorrectly downplay the environment. Someone might look at the rate of Irish dancing in Ireland and Irish dancing in Cyprus and (correctly) conclude that the environment must be the cause. Yet if you took the proposed approach of population geneticists, you might try to adjust for the increased rate of redhead genes and Irish genes in the Irish. This adjustment might leave you thinking that once the increased frequency of Irish genes are accounted for, the rates of Irish dancing are the same in Ireland as in Cyprus. The conclusion would be that the environment isn't important and the difference in Irish dancing is mostly genetics.

So, what will happen when geneticists start adjusting analysis for the harms of poverty, racism, toxin exposure, and so on? If your research has already *biologized* racial disparities—that is, you have already assumed that the differences in health outcomes are biologic or genetic—then any GWAS results you find will be assumed to indicate meaningful differences in biology. If, instead, you are only picking up a marker for coming from a family tree with roots in polluted soil, then adjustments will worsen your environmental assessments rather than improve them. If you know your allele of interest is a true risk allele—like BRCA1 or BRCA2—then adjustments make sense. But if your allele could be real or just a *gene for being marginalized*, then you run the risk of unwittingly concluding that "when you account for the presence of the marginalization genes, the harms of being marginalized are not so bad. Therefore, the real problem in poor communities isn't all the social determinants of health but the fact that they have more 'bad' genes." If a similar genetic *adjustment* was done for gendered outcomes, one might conclude that "when you adjust for the presence of the second X-chromosome, the pregnancy rate between men and women is nearly the same."

Despite using genetic associations to adjust for the environment being dangerously invalid, it is the only "action" the behavioral geneticists propose their work can provide. In a 2023 Cambridge University Press article, Harden and her co-author James Madole attempt to defend the comparison between within-family GWAS studies and randomized drug trials. Despite all the rationalizations the authors admit that GWAS hits do not lend themselves to any direct actions. They state:

> Even if we concede that, at a conceptual level, genes could cause average differences in human behavior, at a practical level, it is not readily apparent what we would do with this knowledge.... [W]e cannot (and should not) readily apply knowledge of genetic causes to change the genomes of large swathes of the population in the hopes of changing their outcomes

Instead, they claim that "The analysis of causes in human genetics is meant to provide us with basic knowledge we require for correct schemes of environmental modification and intervention." So, an approach that was once billed as a means to find cures for diseases through genetics has been redefined as nothing more than a means to tweak environmental science calculations. You would think that the *one and only* actionable outcome that the geneticists outline as a justification for their work would lend itself to an example. You would think even a theoretical example would be in order (like GRIEF from Chapter 17).

The website NIH RePORTER contains a database of all the studies NIH has funded; Since 2007, NIH has spent 17.4 billion dollars on 3,150 research projects that contain "GWAS" as a primary term. PubMed contains over 60,000 papers referencing GWAS. It seems more than a little convenient that, after all this public investment, the only actionable outcome we should expect from population genetics is one that will require continued funding to uncover at some point in the future.

Many of the same geneticists whose work is adored by white supremacists online are suggesting that failure to fund their work would be the true racism because their work will allow for these theoretical (and spurious) adjustments of environmental findings. In fact, the coverage of the Hastings Report indicated that "Daniel Benjamin ... fears such biobank data restrictions will hamper social and behavioral research that could benefit people of African or Hispanic ancestry." It takes a special kind of person to produce the type of work that was cited as *the main inspiration* for a white supremacist to go murder and maim over a dozen Black people and then claim that limiting such work might be bad for minorities.

So, let's revisit Benjamin's "the 74 loci for white people graduating college" paper and see if it would meet the requirements the Hastings Report claims would serve as acceptable standards. Is it possible for a study to be *sufficiently unconfounded* if it focuses on a trait that is immensely impacted by mating preferences, income, and environment, while only partially adjusting for those factors (for example, using only one year of income as a stand in for poverty)? Is a study *carefully conducted* if no negative control is included and the results suggest the impact of genetics is less than 1%, well within the range of background noise that a negative control could account for? Is the study *valid* despite not offering any biologic explanation for how the loci identified could alter metabolism in ways that improved cognition? And as for the risk-benefit profile, it goes without saying that the work was not worth it.

The *Science Magazine* article covering the Hastings Report also quoted both Visscher and Abdellaoui, who unsurprisingly both concluded that their work should go on unencumbered by regulation and that their field can be trusted to perform their research in a beneficial way. All this speaks to their unwillingness to let go of the gene-centric attitude exemplified by the patron saint of modern eugenics-adjacent researchers, the statistician Ronald Fisher. Both the Hastings Report and the

Visscher paper that claimed a one-time hit in *ACE2* was a *success* for COVID-19 research quoted Fisher in their text. The Visscher paper relays that in 1918, Fisher stated, "In general, the hypothesis of cumulative Mendelian factors seems to fit the facts very accurately." This is held up as some sort of proof that Fisher was able to derive the *polygenic* nature of human traits and diseases. However, the *very next line* from the quote they are citing states, "The only marked discrepancy from existing published work lies in the correlation for first cousins," which is in reference to the paragraph before where Fisher opines that "some ambiguity still remains as to the causes of marital correlations."

Fisher was struggling with his finding that people are more phenotypically similar to their spouse than their first cousin. For someone who believed that genetics were the major determinant of who someone was, it would be confusing why people would be more like their partner than a cousin (I searched to see if Fisher had, in fact, married his cousin and was just trying to explain this away, but I could not find any such evidence). Thing is, Fisher absolutely understood that correlation did not equal causation. But he didn't aim his ire at the *curious* reality that phenotypes were more correlated between couples than cousins. No, instead, he took aim at the correlation between smoking and lung cancer. In two successive papers in 1958, Fisher wrote in the journal *Nature*, that:

> The association observable between the practice of cigarette smoking and the incidence of cancer of the lung, to which attention has been actively, or even vehemently, directed by the Medical Research Council Statistical Unit, has been interpreted, by that Unit, almost as though it demonstrated a causal connection between these variables.

> The suggestion, among others that might be made on the present evidence, that without any direct causation being involved, both characteristics might be largely influenced by a common cause, in

this case the individual genotype, was indeed rejected with some contempt by one writer, although I believe that no one doubts the importance of the genotype in predisposing to cancers of all types. I owe to the generous co-operation of Prof. F. von Verschuer and of the Institute of Human Genetics of the University of Munster the results of an inquiry into the smoking habits of adult male twin pairs on their lists.

Fisher then goes on to discuss the data provided to him from a twin study run by von Verschuer. In Fisher's mind, the fact that both identical twins were more likely to smoke than both fraternal twins was seen as evidence that there was "little doubt that the genotype exercises a considerable influence on smoking." As discussed in earlier chapters, this assumes that the pressures for both twins to smoke are equal between the twin groups, which everyone outside of genetics understands to be erroneous (the geneticists themselves admitted as much just after WWII). Fisher continued:

Such results suggest that an error has been made, of an old kind, in arguing from correlation to causation, and that the possibility should be explored that the different smoking classes, non-smokers, cigarette smokers, cigar smokers, pipe smokers, etc., have adopted their habits partly by reason of their personal temperaments and dispositions, and are not lightly to be assumed to be equivalent in their genotypic composition. Such differences in genetic make-up between these classes would naturally be associated with differences of disease incidence without the disease being causally connected with smoking. It would then seem not so paradoxical that the stronger fumes of pipes or cigars should be so much less associated with cancer than those of cigarettes, or that the practice of drawing cigarette smoke in bulk into the lung should have apparently a protective effect.

I should note here that Fisher was himself a lifelong smoker, so maybe he was just sensitive to someone telling him his vice was a health risk. Also know that pipe and cigar smoke are less cancerous than cigarettes because they contain different toxins and are used less frequently. Also, breathing deeply while smoking reduces lung cancer risk because such people die of heart disease before cancer can take hold. Rather than considering any of these possibilities, Fisher implied that the person's genetic make-up will dictate their decision to smoke, what type of smoking they will pick, and their final risk of lung cancer. Fisher couldn't seem to piece together that smoking could lead to lung cancer through effects that could impact every human. Then, Fisher concluded his paper by offering one of science's coldest takes:

> Unfortunately, considerable propaganda is now being developed to convince the public that cigarette smoking is dangerous, and it is perhaps natural that efforts should be made to discredit evidence which suggests a different view. Assumptions are put forward which, if true, would show my inference from von Verschuer's data not indeed to be false but; at least to be inconclusive.

Seeing that Fisher grounded all his claims in the twin study data provided to him by a German researcher in the 1940s made me suspicious enough to look up Professor Otmar Freiherr von Verschuer. He was the director of the Institute for Genetic Biology and Racial Hygiene from 1935 to 1942, who became an official member of the Nazi party in 1940. von Verschuer wrote about how thankful he was that the Nazi connections were able to allow him to continue his work. WWII was making it difficult for him to gain access to samples that he could run his experiments on, but thankfully, the professor had a connection in the form of a former student, Josef Mengele, the newly minted head of research at the Auschwitz concentration camp. Mengele is, unquestionably, the most notorious war criminal in the history of medical research. But thankfully for Fisher, Mengele was able to support the work of von Verschuer in ways that allowed him to supply Fisher with the data needed to conclude that the idea that smoking can cause lung cancer was *propaganda*.

Now, here, some readers are going to be tempted to question whether Fisher knew his buddy was a Nazi (or if *Nature* knew they were publishing praise for one). Yet remember, von Verschuer was a Nazi through the end of the regime in 1945 but Fisher published his work in 1958. The professor was convicted of being a Nazi and made to pay a fine (since the prosecution could not prove he actually killed anyone with his own hands). As an aside, I did not know that Nazi war criminal researchers got off with fines back then, but I think it explains a lot about the direction the field went after the war. von Verschuer paid his fine and then went to work for the University of Munich advocating for liberal eugenics, so he traded his Nazi ideology around "killing the feeble because it was better for society" for the idea that we should "kill the feeble because it was better for them." The professor then went on to help found the quintessential journal peddling scientific racism, *Mankind Quarterly*. That is to say, the idea that Fisher did not know this man was a Nazi with continued Nazi leanings at the time he praised him in the pages of *Nature* is not believable.

Fisher defenders will also then be tempted to say that Fisher was looking at the data handed to him by the Nazi teacher of Josef Mengele and figured that he was going to see the data for what it was, and not be distracted by its origins. Thing is, Fisher was quoted as saying that he thought the Nazis "sincerely wished to benefit the German racial stock, especially by the elimination of manifest defectives" and mentioned that he did not blame the statisticians who calculated that Jews were a dysgenic pressure on society for the fact that the Germans acted on those numbers. So, it certainly doesn't sound like Fisher was admitting the Nazi origins of his data were wrong and was simply trying to look at the numbers objectively. He seemed to think that the origins were just fine, but the Nazis might have handled things the wrong way.

I want to make something clear. I do not care about Fisher the person. I use some of Fisher's equations (but not his data) in my own work and don't care one bit if those equations continue to bear his name. I don't think sweeping the origins of his work under the rug will make the world a better place. But I am harping on Fisher because he seems to be a surreal person for the behavioral geneticists to hold up as some sort of deity as if the founders of population genetics would have trusted any equation that didn't imply the innate superiority of people like them.

After the Buffalo murders, the modern version of population geneticists were accused of being too enamored with their ideas that genes cause everything that they overlooked obvious environmental factors and aligned themselves with Nazi ideology. With a full year to reflect, they published a report that venerated a man who was so enamored with the idea that genes cause everything that he missed an opportunity to identify the harms of smoking through aligning himself with an actual Nazi. If I believed there were such things as *genes for introspection*, I would have to conclude that this group represented a dysgenic force for society.

CHAPTER 19
Genetic Determinism Causes "White Genocide"

BOTTOM LINE UP FRONT: *If there is one word to describe the people who insist that GWAS and PGS are valuable tools, it isn't racist; it is ingenuine.*

The parking lot of the rural Virginia airport was lined with trucks and ambulances. The site was being set up by Remote Area Medical (or RAM). RAM was started by Stan Brock in 1985 as a way of flying medical supplies and professionals into under-resourced areas of Africa, but Brock realized that parts of rural America needed help as well. RAM would bring in semi-trucks that were hauling mobile hospitals. While the outside of the truck looked like any 18-wheeler trailer, the insides were instead a medical doctor's office or a row of dental chairs. Even though the event did not start until the following day, as I pulled up to the event, I saw a long line of cars waiting to get in. Some people had arrived three days earlier and had been camping in their cars just for the *chance* to see a dentist without the often-crushing costs that might befall the uninsured in America. None of the cars in line were luxury vehicles and most were pick-up trucks in a state of disrepair that suggested they were used as a primary work vehicle.

A woman named Susan was one of my first patients. She climbed into the trailer-turned-clinic and sat down with a huff alongside her husband, Jim. Susan had uncontrolled heart failure stemming from a heart attack many years prior (I assure you, I made sure to ask about any signs of Sheehan's syndrome). The medicines she needed to take after her heart attack were too expensive, and so, her heart failure worsened. At the time I was seeing her, Susan fell into the *coverage hole* in the American system. Susan could not work due to her illness. Jim did earn an income, but from an employer that did not offer health coverage. Jim's work did not pay enough for the two to afford health coverage, but paid too much for them to be able to qualify for Medicaid. Living out in the parts of rural Virginia, that struggle to attract health providers also meant that even if they could visit a heart specialist, they would have to make a day trip out of it. Her shirt was the first immediate sign that Susan was one of those patients who seemed to take everything with a laugh. It said something to the effect of, "Where do I go to pick up my White privilege?"

While there are ways in which I think Susan might be failing to recognize some of the advantages of being white in America, no sensible person could look at someone who had camped out to see a dentist while struggling to breathe and think of them as *privileged*. While I might see some missing nuance in Susan's shirt, I agree with the sentiment implied. The idea that white skin would protect against the type of poverty Susan and Jim were facing was, and has always been, a lie. This isn't an argument against cultural superiority. I certainly look at anyone who roots for either the Michigan Wolverines or the Notre Dame Fighting Irish as being from an inferior culture. However, as much as it would be fun to imply otherwise, no one is born innately inferior in such matters. A passing understanding of the origins of genetic determinism reveals that the average white person in the U.S. or England was not the original *audience* of eugenics; they were the original *target*.

Adam Serwer included his work, "The Nationalist Delusion", in his collection of essays titled *The Cruelty is the Point*. Serwer outlines:

> In *Black Reconstruction in America*, W. E. B. Du Bois examined not only the acquiescence of Northern capital to Southern racial hegemony after the Civil War, but also white labor's decision that preserving a privileged spot in the racial hierarchy was more attractive than standing in solidarity with black workers...

> Du Bois wrote. "When Northern and Southern employers agreed that profit was most important and the method of getting it second, the path to understanding was clear. When white laborers were convinced that the degradation of Negro labor was more fundamental than the uplift of white labor, the end was in sight." In exchange, white laborers, "while they received a low wage, were compensated in part by a sort of public and psychological wage."

Rather than be paid in dental access or health coverage, Du Bois posits that Susan and Jim are paid with a psychological wage of their whiteness. However, this psychological wage has the worst exchange rate in the history of finance. Being white *and* poor carried such little benefit for Susan that she noted it on her shirt. Her whiteness did not provide her the drugs she needed to protect her heart nor free her from the indignity of having to camp out for days just to see a dentist. In exchange, the monied interest in the U.S. gets to hoard the resources that could otherwise be spent on meaningfully improving Susan and Jim's lives. Just to be as *absolutely clear* as possible, when I say "monied interest," I *do not* mean Jewish people. This is not code. Jewish people are a group with a shared history and culture but have no magical powers either for good or villainy.

I also don't mean people in dark rooms smoking cigars and twirling their mustaches. It does not take a conspiracy for the system to align with the interest of those for whom the system is working. If enough people are rowing in one direction, then no one needs to actively steer. Monied interests are people with middle to upper class incomes who view the status quo as working well for them. While the white people in poverty are being paid the "psychological wage of whiteness," those with means are suffering what I might call "the moral tax of wealth." Instead of being hit with a literal tax that might allow the state to provide dental and medical services, wealth suffers a moral tax for continuing to tolerate a society that Serwer notes:

> ...like the planter aristocracy that preceded it, impoverished most blacks and whites alike, while concentrating wealth and power in the hands of a white elite. It lasted for decades, through both violence and the acquiescence of those who might have been expected to rise up against it.

While conservatives are more willing to pay the moral tax associated with mistreating poor people, they are not alone. Even my very liberal home state of Maryland allowed Medicaid to expire in 2023 in a way that kicked more than 37,000 people off the rolls. Why was their diabetes a priority for the state one day, but dispensable the next? It was because the pandemic had waned to the point that diabetes being a risk factor for COVID-19 no longer posed a serious threat to those who prefer the pre-pandemic social contract. COVID responses proved that the U.S. could assure health access if it wanted to.

COVID revealed what would happen if the troubles of those whom Jesus termed "the least of these" were to ever creep outside the limits of what the eugenicists have forever considered "proper society." Believing that the American (or I suppose British) way of life is better than other nations can be debated, but believing that Americans or Brits are born superior has killed more white people than any other belief system. The first wave of death came as white Europeans assumed the local infectious diseases of other nations were a sign of innate inferiority of the locals.

An uncountable number of white people died of smallpox, tuberculosis, yellow fever, and more on society's path to understanding microbes *don't see race*. Belief in innate superiority of specific nations led to dozens of wars within Europe from the time of monarchies through WWI. Susan and Jim's son, Jacob, fell victim to one of the more recent examples: the opioid crisis.

As detailed in the book by Beth Macy (and subsequent Hulu TV series) *Dopesick*, starting in the mid-to-late 1990s, several pharmaceutical companies, most notoriously Purdue Pharma, began to target poor white communities with prescription painkillers. The companies researched places with low-income, diseases of despair, and manual labor. These drug dealers with nicer suits hoped to find people likely to need painkillers for work injuries in hopes of getting them hooked. Purdue, run by the billionaire Sackler family, sent speakers into medical schools across the country (including mine) to lie to doctors and tell us that people "could not get addicted if they take the medication for pain." They knew this was a lie but sent their drugs into these communities anyway. As the death toll hit over 500,000 people, including Susan and Jim's son, the executives for these companies cracked jokes in their emails. The penalty they faced for all this death amounted to fines they could easily afford and having the Sackler name removed from some reputation-laundering art museums. That's it.

Contrast this with what happened to Bernie Madoff, who ran a Ponzi scheme to steal money from rich people. He died in prison. The con artist turned fake biotech company CEO Elizabeth Holmes? Also in jail. They didn't pay a fine that barely covered the court fees for those they wronged; they went to jail (albeit I'm sure nicer ones than people who steal wallets). What do you think would happen if the Sacklers had micro-targeted their poison to the children in Beverly Hills, London, or Ivy League towns? Do you think that if the monied interest thought that the people of West Virginia were really of the *good stock* the eugenicists talk about all the time that they would allow anyone to put poison in their water, air, and veins? Had Purdue Pharma centered their scheme outside of Harvard, the Sacklers would have been tarred, feathered, and hung in *the Yard*.

So, what do you think the system would do if rich people could catch heart failure by riding on the same airplane as Susan? Or what if you could contract tooth decay from shopping in the same grocery store? You already know because when COVID upended the social contract between rich and poor, they flew the literal Army into needy towns to set up hospitals and give out vaccines. I should know—I was deployed to a converted basketball arena for the Albany Patroons to assure even the poorest New Yorkers could be safely vaccinated if they wanted to be. Then, when COVID no longer posed a risk to the ability of people like me to take our children to go see an actual basketball game in that arena, what happened?

The willingness of the system to ignore the needs of poor communities is well known among Black and brown Americans. In an interview with then NIAID head, Dr. Tony Fauci, NFL running back and community leader, Marshawn Lynch, wondered why COVID seemed to have a far worse impact in minority communities. Lynch asked, "What is it about our [Black and brown] bodies ...that ...takes a more strenuous toll on us than... a white person?" Dr. Fauci outlined that the decades of racial injustices in social determinants of health had left minoritized communities with higher rates of diabetes, kidney disease, and heart disease—each of which increases the risk for severe COVID. While true, Mr. Lynch quickly pointed out the obvious concern that if the root of the problem with COVID was the underlying disease in minority communities, then addressing COVID alone would be "like putting a band aid on a gunshot wound" especially if people refuse to tackle "the type of shit that is in our communities." Lynch wondered if America was so keen to address the health needs of minority communities for

COVID, "Why hasn't that been the situation and the fight since the beginning of time?"

Dr. Fauci stated, "[When] we get over this outbreak, I hope we don't forget ... that it's the conditions that African Americans have been put into from the day that they are born ... that leads to obesity, ... [and] hypertension." Dr. Fauci is not in charge of the U.S. health response, so it is not his fault that his hopes did not manifest (nor was he in charge of the vaccine or Coronavirus research, for any anti-Fauci readers). However, two years later, the social contract that was willing to let Black Americans, brown Americans, and white people like Susan suffer from preventable and treatable ailments was the only thing that went back to a pre-COVID normal.

Sarah Smarsh echoed this sentiment from the perspective of an impoverished white woman in her book, *Heartland: A Memoir of Working Hard and Being Broke in the Richest Country on Earth*. Smarsh wrote :

> *How can you talk about the poor child without addressing the country that let her be so? It's a relatively new way of thinking for me. I was raised to put all responsibility on the individual, on the bootstraps with which she ought pull herself up. But it's the way of things that environment changes outcomes. Or, to put it in my first language: The crop depends on the weather, dudnit? A good seed'll do 'er job 'n' sprout, but come hail 'n' yer plumb outta luck regardless.*

So, poor communities *of all colors* in America are well aware of the impact the environment plays in their health, but they rarely get a platform to express it. Dr. Winn's grandmother probably put it best when she used to tell him, "I don't need a science paper to tell me what I can see out of my window". Meanwhile, genetics-focused professors write things like this about data suggesting that poverty runs in families:

> *If this interpretation is correct then aspirations that by appropriate social design, rates of social mobility can be substantially increased*

will prove futile. We have to be resigned to living in a world where
social outcomes are substantially determined at birth.

This came from a draft version of a *PNAS* paper by Greg Clark of the University of California Davis as part of his related book *The Son Also Rises*. I will give Clark credit for picking the perfect title for a book filled with hereditarian drivel. Yet, let me spare everyone the time in reading it and just summarize that his point is *the aristocracy is ordained by God through genetics rather than a reflection of societal choice.* His work not only lacked a negative control or adjustments for the environment; he didn't even bother to assess the most obvious cause of generational inequality: inherited wealth.

People like Greg Clark only come in one of two flavors. Either they were born on third base and perceive that they were put there by God. Or they worked hard to climb the social ladder but then failed to see any role for luck, connections, or the kindness of others in their achievements. Clark's webpage makes note that his family consisted of impoverished Scotsmen who came to work in Ireland. Through their toils, he was able to climb the ladder through esteemed colleges on his way to a professorship. I'm sure he and his parents worked hard to make it in life, but I can think of a few non-genetic reasons that the Irish would have a harder time climbing the social ladder in U.K.-occupied Ireland than a Scotsman would.

Then again, maybe I'm just biased by my Irish roots and my base level grasp of Irish history. Clark's entire argument assumes that a nation cannot simultaneously have low social mobility and an aristocracy. In Clark's mind, the fact that Charles Darwin's relatives were also esteemed professors is not a sign that having connections matters but is evidence of genetic superiority of Darwin's biological relatives. Since the average person lacks social mobility, Clark concludes that Darwin's children being successful means that they had special gifts. Can you imagine publicly claiming that because the average person in America speaks English, then data showing that the children of Spanish speaking parents also speak Spanish indicates that there is a gene for *rolling your Rs? Que ridículo.*

There is a running joke among detractors of population genetics daring someone to run a GWAS for nepotism. Clark appears to have answered by implying that one day, a parent might be able to CRISPR their embryo into the aristocracy. When Clark's paper went online, he got push back from those in genetics, including Eric Turkheimer and Graham Coop, who were mentioned earlier. Meanwhile statistical geneticists like Sasha Gusev of Harvard Medical School demonstrated that an environmentally focused model performed better than the gene-obsessed one. However, the population geneticists who each had sworn they denounced discriminatory claims of genetics lined up to defend Clark. Abdellaoui likened criticism of Clark to being "Kardashian." Alexander Young paid Twitter a monthly fee to write prolonged defenses of Clark's analysis. But the bigots truly ate up the implications because they particularly love when people imply innate biological superiority from behind the lectern of a professorship.

But here is the catch: The online bigots think people like Clark are *only* talking about minorities or immigrants. But in reality, much like Darwin before him, Clark was also talking about the average white person and was especially talking about people like Susan and her family.

The vast majority of white people in pre-Civil War America did not own slaves. The vast majority of white people in America today are not rich. The average American may be well off enough to avoid needing to camp out for a dentist, but according to *NBC News*, nearly 60% of white Americans are living paycheck to paycheck.

To those in the *average* white person category, I would like to ask this: Do you think the people espousing white biological supremacy really consider you in the chosen group?

Charles Darwin said society must "bear without complaining the undoubtedly bad effects of the weak surviving and propagating their kind: which he felt would only be limited by the fact that the weaker and inferior members of society not marrying so freely as the sound." Yes, he included non-white races among those "inferior members of society," but he also carried those thoughts about the lower-class white people of his day. Leonard Darwin, the literal *nepo baby* of Charles and mentor to Ronald Fisher, took things further to state that if the U.S. did not implement full scale eugenics "during the next hundred years or so, our Western Civilization is inevitably destined to such a slow and gradual decay."

Leonard Darwin and Fisher championed the eugenics movement. A movement that humanities professor, Alexandra Minna Stern outlined only valued "Anglo Saxons and Nordics, whom they assumed had high IQs. Anyone who did not fit this mold of racial perfection, which included most immigrants, Blacks, Indigenous people, poor whites and people with disabilities, became targets of eugenics programs." Eugenics policies may have disproportionately targeted minorities, but the quintessential Supreme Court case about eugenics, *Buck v. Bell*, was decided against a white woman. In 1927, the plaintiff Carrie Buck sued to block the state of Virginia from forcibly sterilizing her for being "feebleminded" and lacking the financial means to support her child. Justice Oliver Wendell Holmes Jr., along with seven of his eight colleagues, established that the Constitution allowed for involuntary sterilization with his infamous decision that "three generations of imbeciles are enough".

Do you think any of the past eugenicists would defend the people in modern white America who are struggling to make ends meet?

When you hear a professor talk about the innate intellect of the ruling class, do you think he is talking about Susan? When researchers in education genetics swear that they are only talking about how some white people have biologically disadvantaged intellect, why is this okay? Why would it be a career-threatening taboo to build a career claiming innate superiority of the average white person over the average Black person but be acceptable to imply large swaths of white people are innately inferior? These types of genetics researchers routinely claim that "activist scientists" are suppressing the truth about intellect in the name of political correctness. But you want in on some genuinely secret knowledge? Those researchers don't give a crap about people like Susan. They have convinced themselves that people with careers like them were born special, despite those careers being wholly dependent upon science continuing to tolerate their elitist assumptions.

If anyone thinks the researchers in educational genetics are sincere in their research, ask them to name a single discovery of theirs that has helped a child improve their school performance. Ask them to give you even a *theoretical* example of how their work *might* help. They will only ever point to how their work *might* improve future research. But that is like saying, "If you give me more money to pretend to help your child today, I'll have an easier time spending money on pretending to help your child tomorrow." The genetics researchers claiming to be future victims of cancel culture aren't afraid that losing their job will rob the world of "discoveries" that have never helped anyone but themselves. What they fear is losing access to their university's professor's lounge or losing legacy admission eligibility for their children.

Charles Murray wrote a follow up to *The Bell Curve* where he implied that poor children from rural America without any college graduates in their families were *biologically* incapable of being able to catch up to the children of professors. Not culturally or financially disadvantaged, Murray claimed poor white people *grew apart* in a biological sense. He couches his disdain under the cover of Universal Basic Income. His argument is that the Susans and Jims of the world should be given a check from the government to cover their basic needs because he thinks they could never amount to anything more than being able to meet their basic needs.

Do you think people who record podcasts for a living and believe that there are genes causing low back pain understand what it means to complete a truly hard day of work? When was the last time anyone heard the kinds of pundits that rail against non-whites and immigrants demanding justice for the victims of the opioid crisis? Maybe if the Sacklers had something to do with Hunter Biden's laptop, they would care, but it seems the same media outlets that platform genetic superiority of white people are mum when rich white people murder a million poor white people. You think the kinds of people who attend politically connected cocktail parties would be okay switching places with the average American so long as that person were white?

Kathleen Belew, assistant professor of history at the University of Chicago described the conspiracy theory known as "white replacement" as "the idea that somehow, nonwhite people or outsiders or strangers or foreigners will overtake the United States via immigration, reproduction and seizure of political power". Prior to the murder of Heather Heyer in Charlottesville, Virginia, neo-Nazis marched around a confederate statue shouting "Jews will not replace us" to signal their belief in this idea. However, how would anything look different if the pundits who are raking in money shouting about white replacement were secretly being paid by those who want to "replace" white people? While these pundits tell Jim and Susan to get mad about the Jews in Hollywood or schools teaching children that slavery was bad, they are bankrolling politicians who assure Susan won't get lifesaving healthcare and Jacob couldn't get treatment for his addiction. While the pundits tell Jim and Susan to get mad about migrants fleeing war, they are supporting legislation assuring that companies can get away with pouring poison into Susan and Jim's community.

Maybe we should wonder who is making the payments to all the talking heads who claim to be defending the future existence of white men but never talk about how to meaningfully address the toxins, drugs, and poverty harming the white community (and instead just yell about which non-white group to blame). If a truly evil globalist was intent on replacing white people, they would fill their drinking water with lead and hormone disruptors to suppress both their thinking and fertility. Then they would pay some trust fund dweeb to distract them by blaming the woke mind virus on efforts to clean the water or expanding healthcare for everyone. The evil globalist would also make sure that the dweeb was fully vaccinated and working in a TV studio, which required masks to set foot in. Then have him convince white men not to protect themselves against a virus that can cause death, nerve damage, and erectile dysfunction. But that's just my conspiracy theory. Google your favorite pundits' net worth and see if it sparks any theories for you.

Take notice that the pundits who want to defend genetic determinism only ever give their lectures at liberal arts colleges so irrelevant that they don't even have a football team. I'll make this deal; if any of them would like to debate that the people of McDowell County, West Virginia are biologically inferior to the people in Cambridge, Massachusetts, I'll gladly meet them at the West Virginia County fair. Furthermore, I'll challenge them to let their children drink the water from Flint or breathe an air sample from where the trainloads of toxic waste exploded in East Palestine, Ohio. Let's see how interested they are in *genetic predispositions* to toxins when it is their kid in the wake.

The reality is that the one thing the psychological wage of whiteness cannot purchase are protections against pestilence and poison. Poison and pestilence are the only true threats to white privilege—not some trans person drinking a beer on TikTok, not some diversity training class your boss makes you take, and not some Black actress cast as a mermaid in a film that should have never been made live action in the first place. Serwer again notes that there is a "widespread perception that racism is primarily an interpersonal matter—that is, it's about name-calling or rudeness, rather than institutional and political power." This is undoubtedly true, but one could state that it is also true of elitism. The perception being that elitism means speaking ill of the poor rather than being about *institutional and political power*.

But maybe you are reading this, and you *are* part of those institutions of power. Maybe you are a professor or scientist who considers themselves a winner in the genetic lottery. Well, I'd be willing to bet you every penny in your 401k that someone you love has—or maybe you have—a disease of either the neurologic or immune system. I wouldn't make that bet for the renal or cardiovascular system because diseases of metabolism tend to be easier to treat with good access to healthcare and screenings. But just as whiteness did not protect Susan and Jim from their woes, wealth does not protect against diseases caused by inflammatory toxin exposure. I suspect you know at least someone with ADHD, depression, anxiety, Autism, hay fever, eczema, asthma, food allergies, psoriasis, Crohn's disease, or Ulcerative Colitis. I could argue that cancer should be included, as well, given the immune system's role in defending against it. There are only two organ systems in charge of *understanding* the outside world: the brain and the immune system. The brain assesses potential threats and plans the best response while the immune system does the same on a molecular scale. These two systems are the only ones that need to learn from past exposures in order to improve future performance. Is it any wonder that a society that has put these two systems under constant assault has seen a rise in related diseases?

Also, for those who might not see genetic determinism as a threat to your status, I wonder if they are prepared for the future Harvard President to deny their kid entry to the school because they detected a high PGS for committing sexual assault. What if the IRS audited everyone with a high PGS for being a tax cheat? Would the fact that genetics could only *explain* a fraction of a fraction of a percent for tax fraud be a good reason for such a program to be shelved?

The online trolls love using the code words *human biodiversity* to make it sound like they are respecting how genetically different humans can be. But wouldn't the true way to respect genetic diversity be to acknowledge that any allele can show up in any person from any background? If your child had a gene that truly did predict something bad would happen to them, would you want your doctor to ignore it because they are white? If Susan had an allele that could have clued doctors into the best drug for her heart failure, but that gene had been discovered in Koreans, would it be respectful to human biodiversity to let her suffer?

Respecting human biodiversity, however, would also require you to respect how biodiverse human's environments are. A toxin that harms white people will do the same in everyone even if the exposure is uncommon outside Europe. There were plenty of white Americans who were poisoned with lead in Flint, Michigan. So, if science was made aware that a specific gene variant was responsible for prostate cancer in white people, which study should researchers perform next: spend a decade looking to see if the gene is also bad for Black people's prostate cancer or work on making a drug that targets that gene? If scientists found out that a specific toxin was causing Autism in Latinos, should they spend decades trying to see if the toxin also hurts white people, or figure out how to protect all children from this toxin *now*?

When we hear about increased rates of asthma in Black Americans, rather than entertaining the idea that they are innately preprogrammed for the disease, we should ask, "What toxins might they be exposed to?" Even if the toxin might fall *disproportionately* on the heads of minority communities, the wind blows and rivers flow into the rich neighborhoods as well. If we hear about a community overrun with drug abuse, we should ask which powerful interest funneled drugs into those neighborhoods (for Black Americans, it was the U.S. government in the 1980s and for white Americans, it was the large pharmaceutical companies in the 2000s).

If anyone says that a gene *matters*, we must demand to know why the gene didn't matter for our grandparents or great grandparents.

If anyone says that a child has a "genetic predisposition," we must demand to know "predisposed to what?"

If anyone says that genes play only a small role in a disease or trait, we must demand to know what other factors play a role and how they accounted for them in their analysis.

If anyone says that the effect of a gene depends on the environment, we must demand to know which exposure triggered that gene into mattering.

If anyone says that children in affluent schools are all better off than children in impoverished schools, we must demand that they justify spending any future money on research with such obvious results instead of funding improvements in the struggling schools we have now.

If anyone says that the parents' genes make the children's environment, we must demand to know which of the parental genes put toxins into the water or lead into the skies.

If anyone says that genetics can predict someone's race, we must ask them if gene therapy could turn a Norwegian into a Mexican. We should inform them that with the notable exception of English, research has shown genetics can predict what language someone speaks. Then we must ask if someone could mutate their way into speaking Mandarin. We must insist on a better understanding of the difference between genes *associated with* groupings and genes *causing* groupings.

I don't think the geneticists who think they are going to find genes that explain income or education are racist, despite how much the racists think they are racist. I believe them when they say they do not want to harm anyone with their work. I do not, however, think they are genuine in their claims of wanting to improve the lives of others. We must demand better.

The belief of white genetic superiority is rooted in the claim that God gave Europeans genes for supremacy, but the expected natural order has been lost due to multicultural democracy. However, those longing for a racial hierarchy invariably see a class-based hierarchy as equally foundational to society. The results of using racism to distract from elitist policies was well articulated in the book *Dying of Whiteness: How the politics of racial resentment is killing America's Heartland*, by Vanderbilt Professor Jonathan Metzi. Professor Metzi interviewed white Americans who opposed expanding government healthcare services despite their lives depending on those very government programs. The interviewees feared that minorities would "use up" the resources and leave them with nothing.

However, these concerns do not match the ground truth I gained while driving past the long line of white Americans camping in their cars in hopes of accessing healthcare. The truth is that the policies which stole nearly all of Susan's white privilege *conspire with* anti-Black racism rather than oppose it. The laws aimed at denying services to those deemed to be of lesser stock have never been opposed to including impoverished whites among the targets. Susan didn't lack for care because all the dentists went to work in minority communities that are in just as much need as her town. Teaching critical race theory didn't spew the particulate matter into Susan's air or drip the heavy metals into her water either. A 2021 *New England Journal of Medicine* report identified air and water pollution as major contributors to the types of heart conditions Susan faced. The true paradigm of eugenics isn't just about race, it is that the hierarchies of *both* race *and* class are biologically encoded; this is the paradigm whose shift is nearly upon us.

PART IV

Re-Tilling the Soil

CHAPTER 20
How Paradigms Shift

BOTTOM LINE UP FRONT: *The Structure of Scientific Revolutions provides insights into how we can get away from the gene-obsessed paradigm and why it is so hard to do so.*

One of the most cited academic books of all time is Thomas Kuhn's *The Structure of Scientific Revolutions*. The field of science, Kuhn pointed out, presents itself as a slow but steady process of learning about the world. With each experiment or each new published discovery, scientists get a little smarter. Slowly, the collective knowledge grows in a day-to-day process. Kuhn pushed back against this idea by introducing the concept of *paradigms*. Kuhn argued that scientists first need a way of thinking about a puzzle that allows them to attempt to solve it. For example, early investigations into how electricity works constructed a paradigm that supposed electricity flowed like a liquid. The researchers conducted experiments under this set of assumptions. In so doing, they ran experiments in which they already had established expectations of what the outcomes would be. If none of their results made any sense, they would return to the drawing board and construct a new paradigm. But if enough of the findings fit their priors, they would keep that paradigm as the foundation for future experimentation.

Textbooks do not highlight the process of trial and error, and present only the successful outcomes, which gives the misimpression that scientists spend their time shouting, "Eureka!" rather than convincing themselves and then others of their conclusions. Kuhn argued there were only three main types of research phenomenon. The first type is when researchers fully expect the results and are just trying to reproduce findings from peers or predecessors. The second was the most common and represented when the findings fit the paradigm, but the exact mechanism is unknown (e.g., if you know that stimulus A reliably leads to result Z, but the steps in between are unknown). The third phenomenon was rare and represented when results were anomalous (meaning unexpected but still reliable) but could not be assimilated into the standing paradigm.

Since no paradigm is ever perfect, some findings won't quite meet expectations. Those working on the flow of electricity found some evidence in support of their theory, but since electricity does not behave exactly like water, many of their findings were *off*. Kuhn referred to these displaced results as *anomalies*. If later experiments could explain the anomalies, the foundational paradigm would strengthen.

Those who had built their careers on the existing paradigm were often quick to dismiss anomalies generated by other researchers. Any anomaly that was unable to be brought back under the paradigm might be dismissed as a failure of the scientist's methods of analysis rather than a weakness in the paradigm. But if anomalies began to accumulate, the field might go into a *crisis*.

During the crisis phase, paradigms would compete against each other with rival factions. Eventually, the controversy would make its way into the classroom and textbooks. Younger scientists would be trained in such a way that they would simultaneously learn the strengths and weaknesses of any paradigm. Whereas the old generation used the paradigm as a tool to solve their chosen puzzle, the younger scientists saw the anomaly-filled paradigm itself as the puzzle that needed to be solved. Kuhn wrote:

Probably the single most prevalent claim advanced by the proponents
of a new paradigm is that they can solve the problems that have led
the old one to a crisis. When it can legitimately be made, this claim
is often the most effective one possible.

When the new paradigm was seen to be capable of solving the anomalies of the prior paradigm, the field would undergo what Kuhn coined as *a paradigm shift*. But importantly, Kuhn noted that while the field of research may shift paradigms, the established researchers rarely did. He quoted Max Planck's observation that:

a new scientific truth does not triumph by convincing its opponents
and making them see the light, but rather because its opponents
eventually die, and a new generation grows up that is familiar with
it.

Kuhn cited numerous examples of researchers who had risen to prominence in a particular field refusing to move away from the paradigms that had earned them prestige, professorships, and profit. So, it was up to the new generation, or an *outsider* that was not embedded with the prior paradigm, to provide the proof needed for a field to progress. As the new generation aged, they too would fall prey to solidified dogma and the cycle would repeat until, once again, the anomalies overran the paradigm. Kuhn stressed that paradigms were never abandoned so much as they were exchanged. Scientists needed to have some paradigm that could guide their experimental design, and so, even if the old one was flawed, they would cling to it until a newer one allowed for better studies. Thus, for a paradigm shift to occur, the new paradigm must not only account for past anomalies but also provide a novel means of studying the problem. If the new paradigm could not be easily tested, then it would not be adopted, given that the primary value of a paradigm is in its ability to guide experimentation. Kuhn's write up focused mostly on the fields of chemistry and physics as he felt that the history of biological science was not yet long enough to make the same detailed observations.

As we sit today, I would propose that biology has three dominant paradigms. The first is germ theory, posthumously identified by Dr. Semmelweis but brought into acceptance by Louis Pasteur and others. The COVID-deniers notwithstanding, the idea that microorganisms can cause disease seems to be on solid footing. The second is toxicology: the idea that a poison can cause harm. These two are fairly related given that researchers would assess them using similar experimental designs. Had the wells tested by John Snow been poisoned with toxins rather than infected with cholera, his analysis would have been the same. The third paradigm is the one that has proven to be so badly in need of being exchanged: the idea that any disease that cannot be squarely put into one of the first two categories should be assumed to be genetically caused until proven otherwise.

Cadaverous poisoning was assumed to be an innate property of the women dying in labor until it was proven to be caused by infection. Fisher believed lung cancer and smoking were both innate properties of certain people until their causal relationship could be uncovered. Until someone can prove an environmental factor causes a disease, biomedicine tends to *poke the dead fish* and focus the research efforts on the possible innate contributions. So, why has it been so hard for this paradigm to shift? How many GWAS studies with claims of *explaining* fractions of fractions of percentage points for diseases would we need to see before we can move on? With each round of modification, the estimated contribution of genetics has dwindled; what was presumed to be 85% in twin studies became 15% in GWAS, then 3% in within-family PGS, to now under 1% when accounting for assortative mating. How much lower should things slide before the paradigm of assumed innate encoding can shift?

Kuhn's work provides valuable insights into why the gene-centric paradigm has been so difficult to shift away from. The first of which is what was outlined by Plank when he articulated that the old guard needs to die off. When Kuhn published *Structure* in 1962, U.S. life expectancy was 5-8 years less than pre-pandemic America, meaning that the scientists who build their careers on flawed paradigms are capable of being *thought leaders* for a lot longer. Science also has a soft spot for pedigrees. While success in science may be more about what you discover than who you know than, say, in big business or politics, science is far from a meritocracy. If you worked in the lab of someone with a big name, then you are inherently viewed as being more capable that others, even if they accomplished comparable work with fewer resources. Dr. George Santangelo's group, in the Office of Portfolio Analysis at NIH, has several publications detailing that women scientists make larger impacts on the field of medicine despite fewer resources. They also receive fewer citations despite their more impactful work. So, bad dogma can hang around longer because the scientists who made their careers on the flawed paradigm live longer, get more resources, and then assure the careers of their trainees are also built upon that flawed paradigm, thus repeating the cycle. But there are several other ways in which Kuhn's work, and the work of others, can inform us as to why genetics continues to be the default biologic assumption despite generations of failure to improve the lives of anyone outside the academic in-crowd.

Haters Gonna Hate

Perhaps the most obvious blockade to a new biomedical paradigm is that the racists and bigots have always seen genetics as their savior. They are not interested in a biologic mechanism for racial disparities; they only want an immutable biologic explanation. If the achievement gaps in race, gender, or immigration status could be explained by toxins and resources, then they could be solved by cleaning up the environment and providing services. This is why they cling to twin study estimates with such zeal. Telling people that a trait is 2% genetic based on within-family GWAS just isn't as convincing as telling them that any day now the "missing heritability" will be found to prove that 90% of the trait is genetically encoded. Thus, the claim that the *proof* of traits being 80-90% genetic being constantly five years away is one they cannot let go of.

One of the most frustrating practices of the behavioral geneticists is that they will engage with these bigots as if they are operating in good faith. Behavioral geneticists are critical of overtly bigoted claims but seem to believe that if they present just the right graph or articulate their argument in just the right way, then the overt eugenicists will suddenly realize that they are wrong in their assessment that "the Jews" are trying to "replace" them. The behavioral geneticists may think this indicates their objectivity, but in reality, it only shows their willingness to provide a façade of scholarly validity to people who are nothing more than racist conclusions in search of supportive data. *The Atlantic*'s Adam Serwer put it best when he wrote, "They don't want civil-rights laws that would level the playing field, or policies that would erode the privileges they've inherited. Of course if you are really genetically superior, you do not require a society built on that premise. You seek to create one because, on some level, you know you are not."

The Heresy of the Prosperity Gospel

Perhaps the most heretical statement about Jesus Christ is that he hated the poor and loved the rich. Google, "Bible quotes about the rich," and see if any wouldn't be labeled as *class warfare*. The Prosperity Gospel is the distillation of this heresy, which states that God shows his favor by providing you with material wealth. Any rational person who has read more than five sentences of the New Testament can see that this was nothing more than an excuse for rich pastors to live a life of opulence on the donations provided to their churches. However, the core rot in this belief is the inverse, that the poor are without God's favor. The commonly used phrase "there but for the grace of God go I" is intended as a reminder that being "better off" does not make one "better than". However, Reverend Deacon Terri Murphy of The Church of the Ascension preaches that while this phrase is rarely said with malice, the saying blasphemously insinuates that those who have been unlucky have lost God's grace.

The notion of the *undeserving poor* is one that rears up in all Western societies but is particularly prevalent in the U.S. and U.K. Its tacit acceptance makes it easier for researchers to imply the innate inferiority of poor white people, so long as they steer away from the far less politically correct subject of implying the innate inferiority of minorities. The genetic implication of this belief persists in politics because those who do not want to provide welfare to the needy find it easier to paint the poor as so biologically inferior that all welfare programs are doomed to fail. Not every welfare program is valuable, but debating any specific policy flaws takes more time and political skill than advocating for genetic determinism. Thus, because society continues to be more tolerant of statements chastising the poor than those demonizing specific ethnicities, genetic claims that are limited to impoverished whites will find greater acceptance. Even as every racist in the world infers the predicably bigoted extrapolations.

The Sunk Career Fallacy

The first guidepost for being *anti-eugenics* offered by *The Genetic Lottery* was "stop wasting time, money, talent, and tools that could be used to improve people's lives." Reading this written by a population geneticist genuinely made me wonder whether there may be a causal gene for psychological projection. The reality is that entire departments at big name schools have spent over two decades searching for genetic contributions to diseases and traits. The behavioral geneticists in particular have abandoned any pretense of performing biological validation for their claims. Thus, the only skill they have cultivated is combing the U.K. Biobank for some new association that they can mathematically gerrymander into a novel finding.

If genetics were seen as the minor modifier of environmental exposures, a lot of academics who lack the skills to do anything else would be out of work. Next to them in the unemployment line would be the numerous members of the editorial staff who spearhead publications of journals that would be more properly titled *Nature Correlations*. They are all in the sinking vessel together, with each manuscript serving to wade just enough of the rising tide to allow the editor or researcher to make it to retirement. As failures mount, people have no choice but to come up with more and more absurd rationalizations for their work including racializing the methods, which is exactly what Kuhn pointed out happens to dying paradigms. One recent example was published in the *British Journal of Dermatology*, trying to rationalize the continued interest in *clinically* useless *FLG* gene by connecting it to palm reading. No, really, the lead author tweeted out the following:

Edel O'Toole
@EdelOToole

In June issue @BrJDermatol we published our first phenotyping paper in 506 Bangladeshi children and young adults with eczema. We sequenced the filaggrin gene and examined the palms of 506 individuals. #dermtwitter #eczema

academic.oup.com
Deep palmar phenotyping in atopic eczema: patterns associated with filagg...
Loss-of-function (LoF) variants in filaggrin (FLG) are associated with hyperlinear palms (HLP). It is not clear what defines HLP and predictive ...

2:11 PM · Jun 14, 2023 · **6,775** Views

I'm sure the authors are good people who want to improve care for patients. But when your skill set is poking dead fish, turning to look at the lake is a nonstarter no matter whether you are researching eczema, Schizophrenia, or cheese flavor preference. Upton Sinclair once wrote, "It is difficult to get a man to understand something, when his salary depends on his not understanding it." I would extend that it is equally difficult to get someone to understand something when his or her scientific reputation and/or research funding depends on not understanding it.

If A Result Topples Genetics And No Geneticist Is Around to Read It, Does It Make An Anomaly?

Kuhn notes that the key to a paradigm shift is the emergence of crises caused by mounting anomalies. But what if each scientific specialty were so isolated from the larger scientific field that anomalous results went undetected? Population genetics judges itself on the fit of its mathematical models, with no regard for how others see their flawed assumptions. With the silo-ing of niche disciplines, a result that would be anomalous for genetics may never penetrate the geneticists' bubble. Epidemiologic data showing the impact of migration on allergy or the rapid uptick in rates of disease would completely go against the idea that genes are the central driver of the disorders. Studies in the fields of social determinants or environmental sciences may find variables for which any genetics study should adjust, but if the geneticists never read such literature, how would they know? Academic publishers assessing a paper on the genetics of eczema would never send the paper to someone like me to review (especially not after this book). They would assume (correctly) that I do not have expertise in the statistical nuances presented by the geneticists. But even if they did send me the paper for review anyway, they probably wouldn't appreciate critiques that apply to the entire field of population genetics. Each field needs a paradigm along with a base level of assumptions. When the work that challenges gene-centric thinking goes unseen by population geneticists, whether because it was ignored or just missed, anomalies can't mount.

Cash Rules Everything Around Me (C.R.E.A.M)

A lot of people make a lot of money off genetic determinism. I don't even include here those who profit from framing the poor as innately defective. Here, I mean directly cashing in on people believing that DNA is fate. Sites like *23andMe* sell people reports stating "you probably wake up at 8:57 a.m." and "there is a 65% chance you prefer chocolate to vanilla ice cream." Convincing the public that DNA has this kind of predictive power is profitable for those who pedal pseudoscience. These sites do offer legitimate assessments of proven risk alleles (such as for cancer), but the vast majority of people don't have such alleles and those that do can get tested by a clinician. Ancestry sites do offer some insights into people's family trees, even if it is suspect to claim that their approach possesses the precision to state you are 12.7% Irish. As the artist Canibus put it nicely in his song titled *Chaos*, "The best place to hide a lie, is between two truths."

The next wave of profiteers will be those selling embryo selection for traits to couples who have more money than sense. Those who can afford to pay tens of thousands of dollars to custom order a baby with a PGS for high achievement are already lining the pockets of grifters.

Genetic determinism can also generate profit by protecting polluters. As outlined in *The Tyranny of the Gene*, whenever you hear about a pollutant causing a disease, the overwhelming majority of the time, there will be a company that is polluting. An old adage is "one man's cost is another man's income." Well, the same goes for medical bills. If eczema is part of the cost of isocyanates and xylene, then the only way to shift those costs is to reduce the profit of the companies that use these chemicals in their products. Even if banning the chemical is too extreme, then requiring increased money be spent on filters or disposal still impacts the bottom line.

More recently, biologic determinism has been used to segregate the population to aid pharmaceutical companies. In her book *Fatal Invention*, Dorothy Roberts details the story of the drug known as BiDil. This drug was about to see its patent expire. The company conducted a study in only African American patients with heart failure. The drug was successful in improving outcomes but that alone would not have extended the patent. So, the maker of BiDil claimed that the drug worked *better* in Black people. Their study lacked any comparison group and there was no reason to suspect the drugs would fail to help non-Black people with heart failure. But the FDA approved the drug for patent extension under the claim that it would help close the racial disparities gap. BiDil was given 13 more years on the market but ironically still failed due to the same social determinants that limit care in Black communities. Communities distrusted a drug labeled as *for Blacks* due to the lasting stain of the Tuskegee experiments. Even if a patient was open to using the drug, they may lack access to providers, and lower incomes made it less viable to buy BiDil instead of combining two far cheaper generic versions of its main components. To be fair, BiDil might have failed for reasons other than social determinants, but it seems safe to say that being the first drug marketed under the claim that Black people have differing biology was *probabilistic but not deterministic* of failure.

The Troubling Persistence of Ableism

Dr. Christopher Donohue is an historian with a mathematics background with the National Human Genome Research Institute at the NIH. He is also an advocate against ableism in all its many persistent forms. Dr. Donohue has Cerebral Palsy, which is a disorder impacting the ability of the brain to control muscle movements. Cerebral Palsy is wholly caused by environmental factors, such as intrauterine infections or trauma during or after delivery. Donohue's work has outlined the lack of mathematical sophistication in population genetics approaches like GWAS and the PCA plots discussed in Chapter 4. He described current GWAS calculations as "first semester linear algebra with some basic regression analyses" and outlined the numerous ways that population genetics has failed to take on the modern statistical approaches identified by other *big data* fields.

In the discussion around eugenics, typically racism and antisemitism take center stage. However, the eugenicists were equally keen on purging society of the *stock* they blamed for any disability. Donohue told me that "ableism needs to be viewed through the lens of each impacted community." Ableism is at its worst when it fails to make a distinction for the scope and range of abilities within a community. For example, while Dr. Donohue has some specific needs, he has been extremely successful in his academic career; whereas others with Cerebral Palsy may be afflicted to the point of debilitation. Treating every disability as if it were experienced in the same negative way in all people allows the festering idea that disabilities should or can be purged rather than treated. Some people with Autism see their diagnosis as part of their identity and would never consider any offer to "cure" them. Others suffer from a level of disruption that causes them to yearn for an opportunity to treat their symptoms. Donohue has been incredibly successful and yet has been denied care at clinics that see *Cerebral Palsy* on his chart and respond by claiming "we can't handle your needs and you should see a specialist."

In the book, *Beyond Bioethics*, bioethicist Adrienne Asch wrote about prenatal testing:

> *For people with disabilities to work each day against the societally imposed hardships can be exhausting; learning that the world one lives in considers it better to 'solve' problems of disability by prenatal detection and abortion, rather than by expending those resources in improving society so that everyone—including those people who have disabilities—could participate more easily, is demoralizing. It invalidates the effort to lead a life in an inhospitable world.*

Some disabilities undergo *medicalization* by the research community. For example, cystic fibrosis could be considered a *pulmonary disability* while diabetes might be thought of as a *metabolic disability*. But no one would ever use these terms for medical diseases. Instead, patients with these diagnoses are seldom addressed in the same way as individuals who require a wheelchair or those experiencing cognitive impairments. Donohue outlines that disabilities get stigmatized under the influence of three main factors. First, the rarer the disorder, the more likely it is to be stigmatized. Second, any disability of movement, speech, or cognition will feed into concerns of whether the person afflicted will be able to provide for themselves and/or how much support they will need. This is particularly true for any disorder that would be noticeable to those attending a eugenics society cocktail party. The person who depends on insulin for survival can go unnoticed whereas the person who needs a wheelchair for mobility disrupts the eugenicists' views of a perfected society. Finally, adult patients with disabilities are more likely to be mistreated than children, perhaps due to the general expectations that children will not be self-reliant and need an additional level of care and support.

Professor Ijeoma Ononuju expands on this concept when he discusses the use of the term "dis/ability" (employing a slash between dis and ability) instead of disability. He teaches that the small change in language is meant to "(1) counter the emphasis on having a whole person be represented by what they cannot do rather than what they can and (2) disrupt the notions of permanency of the concept of disability, seeking rather to analyze the entire context of which a person functions". Stevie Wonder is one of the world's most famous and successful music artist, while I can neither play piano nor sing to save my life.

Stevie Wonder is legally blind, whereas I have full sight. To label Mr. Wonder as "disabled" suggests musical talent isn't an ability. The term dis/abled is meant to convey that he has parts of him that are "dis" but parts that are highly able. Dr. IJ outlines that the traits that are traditionally thought of as a disability are often things that society perceives as potentially inconvenient for others. People requiring wheelchairs need ramps, wider aisles, and up-front parking spots. None of these are cumbersome but society views them negatively, often because of how they look through an economic lens. Whereas my lack of musical talent does not inconvenience anyone other than my family who must suffer listening to my off tune singing at home. Someone who is an exceptional mechanic but reads slower than average might be told they are *learning disabled*, while someone who can read quickly but couldn't change a tire let alone repair an engine might be eligible for a *gifted and talented* program.

Everyone has things they can do and things they can't do. That isn't meant to minimize the challenges of those who can't see in a world built for those who can, but this realization stresses that society defines "disability" on societal terms, not on individual terms. Eugenics takes the idea of inconvenience to its purest economic form in deciding that rather than focusing on the diversity of talents and abilities in a population (and individuals), those who present any perceived inconvenience to *the system* are to be eliminated rather than nurtured.

Donohue has seen how the genetics era has worsened ableism by continuing to suggest that disabilities could be eradicated by selective breeding or embryo screening. Starting in 2017, papers sprang up claiming to have found genetic *risk alleles* for Cerebral Palsy. These have progressed to now being part of risk scores offered to pregnant mothers despite the fact that genetics risk in Cerebral Palsy is negligible compared to environmental risk or potential malpractice during delivery. There is a troubling and continued willingness of society to see dis/abilities as something to be purged, irrespective of whether the person afflicted actually considers themselves worse off. Prospective parents want to maximize their chances of having the healthiest child possible. Thus, they may take vitamins, exercise, and give up smoking or drinking. While modern genetics can accurately screen embryos for the kinds of monogenic disorders that tend to be medicalized, the continued desire to prevent dis/ability fosters tolerance for the idea that genetics might *cure* other dis/abilities as well.

One other aspect is the aversion of the medical community to perceive statements about medicalized disabilities as ableism. Harden's book spent quarts of ink on the idea that people with a disease should be seen as victims of bad luck and thus never viewed as innately inferior. Viewing all humans as innately worthy of respect is a sentiment that I think all decent people can agree on, but I don't think we need to define all diseases as genetically based to do so. The problem comes in separating the patient from their circumstances. *The Genetic Lottery*'s kumbaya moments assume that if science could prove that mental health disorders or other dis/abilities were nothing but flukes of genetics, people would be more caring.

If depression were no more circumstantial than sickle cell or if unemployment were akin to cystic fibrosis, then who could blame anyone who was down on their luck? Once again, this completely misses the point of genetic determinism. The goal of eugenics was to rid *society* of these traits with little regard to how they did so. Furthermore, you don't have to be discriminatory against people diagnosed with a disease to realize that certain *diagnoses*, especially if they can be ordained before birth, will always be viewed as inferior.

If educational attainment really had been just one gene, then it would have meant that someone might be able to make a drug out of the one protein it encoded. If you could take this hypothetical medication during pregnancy and assure your kid had the cognitive ability to graduate from college, would you take it? What if there were a supplement that could assure your child would never suffer from an inflammatory or neurologic disease? Would you consider people who had access to that medication to be in a *superior position* in terms of health? This is not to say that professors are innately superior to blue-collar workers, nor do I suggest that they are. But even if there were no perceived superiority related to career or income in society, the question would still persist around *inherited advantages*.

You can simultaneously understand that inheriting gifts does not make you superior, but the *situation* of inheriting gifts is still *superior* to not inheriting those gifts. According to the March of Dimes, a nonprofit organization, 97% of women take prenatal vitamins at some point during their pregnancy (although only 37% take them before pregnancy as recommended). These medications provide protection against spina bifida. You can recognize that mothers taking these vitamins are trying to *pass down* superior health to their children without implying that those with spina bifida are inferior. The statement, "People with diabetes are not innately inferior," is correct, but that does not mean diabetes itself is just as good as not having diabetes. Jesus both loved the blind and tried to cure them. Just as you can "hate the sin but love the sinner," one can "hate the disease but love the patient." This blind spot causes those working on the genetics of disease or dis/ability to fail to recognize how their claims of innate encoding of these disorders fuels the notion that *societal outcomes* could be perfected through genetics.

This is not ableism in the traditional sense, especially since most people with medicalized dis/abilities are more likely to want to find a cure. For example, I can assure you that no one in the eczema community or families with food allergies are hoping to see their children and grandkids impacted. Yet, personification of the nuance around disease versus individual can be seen in former NFL defensive back Ryan Clark. Although he was undrafted, he made the New York Giants roster. He would progress to earn a spot with the Pittsburgh Steelers with whom he made his name. After retiring from football, Clark would go on to be an NFL pundit and co-host of the popular podcast, *The Pivot* (with his fellow former NFL players Fred Taylor and Channing Crowder). In one episode, Clark spoke with the Steelers' Head Coach, Mike Tomlin.

Mr. Clark and Coach Tomlin revisited the time when Coach Tomlin (correctly) refused to let Clark play a game due to his genetics.

Clark knew that he had sickle cell trait long before his Steelers were set to play against the Broncos at the high-altitude stadium in Denver, Colorado. The low oxygen environment in Denver puts people with sickle cell at risk for having a sickle crisis, but Clark had never had any issues previously and so he was cleared to play. After the game, Clark became extremely ill due to what was later known to be a sickling event brought on by strenuous exercise in the relatively low-oxygen environment of Mile High Stadium (I refuse to call it by whatever corporate name it goes by these days). Clark had to have both his spleen and gallbladder removed as a consequence of the sickling crisis. After intense rehabilitation, he worked his way back onto a professional football field. The NFL scheduling rarely pits Clark's Steelers against the Broncos in Denver. Thus it took a full five years after Clark's crisis before his team was slated to return to Mile High. In his podcast interview with Coach Tomlin, Clark relayed that he hoped to convince his Coach to let him play in the game. Coach Tomlin, I think correctly, refused.

The Genetic Lottery frequently uses the example of Shawn Bradley, a seven-foot-six-inch-tall center who enjoyed a 12-year NBA career with career earnings of 27 million dollars. Harden presents Bradley as walking evidence of how *winning the genetic lottery for height* made Bradley rich. In comparison, Clark enjoyed a 13-year career in a sport that is far more brutal and prone to early retirements from injury. Clark played seven years *after losing two internal organs*. In total, Clark made 22 million on the field in a sport where the average salary is 2.5 times less than the NBA and left with millions more in prospects than Bradley did. So, it seems important to highlight the level of Clark's success despite his genetic *losses* compared to Bradley's career built on whatever genetic wins he might have had.

Despite Clark's clear physical abilities, his story is relevant in the ableism discussion because of one central question: "If you were in charge of the Denver Broncos, would you sign Clark to a contract?" Coach Tomlin made the correct decision to keep Clark from risking his life over a single game. But what if the team played half its games at high altitude like the Broncos do? What about teams in the Broncos' division that must play in Denver every year? Clark played for the respected LSU football program in a sport where fewer than one in 33 high school players make a top division roster. Clark became a professional in a sport where only 1.6% of college athletes make the NFL. He had a career nearly four times longer than the NFL average of 3.3 seasons, all while taking home nearly 10-times the average career earnings. Clark may never be viewable as innately inferior, but his medical condition can still be viewed as a "disability" worth discriminating against in the right context. A society that continues to desire removing *all* such risk will remain open to the idea of using genetics to do so.

Moral judgments are also inherent in any claims of innate encoding of an undesirable behavioral trait. You can't say "the Irish have an increased genetic predisposition to alcoholism" in a vacuum. There is an inherent judgment to such a statement and an inherent connotation that the stereotype of the Irish is validated by innate biology. It is also possible to love and respect people struggling with addiction and know that they are not *less than*. But it is unmitigated denialism to say that you think that being alcoholic is interchangeable with not suffering from the affliction. There is no morally neutral way to claim that minorities have *genetic predispositions* to diabetes, asthma, cancer, drug use, crime, poor education, poverty, mental health disorders, and everything else in the medical textbooks. Furthermore, why would there be a demand for biomedicine to invest billions of dollars trying to treat diseases if health were not superior to illness? This is especially true for situations in which one might be able to protect their children. It is completely possible to both treat every person diagnosed with Autism with dignity and grace while also being hopeful that, one day, *the option of a treatment* might be presented to those diagnosed or a reliable prevention will be discovered.

In the *Journal of Sexual Medicine*, a group in Finland reported their claim to have found alleles linked to *male pedophilic sexual interest*. If the average population geneticists were found to be more likely to carry those alleles, do you think they would find it morally neutral to write a headline saying, "GWAS researchers have higher rates of alleles linked to being a child molester"? As long as people believe that *society* would be better off without the challenges caused by even the most contextual of dis/abilities, genetic determinist thinking will find a place in the conversation.

Liberalized Genetic Determinism

In his famous 1963 Letter from Birmingham Jail, Reverend Doctor Martin Luther King Jr. wrote:

> *I must confess that over the past few years I have been gravely disappointed with the white moderate. I have almost reached the regrettable conclusion that the Negro's great stumbling block in his stride toward freedom is not the White Citizens' Councilor or the Ku Klux Klanner, but the white moderate, who is more devoted to "order" than to justice; who prefers a negative peace which is the absence of tension to a positive peace which is the presence of justice; who constantly says: "I agree with you in the goal you seek, but I cannot agree with your methods of direct action"; who paternalistically believes he can set the timetable for another man's freedom; who lives by a mythical concept of time and who constantly advises the Negro to wait for a "more convenient season".*

If I were writing this about the hurdles to overcoming genetic determinism, the only thing I might edit would be to remove "white" from the "white moderate" comment. While the field of population genetics is white, even for a scientific niche, there are plenty of researchers who give the impression that the genes for being *skinfolk* and the genes for being *kinfolk* are not a perfect overlap.

To its credit, *The Genetic Lottery* points out some of the liberal reasoning for believing in genetic determinism. For example, Benjamin Neale (another Hastings Report member) wrote a paper in *Science* in 2020 claiming, "Large Scale GWAS reveals insights into the genetic architecture of same-sex sexual behavior." Their findings claimed five loci were linked with same sex behavior, each just barely over the arbitrary cut off for significance. The work was devoid of negative controls and carried the standard *whites only* caveats on the analysis.

Yet the idea that there could be a *gay gene* is not new. A twin study paper from 1993 by Whitam and Martin claimed that sexuality had a heritability of over 70%. Since those identifying as homosexual are constantly defending themselves against the idea that their sexuality is a *choice*, it is tempting to believe that any such feelings were immutably encoded into the genome. Research into possible environmental causes of homosexuality is more often run by homophobes who want to prove intrauterine exposure to hormones or toxins *cause* the children to be gay. So, we have competing goals: one side that sees biological determinism as protection against persecution and the other that sees environmental causes as justification for discrimination and/or a means of preventing homosexuality through environmental mitigations.

However, the idea that persecution of traits would diminish if only we could prove the immutable nature of those traits requires that we pretend bigots argue in good faith. In some U.S. states, it is now a fireable offense for a gay schoolteacher to have a picture of their partner displayed on their desk. Are we really supposed to believe that proof of a *gay gene* would cause the legislators responsible for such laws to change their ways? If there was a gay gene, gay microbiome, or gay epigenetic change, what might people who already want to get rid of homosexuality do with that information? Within a week, they would be researching "cures" for homosexuality not embracing human diversity. For people who oppose LGBTQ rights, a gay gene would not be viewed as proof God created homosexuality; it would be viewed as a mutation sent from Satan and likened to cancer risk alleles. The only truly egalitarian answer is to forgo debating how biological someone's existence is and instead point out that it doesn't matter *how* sexuality is ultimately determined. *If* someone is gay matters because their sexuality represents a part of their identity that might need support in a heteronormative world; but I cannot think of anything good resulting from caring about *why* someone is gay, especially on a molecular level.

Harden also frequently states that discovering genes associated with drug use improved care for people struggling with drug or alcohol addiction. In 2023 she claimed: "genetic research on substance use disorders (SUDs) contributed to a paradigm shift in conceptualizing addiction as a chronic disease that resides within the body of the individual rather than as a moral deficit that resides within their will". This would be a great claim if not for how incompatible with history it is. Metabolic biomarkers associated with drug addiction were discovered before GWAS was even conceptualized. Similarly, the American Medical Association recognized addiction as a 'disease that resides within the body of the individual' in 1956-over two decades before DNA sequencing was invented. Furthermore, the belief in inborn predispositions to drug dependance was used to justify ignoring the early stages of the opioid epidemic. The poor white neighborhoods where the first evidence of oxycontin addictions arose were dismissed as innately inclined to addiction rather than seen as a sign of what was to come. The conclusions are clear: People obsessed with *inborn* morality (as they define morality) will never be allies in the fight for a fairer society.

A Perceived Lack of Simplicity

By far the most limiting factor is that genetic determinism represents the quote, "For every problem there is a solution that is simple, neat—and wrong" (attributed to Mark Twain, H. L. Mencken, and others). Lewontin commented on the simplicity of genetic determinism in his work, *Not In Our Genes*, stating:

> *Critics of biological determinism are like members of the fire brigade, constantly being called out in the middle of the night to put out the latest conflagration, always responding to immediate emergencies, but never with the leisure to draw up plans for a truly fireproof building. Now it is IQ and race, now criminal genes, now the biological inferiority of women, now the genetic fixity of human nature. All of these deterministic fires need to be doused with the cold water of reason before the entire neighborhood goes up in flames. Critics of determinism seem, then, to be doomed to constant nay-saying, while readers, audiences, and students react with impatience to the perpetual negativity. "You keep telling us about the errors and misrepresentations of determinists," they say, "but you never have any positive program for understanding human life.*

Kuhn outlined that in times of crisis, the simpler paradigm tends to prevail in the competition between two different paradigms. Recall that a sign of a dying paradigm is an escalation of complexity and caveats. Population genetics went from claiming that one gene will explain why Michael Jordan was a great basketball player to claiming that many genes combined to determine his skills to claiming that all his genes were needed to account for his abilities. Now, they make claims akin to stating that $1/14^{th}$ of an inch of his vertical leap can be explained by genetics but only when he is playing in Charlotte on a Wednesday. For so-called complex traits, simply saying "it's genetic" is easier—not only because of its outright simplicity but because the *fire brigade* that Lewontin mentioned does not busy itself trying to construct models that are equally simplistic and equally false. If you want to go around claiming that an Irish-Italian lady from Ohio loves tikka masala because of her genetics, you are an arsonist, but are likely an unopposed arsonist. No one else would waste their time trying to calculate all the factors that might play a role in a phenotype that is so complex.

However, I would argue that for biomarker diseases, genetic determinism is the *more complicated* answer. You have two choices to pick from here: Option 1 is that some 10,000 genetic alleles combine with some unspecified environmental factors in just the perfect way to generate eczema in a child. These alleles existed in every previous generation since the dawn of humanity but were never combined in just the perfect way until after 1970. Option 2 is that a disease that has presented the exact same way in humans since antiquity represents a specific biologic process. That process is initiated by a small number of toxins that cause the disease through predictable mechanisms. The only thing that has changed throughout history is the intensity and frequency of toxic exposures. The reason that our ancestors didn't suffer this disease with any regularity was that exposure to these toxins was rare, and even when exposed, levels were often low enough for our bodies to handle them. But we have found a way to greatly increase the rate of exposure and assured exposures are more concentrated than our bodies can deal with.

Again, option 1 is that genes that cause 80-90% of diseases just sat silently in the human genome until after modern industrial exposures took off. Option 2 is that once upon a time, inflammatory diseases were caused by chemicals that were natural but rare; today those same diseases are caused by those same chemicals, but we have turned those exposures into common and synthetic. Scientists seem to understand this dynamic for the paradigms of infection and toxins. We know that the biologic processes governing Tuberculosis infection unfolded in the same way for pharaohs and ancient Greeks as would unfold for someone infected today. Anyone breathing in the volcanic ash from Mount Vesuvius in 79 A.D. would suffer the same pulmonary damage from sulfur compounds as people today living near an oil refinery. The biology is identical; only the exposures differ.

Numerous exposures linked to modern diseases have been shared by people throughout human history. Believing in genetic determinism requires an elaborate narrative to explain why the impacts of these exposures are restricted to certain times and locations. Believing in an environmental cause requires no such complexity.

The Timeless Hubris of Inherited Wealth

Most cultures have some form of an adage saying "We borrow the Earth from our grandchildren". The notion is meant to encourage people to think about the long-term consequences of their actions. A noble goal indeed, but one that is predicated on the idea that there will always be future grandkids. In his book, *Human Extinction: A History of the Science and Ethics of Annihilation*, Émile Torres teaches us that, for most of human history, the notion that humans would live forever was a given. Even though *The Book of Revelation* in *The Bible* discusses *the end times*, some chosen few humans were foretold to return to Earth and continue humanity's life here. Torres tells us that it wasn't until humans discovered that all the dinosaurs were wiped out by a giant meteor did the possibility of total human extinction become a topic for debate. Darwin, for his part, was so convinced that species change happens at the slow pace of natural selection, he refused to believe the fossil evidence for mass-scale extinction was anything more than a sign that anthropologists needed to dig deeper.

When total human extinction became a consideration, the noble idea of considering future generations when deciding policy became geared towards saving *future* humanity in the form of what is called *longtermism*. Since it was born in the same social circles that fancied eugenics, longtermism attempted to justify incentivizing the breeding of *certain* people while limiting breeding of *certain* others as a *service* to future generations. You probably guessed that those who would be allowed to breed just so happened to live in the same countries and look exactly like those espousing long-termism. This version of repacked liberal eugenics went into hyperdrive when it was combined with ideologies that Torres has summarized under the acronym TESCREAL (pronounced tess-cree-all). The letters stand for transhumanism, extropianism, singularitianism, cosmism, rationalism, effective altruism, and longtermism. Other than longtermism, the term most relevant to what has become super-eugenics is transhumanism.

Torres offers a professional academic definition, but I will describe transhumanism as the idea born from playing games like *The Sims* too often. These kinds of video games were similar to many others that allow you to make your own digital avatar. However, in these games, you do digital chores and go to digital work instead of fun things like shooting Nazis or solving puzzles. Transhumanism as it exists today comes from guys who moved to Silicon Valley and dedicated their lives to assuring that not only would everyone play *The Sims*, but all future people would literally live in a digital world like *The Sims*.

I neither know nor care how they came up with this number, but the TESCREALists calculate that if biologic humans reach our maximal intellectual potential, then we will be able to "create" up to 1×10^{52} digital future humans. They then argue that actions taken today must be viewed on that *1 with 52 zeroes behind it* scale. To them, the lives of trillions upon trillions of future humans rest on whether current humans reach their full intellectual capacity. This has allowed eugenicists to justify the miniscule impact of genetics by inflating the impact. If two rich parents spend their money to select an embryo with a PGS that might enhance the baby's intellect by 0.0001%, most would laugh at them. But they can now counter that improving human intellect by 0.0001% is akin to saving the lives of a *1 with 48 zeroes* worth of future digital humans that will not be "born" if we don't max-out. Torres summarized that the TESCREALists see "eugenics as a great way to increase the probability that we're smart enough to build the tech needed to colonize space, maximize value, etc.". This is digitizing the endless tradition of rich children becoming rich adults who just want to pretend their perceived innate superiority has left them "burdened with glorious purpose" as said by Loki, who is also a digital villain residing only in an imagined universe.

Here again, their lack of interest in lead and other neurotoxins shows us that they are full of crap (or, I suppose given they are mostly born in the same era, their brains are full of lead). If an atrocity can be justified on the claim that you are protecting trillions of future humans from a modern dysgenic pressure impacting only a theoretical fraction of a percent of IQ, then a neurotoxin that is proven to be able to rob people of 10% of their IQ would be the greatest threat in eHuman pre-history.

I'm not sure how much staying power TESCREALists will have. Mostly because many of the big names in that space are being uncovered as overt racists or alleged financial criminals. But considering how easily past eugenicists justified crimes against humanity, I would not underestimate the potential harms of the new crop of old-money, politically connected men who have created a belief system that allows them to rationalize any action under the guise of saving the future.

The Environment is Harder to Study

In addition to improved simplicity, Kuhn noted that a new paradigm must have a way to easily test the remaining puzzles in the field. Genes are easier to study than the environment. For one, genes are finite. As much as geneticists may like to talk about the 3.2 billion base pairs in the human genome, the vast majority of those are identical among people. So, even though the number of potential variations is large, they are still limited in number. By comparison, the potential environmental contributions to diseases are neither known nor finite. In our analysis in which we made the *poly-pollutant score*, we measured the effects of hundreds of tracked chemicals but thousands more float untracked through the environment. A report in 2020 by Wang and colleagues identified that *over 350,000 total chemicals including over 70,000 new synthetic chemicals* had been added to the chemical databases since 1960. During that time, at least 22,000 of those chemicals have been added to the EPA Toxic Substance Control Act registry – meaning that they are chemicals with known or suspected harms that are actively being used in US products. Each of these may have interactions with any or all the chemicals that they come into contact with.

Adding complexity to the research plan is that environmental exposures are not static. The lead contamination in the Flint water system was not the same minute-to-minute. So, one person may have suffered dangerous inhalation of lead by taking a shower when lead contamination was at its peak, whereas their neighbor may have been exposed in a less dangerous manner by a bath hours later when lead levels were far lower. Toxins can deposit in the body and carry effects for years. Microbiome changes can last a lifetime and be passed down across generations. The geneticists may love to claim that within-family genetics can inform generational inheritance, but epigenetics, toxin deposition, and the microbiome can each transmit harms in heritable ways. The difference is that harmful changes in epigenetics, toxin exposures, or the microbiome can wreak havoc and then disappear below the level of detection. A 45-year-old woman with newly developed rheumatoid arthritis might have been set on her disease path by an exposure she had in infancy. Without longitudinal studies spanning lifetimes and even generations, revealing the environmental causes of disorders will be more challenging than just searching for associations in a finite set of static genetic differences.

The National Oceanic and Atmospheric Administration (or NOAA for short) currently runs field campaigns under the Chemical Sciences

Laboratory. This program uses mass spectrometry mounted to planes, boats, and vans to collect samples of all the air pollutants in an area at the same time. Much like you can go to nearly any city in the world and get a day-to-day read out on the types of pollen in the air, if this program were expanded and the data formatted for the broader community, perhaps in the future researchers could track all the chemicals in the air and water in real time. This would be a giant step up from the current *air quality index* reported each day.

The *air quality index* is limited to only a handful of chemicals that were more relevant in the 1970s than today. Having the ability to track all potential toxins in real time could be paired with symptom trackers in patient populations. For sufferers of eczema, the flares they perceive to have no obvious trigger might be shown to be preceded by a large factory output in ways that are less obvious than when air pollution is coming from a massive forest fire. Maybe toxin data could explain why a person who has had blocked arteries for a decade suddenly suffers a heart attack on one particular day. We can't readily look for environmental triggers for diseases until we can reliably measure all the possible environmental triggers.

Therefore, the new paradigm needed in biomedicine is one that recognizes that if a disease: (a) becomes more common as you move from less industrialized eras or areas to more industrialized, (b) has inconsistent or weak genetic search results, and (c) can be created in mice or cells using chemical exposures, then the disease should be considered environmentally induced and our energy should be dedicated to finding the offending toxin.

So, given all these hurdles, how do I predict this will play out?

As Professor Dave Curtis of University College London pointed out, as whole genome sequencing becomes cheaper and easier to analyze, the loci-type GWAS will disappear. There will still be researchers dedicated to combing through the U.K. Biobank, of course, but no one will fund a study looking for a polygene-*ish* score when they could look for the specific sequences of actual genes. As whole genome evaluations proliferate and their databases build, there will be a rush of new discoveries. There will be single-gene disorders that are uncovered that will mimic the more common version but be extremely rare. Each of these discoveries will be held up as evidence that continued investment in genetic determinism will provide benefit to those suffering with the more common, non-genetic, versions of these diseases.

However, the new generation of students interested in genetics will be drawn in by technologies like CRISPR instead of GWAS. It may be appealing to aspire to be the first person to run a GWAS on hay fever, but no one enters science to be the 118th person to do so. Using CRISPR to assess the value of any gene modification requires having a specific sequence to target, and thus, to have a biologic model. As the whole genome databases fill up, the possibility to find spurious associations will increase. Some researchers who have invested their career in finding the *gene for* Schizophrenia will claim that if you limit the population to redheaded Mongolians with linear triangular palm creases, a single base pair change in a gene that is expressed in the brain *explains* 5% of the risk. However, this era will also see an erosion of many of the prior genetic claims made under GWAS. In a way, this has already begun. A paper by Shekari and colleagues in *Nature Medicine* reported that the loci that had been associated with premature ovarian insufficiency via GWAS could not be found when looking for the exact base pair change in the genes at those loci.

The gene-specific spurious hits will be far more damaging than the benefit provided by the whole genome revisions of GWAS. Furthermore, the flawed study design of population genetics won't suddenly improve just because of an upgrade to whole genomic sequencing. Instead, they will continue to find spurious results, especially with racialized populations. One researcher will find a base pair change in gene A is associated with asthma or education in European Americans while another claims those diseases or traits are linked with gene B in African Americans. Whereas the polygene-*ish* score issues around loci might serve as a fig leaf against claiming different biology, here, they will have to either admit the entire exercise is inconsistent with known human biology or go full-on race realism.

This will deepen the schism in the field that already exists today. The geneticists who demand genes be connected to the rest of the biological processes in life will see racialized genetic claims as becoming increasingly disconnected from reality. Those who work to connect the environment to genetics will also try—and fail—to identify a connection between known exposure triggers and the proposed genetic susceptibility. It is easier today to passively hand wave the *gene-environment* interaction claim when neither variable is specific. But once people start making specific genetic claims that can be tested via CRISPR, then those researching how specific toxins might impact diseases will expect to see those toxins impact the genes being invoked. If cigarette smoke made no impact on *Wnt*, then the claims of *Wnt*'s importance might not be immediately invalidated, but they would certainly be greatly diminished.

A few pharmaceutical companies will become enamored with a whole genome association and skip the modeling phase of drug design in a race to try some drug that alters that gene's function. Their results will likely be a let down because the originally reported association was devoid of any validation in transcriptomics, metabolomics, or proteomics. The pattern of geneticists claiming retroactive victory whenever they identify a gene already known as a viable drug target will continue, even if there is insufficient evidence to confirm that the specific sequence is a problem. Even if any genetic hits do translate to the rest of biology, the trivial impact of the gene sequence will indicate a trivial role for the protein it encodes and thus won't translate into any pronounced therapeutic value.

The population geneticists who are already comfortable with racializing their work will continue to find favor with the eugenics audience. Meanwhile, the academic audience will shrink. Students will be educated with an awareness of these successes and failures. Those interested in mental health will be taught the 20-30 years' worth of failures to find a genetic cause and will thus be more open to alternate causes. This will only accelerate as examples of successful interventions targeting the microbiome or mitigating a toxin make it into the scientific literature and the news. Each successful CRISPR-based cure might inspire more burgeoning geneticists, but will also make the work ascribing mere associations seem all the more unimpressive.

Further damage to the legitimacy of population genetics will be done as those in the field attempt to *adjust for* genes they think are causal that represent mere reverse causation. Much like the idea of having a redhead gene *causes* any harm in an anti-ginger society, adjusting for that gene in their analysis will only serve to downplay the role of the environment. In essence, the population geneticists will report, "When we adjusted for the causal alleles for being marginalized, the effects of marginalization were no longer evident." They will hint at the implication that the impact of marginalization is primarily driven by genetics instead of the environment. The disconnect between researchers and marginalized communities will widen further when people in those communities realize that the DNA samples they provided under the promise of improving the health of their people was instead used to minimize the role of the causal environmental factors that they could "see outside their windows".

The population geneticists working on medical diseases will be wise enough to avoid framing genetics as the main factor in health disparities.

Instead of highlighting genetics in ways that seem stigmatizing, they will frame their findings as good news for marginalized communities. They will claim that all the worry about the hard work needed to correct environmental injustices or address social determinants of health can be circumvented by targeting their proposed allele. It will be BiDil redux as they claim that deploying a *Black kid inhaler* will be better for asthmatics than addressing pollution. While these claims may find favor with those "more devoted to order than to justice," the failures of these medications to offer anything beyond what they would to any human population will be viewed with distrust by academics and the public alike.

As the population association-ists receive fewer speaking invitations and see their work increasingly deprioritized by even the U.K.-based journals, they will blame *cancel culture* for *suppressing* their work. This will only further endear them to their neo-Nazi audiences who love a good story about a *controversial* speaker being shunned by the *illiberal elites*. Yet, because many in the media are so dedicated to faux centrism that they are willing to be Nazi apologists, the types of geneticists who claim academia is suppressing their findings will always receive a larger platform than the geneticists who serve on the fire brigade Lewontin described. The supposed discovery may be nothing more than an eighth of a percent of college attendance correlating with a handful of base pair changes devoid of biologic validation, after accounting for several key factors, and limiting the analysis only to Norwegians named Magnus. Yet, as it is now, the story will be about the perception of censorship rather than about litigating whether behavioral genetics provides data is worth paying attention to. At some point, a few population geneticists will test the legal protections of tenure and make some comment about how their data proves Jews, minorities, or people with a dis/ability are dysgenic for society.

Here is where population genetics will live out its destiny to repeat the arc of phrenology. Professor Samuel Greenblatt of Brown University wrote an excellent summary of the history of phrenology for the journal titled *Neurosurgery*. Like population genetics, phrenology was designed by people more interested in supporting status quo political policies than truly unlocking medical mysteries. Phrenologists began to combine their claims into a poly-phrenology score. Practitioners asserted that sloping brow may indicate one problem, cranial bumpage another, but both held even more insight than either in isolation. Long before population stratification dismantled correlations for geneticists, phrenology failed to appreciate that physical differences between groups were secondary to group mating preferences rather than the cause of group difference. Like population genetics, phrenology could be split into *gentlemanly* versions hoping to solve Schizophrenia and the bigoted versions trying to blame Irish poverty on sloping brows. Despite its link to discrimination—or because of (depending on your views)—phrenology became wildly popular between 1810 and 1840. Several academic journals were dedicated to phrenology and the field frequently published in the *Boston Medicine and Surgical Journal*, which became the *New England Journal of Medicine*.

Ultimately, phrenology fell apart on two fronts. First was the work of researchers like Dr. Pierre Flourens, who tested the phrenologists' claims that certain bumps on the skull overlaid brain regions controlling a specific function. Flourens ablated parts of animal brains to test whether the associations claimed by phrenologists had real biologic meaning. Flourens did find parts of the brain with specific functions; for example, he found what we now know to be the motor cortex that controls movements of the limbs, or parts of the brain that govern vision. However, Flourens could not find any surface spot that controlled behavior in the ways claimed by phrenologists and joked that he certainly could not find the part of the brain that contained the animal's soul. In effect, Flourens both started and ended the *candidate skull lump* era of phrenology. He took each proposed association, tested its validity in biology, and found it was erroneous.

However, phrenology ultimately died at the hands of public mistrust after high profile failures found their way into the news. Centuries past, wealthy couples paid phrenologists to read their kid's skull and tell them what special gifts their kid had. People paid good money to be told their kid had the physical features of a future music star, only to find their child was tone deaf or hated music. Couples feeling that they had been duped demanded recourse and their wealth and connections afforded them an audience with other rich people who worried they too could get suckered. Stealing money from rich people will get you in trouble every time and phrenology became a grift targeted at the rich. Since GWAS and PGS seem pre-ordained to repeat phrenology's mistakes, we will see rich people pay money for an embryo they are told will have advanced music skills, generational sports talent, or heightened intellect. Children will be aborted based on PGS for diseases with vanishingly weak genetic signals while embryos are implanted based on PGS that are even weaker. Some of these children will succeed in life just by the sheer value of being born to a couple with enough money to pay for such services. However, stories will eventually arise of a *designer embryo* who not only fails to prove to be a savant but is also diagnosed with a behavioral disorder, Autism, Cerebral Palsy, autoimmune disease, leukemia, or some other diagnosis. Since the PGS embryo selection makes a near zero impact on disease risk, we can expect the same risk of diseases for rich peoples' designer babies as we would for rich peoples' babies in general. When one of these couples sues for millions in damages from the *trauma* of learning their kid wasn't the Über-mensch they paid for, the final phase of the GWAS phrenology arc will be upon us.

To be fair, my field of the microbiome will also repeat many of the same errors of population genetics. As the microbiome databases expand to include behavioral traits, someone will claim that there is a microbiome for graduating college just as Harden has for genetics. Their work will lack a negative control, their work will lack a molecular mechanism, and their work will fail to account for how shared environment and assortative mating can also beget more similar microbiomes they way they do for genetics. The only protection here is that a *polymicrobial score* can still be experimentally tested through stool transplants or probiotics. So, the expectation for providing a model may protect against things going too far into the abyss.

Whereas students in the past would have opted for just one field (e.g., genetics or microbiome), students in the future will learn a more integrated analysis. Courses exemplified by the Systems Medicine curriculum led by Dr. Sona Vasudevan at Georgetown University teach students to look at problems from numerous databases. If you only understand genetics or the microbiome analysis, claims of an *Autism microbiome* are no easier to assess than claims of an *Autism gene*. For students trained in multiple data sources, claims of a disease association can be cross-checked across disciplines. If a particular gut microbe really did cause Autism, then one could check: if foods linked to increased rates of Autism impacted that microbe; if populations with higher likelihood of having that microbe had higher rates of Autism; if drugs linked to Autism impact that microbe; and if the reported *genetic predispositions* might influence how humans interact with that microbe. Then, when the associations derived from a systems medicine approach are modeled, Dr. Vasudevan's students will be able to evaluate the RNAseq and metabolomics data to determine if their identified associations are consistent with biologic experiments.

Every young biomedical scientist enters their field with hopes of making the world a healthier place. Some work to improve our molecular understanding through basic research. Some work to improve our medical understanding through epidemiology, statistics, or clinical research. Some work to translate various findings into clinical practice.

The next generation of scientists will learn about the anomalies in population genetics *before* they commit themselves to any specific approach. Thus, they will be less likely to join the ranks of the established population geneticists who, rather than shift their paradigm, have decided to prove correlation coefficients are subject to Goodhart's law. Yet, as Kuhn promised, eventually the old guard will die, and the next generation will ascend. It remains to be seen how much damage the vestigial researchers will do on their transition from professorship to irrelevance.

CHAPTER 21
An Actual Ideas Chapter

Typically, one of the final chapters of any nonfiction book is supposed to include the various ideas the author proposes for addressing the problem outlined in the rest of the book. As stated, the *ideas chapters* for the genetic fanatic books are more than a little thin on policy proposals. While I have been backhanded to many in the population genetics field, I will note that the defenders of population genetics should be offered a soft landing should any of the GWAS fanatics change their minds. While most critiques about *cancel culture* are deflections, it is a valid point that changing your mind should not be met with an Internet pile up. Mind you, I have little illusion that the pro-eugenics crowd would ever move away from their eugenical beliefs, but should any of them do so, I think we can all guess that most online content will be aimed at mocking them for past wrongs rather than praising them for coming around. We may be skeptical initially of their changes in words until they put actions behind them. Yet, if they do change their ways, we should celebrate. So, for those who might be more open to trading the *Selfish Gene* nonsense for a *Collegial Cell* mindset, here are some ideas for how we can speed up the process of shifting the paradigm away from genetic determinism.

For Researchers

As I completed this writing, NIH's campus became dotted with signs celebrating the 20th anniversary of the Rose Garden announcement of the Human Genome Project. For the next generation of medical researchers who be at home watching on TV as I was two decades ago, be sure to ask yourself how you think we should define a successful scientist. Is success in science measured by tenure? Publications? Impact factor? Not everyone will go into translational medicine, nor should they. But everyone should work towards a meaningful impact on improving health for others.

~ The study of genetics will continue to offer immense benefit as long as it maintains connection with the rest of biochemistry. Identification of rare, monogenic disorders will occur at a progressively faster pace. The advent of whole genome sequencing will flag potential gene candidates which can be tested in models using CRISPR. This may lead to direct gene therapies or targeted drugs, but know that the incidence of patients with monogenically derived versions of common disorders will be exceedingly small. Furthermore, know that genome associations are not infallible. Some will inevitably be spurious, and so, when there is no hint of impact on biochemical profiles or when modeling fails, move on. Don't invest your career in chasing *missing heritability* under the belief that technology will compensate for flawed assumptions and poor study design. Ponce de León didn't fail in his quest to find the fountain of youth because he lacked a GPS guidance system.

Cancer genetics will continue to be a thriving area of research, and among various genomic sciences, it holds the most potential. CRISPR technologies will one day allow researchers to identify the exact mutation causing their cancer and then custom design a drug to replace the mutation in a patient's tumor. The healthy tissue would be left unaltered, since it would not contain the mutated sequence. Cancer genetics have already been used to design custom T-cells that can attack the specific tumor in an individual patient. Thus, if you have an interest in genetics, strongly consider cancer research. Just be warned, the oncologists don't seem as excited by mouse models and certainly don't care about mere associations. The expectations in the field of cancer are that you either add time or enjoyment to the life of patients, or you step aside for someone who can. However, one notable caveat is that the definition of success for pharmaceutical companies selling cancer treatments is strictly related to profitability. Thus, many of the new drugs that will be celebrated as precision medicine will offer little more than standard approaches.

~ Consider using the polygene-*ish* results as a tool to sort true genetic associations from the spurious. Take this rudimentary example: If Loci A, which contains genes 1-5, is associated with eczema in Europeans, Loci B which contains genes 2-6 and is associated with eczema in Africans, and Loci C which contains genes 4-8 and is associated in Asians. Rather than shrugging off the lack of portability, combine the analyses so that you can narrow the potential genes of concern to just genes 4 and 5. If you focused only on Europeans, you couldn't really know what gene linked with Loci A was the one generating your signal. But if the gene really does play a role, then wherever it is located in the genome of other populations, it too, should make a difference. If you claim that your results are unique to one population because of allelic frequency differences, prove it.

~ If you work in population genetics, demand use of negative controls for your own work and the work of others. Measuring loci associations versus chance alone is an anti-intellectual stance in a world where assortative mating shuffles genes by chromosomes. Dr. Steve Pittelli, author of the blog, *Unwashed Genes*, and a long-time critic of GWAS for psychiatric disorders, proposed running a GWAS for the last digit of the cohort's social security number (or enrollment number for the U.K. Biobank or other databases). One could also run a GWAS using the first digit of the social security number (which are regionally assigned in the U.S.). A GWAS for the last digit of the social security number would be a true randomization; a GWAS for the first digit would allow for assessing how the regions people live in might impact the surrounding gene pool.

Perhaps consider running two analyses in the same cohort: one as a negative control and the other for the trait or disease you are looking for. Take a cohort and run a GWAS on the language spoken, then rerunning the same sample for Schizophrenia, eczema, or any other trait. This will allow direct comparison between the strength of the genetic associations with diagnoses and the strength of the genetic associations with assortative mating. This would provide you with a meaningful p-value for the difference *between* your trait and your negative control, rather than the meaningless p-value for the difference between your trait and chance.

~ Young researchers should look to environmental causes. While measuring an ever-changing environment is difficult, the microbiome gives some ability to screen the residual impacts of environmental exposure. Although not discovered in this order, the gut microbiome that was linked to obesity was taken into the lab and shown to be unable to digest fiber but able to live on refined sugar or saturated fat. This was then modeled with mice. Had the link between diet and obesity not already been established, this realization that *dysbiosis recapitulates environment* could have pointed researchers to the key macronutrients important in healthy diets. People would have seen that the gut microbiome in obesity appeared to be related to the balance of fiber and refined sugars and fats, modeled their effects, and then could have looked to see if those macronutrients were over consumed.

Our work used the microbiome as a *canary in the coalmine* to look for toxins related to eczema and this could be copied for any number of diseases associated with industrial living. The microbes associated with Autism, for example, might not cause Autism (in the real sense of the term, not the GWAS sense). But if those microbes seem especially adapted to particular toxins, it might signal that the gut microbiome has been able to adapt to toxins in ways that human biology cannot. Our microbes will adapt to their environment and those adaptations might provide clues into which toxin they were forced to adjust to. Find that toxin and maybe you have the key trigger for your disease.

Aspiring researchers could even make good use out of the mountain of failed GWAS studies by pairing them with the environmental analyses in ways that would illuminate, rather than distract from, the true causes of disease. What we need in science is a *Human Exposome Project* where dedicated staff expose different types of human cells to various chemicals. They could then observe which genetic and metabolic pathways were disrupted by those chemicals, as was done to show that cigarette smoke impacted the *Wnt* pathway. If a database were created that could match toxins with the signaling pathways they disrupt, you could then fold in the prior GWAS studies. To do so, you would need to look at GWAS in terms of pathways, which is currently not the norm. But if the *genetic predispositions* in a certain disease tend to cluster in certain pathways, then any chemical that influenced those pathways may also be involved.

If your claim is that a 2% reduction in *STAT3* activity *caused* eczema, then a toxin that reduces *STAT3* activity by 2% in the skin could be expected to do the same. This approach could re-purpose the GWAS results with correlations in the fraction of a fraction of a percent range. It wouldn't matter if an allele only rarely impacted asthma; that allele could clue researchers into which toxins would impact the genes that are supposedly so *causal*. After all, if the geneticists are going to claim a gene's activity could be 100% contextual, then an allele that is only activated in the presence of a toxin would make perfect sense. This would be vastly different from the idea of getting the genes "out of the way" proposed by Dr. Harden. If geneticists want to help tease apart the environment, the tools are all there to do so.

~ If anyone outside of my family, friends, and lab mates read this, I suspect the largest pushback will be along the lines of "not all population geneticists". I will be accused of having constructed a strawman of the field based only on the most visible (and risible) geneticists. To this I would say, "prove it". If you are a population geneticist who demands your work be validated in biology, requires falsifiability of your claims, and refuses to racialize your analysis, then show me where you have demanded such standards be placed on publishers and funders. Suppose you get a publication into *Nature Genetics* that includes biologic validation, models, and genuine adjustments for environmental factors, do you object if it is printed adjacent to a paper claiming Mexicans are genetically inclined to diabetes or a paper using the national IQ dataset that implies the average Nigerian has an IQ of 70? Do you even notice? If you are not, as the fabled adage goes, among the "good people doing nothing" as evil triumphs, show your work. If I have unfairly presented population genetics as more deterministic than it is, the field should denounce twin studies for their frequent insinuation of traits being 70-90% genetic. The field should openly state precisely how genetically advantaged they think some groups are—if less than one or two percent, then they should explain the value of continued funding. And importantly, the field should measure the environmental variables in their models, rather than divining them.

For the Journals

~ Every journal should have a moment of reflection on their tolerance for genetic determinism, especially those in nations that still bend the knee to a hereditary monarchy. Any publication that makes claims about rich Europeans contrasted with poor Europeans should be assessed as if it were claims of Black versus white. The Buffalo murders show clearly that journals cannot pretend that claims of innate advantage of one set of white people over another will go unnoticed by the bigots just because the paper came with a 17-page FAQ section. If a journal believes papers implying innate classism are okay while those implying innate racism are not, they should state this outright. If it helps, anywhere an author talks about an allele or PGS *causing* a trait or disease, force them to change the phrasing to *advantage* or *disadvantage* and see if they are still comfortable with their message.

I understand why journals and geneticists would shy away from saying someone has genetic superiority in intellect, but if intellect is a desirable trait, then genes that gift it to a child would provide an *advantage*. Alleles that greatly increase the risk of a disease would provide a *disadvantage*. If a journal would be unwilling to publish a paper claiming that "Black Americans are genetically disadvantaged for development of asthma," then don't write "Americans of African descent had increased frequency of alleles causally linked to pulmonary disease." If a journal would not publish the statement that "people from Nottingham are genetically disadvantaged for developing cognition compared to people from London," then don't write "select U.K. populations were enriched for alleles causally related to cognitive development."

~ Prominent journals should demand papers have falsifiability. Note that I said *prominent* and not *all* journals for those wishing to construct a strawman about *bans*. A claim that 10,000 alleles combine in some perfect way to create asthma can't be experimentally disproven. Since the alternative is that the 10,000 alleles are just mere association, then unless the alleles are put into pathways, you can't test whether the association is biologic or sociologic. This would be true for any evolutionary psychology claims, all PGS claims, most behavioral genetics claims, and many other situations discussed. High profile journals should not platform claims that lack falsifiability. Those who want to publish it can do so in lower impact journals.

~ Journals should hold genetic claims to the same standards as they would for toxicology. Using the terms *causal*, *determines*, *contributes to*, *drives*, and so on for a toxin would require modeled proof of direct connections. One massive reason for this is the legal risk for blaming an environmental exposure without solid evidence. If someone published a paper claiming that children's cereal caused Autism, there is no correlation coefficient high enough to protect them from the cereal company's lawyers. Even cigarettes underwent biological testing prior to Surgeon General Luther Terry issuing the now famous warning label. These studies were demanded by health officers despite cigarettes having far stronger associations with diseases than the GWAS and PGS could ever dream about.

At minimum, demand that any genetic claim have some biochemically correlated measure. Don't continue to allow papers to make astrology style claims when a loci is near seven genes but only one make sense with their disease. If that gene is truly differentiating for that disease, then the protein it encodes as well as the metabolism that protein is involved with should also be askew. Failing this, the journal editors should reflect on their own internal biases that pre-filter papers on environmental causes of disease. Editors should ask themselves why papers detailing a community intervention that lowers average diabetes measurements by 15% or a toxin that might increase the risk of diabetes by 20% are not prioritized, while papers pointing to an association between genetics and a fraction of a percent's worth of diabetes (in only white people) are given priority.

~ Require genetics papers to show both their adjusted and unadjusted correlation values. Do not allow people to say "genetics explained 15% of the variation" when the correct statement is that genetics explained 0.01% of the total variation but 15% if you adjusted for a slew of impactful variables.

~ Publisher should take Jedidiah Carlson's approach to track papers through neo-Nazi circles online and tally the results by individual researchers, institutions, and the journals within their publishing family. Each year, publishing companies should publicly report which of their journals, and which authors within those journals, were the most likely to be signal boosted by bigots. Since any racist can publish their trash work in predatory journals, it is essential that the journals that have earned respect in science live up to their own standards. Dr. Carlson told me that when he has privately shown behavioral geneticists how their work percolates through bigoted online circles, few seem to be concerned enough to change their ways. I have little doubt that if this program were implemented the population geneticists would recoil. However, I wonder how many will see the hypocrisy of their field protesting prestigious journals publishing stigmatizing associations devoid of mechanistic validation.

~ In science, significance is not merely synonymous with notable or important. It has a rigid mathematical definition. Therefore, journals should take a similar stance to terms like *unlikely* or *seldom*. The field of academic publishing should establish formal definitions for these terms. For example, "likely" could mean that the event has a 51% or more chance of occurring, "unlikely" could mean 10%, and "negligible" could mean less than 1%. Certainly, other words and cut offs are possible, but scientists stressing the importance of clearly communicating information to the public should not *constantly* (defined by me as >75% of the time) use phrases that could mean different things to different readers.

~ This one is a little technical, but journals need to share more data about themselves other than impact factor. Impact factor is a metric that indicates how frequently the papers published in that journal are cited by other papers and journals. It is a metric of how much attention the papers get. It is supposed to be a proxy for quality and, like all proxies, is imperfect. We need not dispense with the impact factor but today, there are additional metrics that paint a better picture. For example, the relative citation rate (RCR) captures how frequently a paper is cited relative to its age; papers like the one describing CRISPR or mRNA vaccines have been cited far more frequently than the age of those papers suggest. There are also clinical citations metrics run by the Office of Portfolio Analysis at NIH. This metric captures how frequently publications are cited by treatment guidelines. The metric is meant to indicate whether a discovery is considered valuable enough to make it into clinical practice. The clinical metrics are not just for translational or clinical work. The clinical metrics also capture when a basic science discovery eventually leads to a new drug, even if it takes decades to do so. When a paper changes the standard of care, it should be properly credited even if other basic-science researchers do not cite it. Furthermore, the RCR and clinical impact metric provide a more equitable assignment of praise both in the sense that it would value the diversity among scientists and value the diversity of ways science impacts society.

For Funders and Regulators

It has often been said that "there are no stupid questions." I would like to add one provision to the end that states, "There are no stupid questions...that are freely asked and freely answered." If a kindergartener asks, "Why didn't I see unicorns at the zoo?" this is adorable. When an adult asks the same thing, it is a little odd, but tolerable. But when an adult asks the federal government why they won't invest large sums of money to evaluate the reasons he didn't see any unicorns at the zoo, that is a stupid question. When an adult asks that we invest the 500 millionth dollar to look for the 110th time into why he didn't see any unicorns at the zoo, this would be pathologically stupid. But it would be unfair to ask journals to deprioritize eugenics fandom studies if funders are going to keep sending money to the authors of the studies searching for helical unicorns. Thus, the major biomedical funders should follow the same rules outlined for the journals and demand grant proposals present the same level of supporting evidence for genetics claims outlined above.

~ Similarly, no more state or federal money should go to twin studies. If people somehow insist on funding these studies, then the amount of money provided should be pegged to the era that twin studies still held promise. If you want to run a twin study today, funders should not offer more than $100 worth of doubloons since that would have been a lot of money in the 1800s.

~ Funders (and the journals mentioned before) need to respect the work Kuhn performed on paradigm shifts. When a paper is submitted to a high-profile journal, or a grant is submitted to a funding study section, it gets reviewed for acceptance or rejection. Kuhn's work outlined that those who built their careers on the prior paradigm will die before they change their ways, especially if that change would undermine their previous discoveries. Kuhn's insights have proven to be accurate time and again. So, when someone comes along with a new idea that challenges the current paradigm, why would we exclusively turn to those same paradigm-entrenched scientists for their opinions? Consideration should be given to multidisciplinary reviews for grants or high impact papers to assure big decisions are not made only by those most established in a scientific niche. If funders and journals are going to claim that they want paradigm-shifting work, then they should act accordingly and purposefully seek diverse opinions.

~ The NOAA Chemical Science Lab program which is capable of measuring all of the various molecules in our air should be coupled with databases on disease rates in the surrounding areas. Patterns may emerge linking specific pollutants to disease outcomes that can be further validated with biologic models.

~ As detailed in *Fatal Invention*, the requirement that all studies be sub-analyzed by race should be removed. I fully understand that the purpose of this policy was to assure that study participants were not all white. Asking researchers to show they had a diverse cohort did incentivize them to recruit populations that had previously been ignored. However, the unintended consequences have become worse than the original harm. When every biologic medical study reports differences based on race, despite race having no biological validity, then it perpetuates the misconception that racial groups are biologically and/or medically valid.

~ The U.K. Biobank should either go on record that it adheres to every claim made with its data or enforce its own stated standard that "the purpose of U.K. Biobank is to set up a resource that can support a diverse range of research intended to improve the prevention, diagnosis and treatment of illness, and the promotion of health throughout society." How does a study claiming that people inherit their parents' occupation *through genetics* rather than learned behavior meet this standard? If the U.K. Biobank is fine being the neo-Nazis' favorite database, they need to articulate as much in their mission statement. The U.K. Biobank should consider mimicking the *All of Us* program's restrictions against use of its data. Some of the more laughable and disturbing examples of genetic associations presented in this book could be prevented by the Biobank committing to its stated ethics.

In a lecture at NIH the current head of the All of Us program, Dr. Joshua Denny, outlined the protections the system relies on to prevent some of the nonsense seen with the UK Biobank. First, the system is dashboard based, meaning that researchers must prompt evaluation from the All of Us server, rather than use the data directly. So, if a researcher requests an evaluation for genes associated with alcoholism in the Irish, the dashboard reviewer would deny the request and inform that researcher that such work would be stigmatizing. The All of Us team would be willing to work with the researchers to find a way to answer the researcher's questions in the most valid way possible; since segregated genetics is rarely the best approach, the dashboard approach also reduces the risk for discriminatory studies. The program enshrined further protections against abuse in the patient consent form as well. Even if someone were to download the entire database and decide to run the Irish drinking gene study by circumventing the dashboard, they would be in violation of the terms of consent. Said another way, the researcher using the data for harmful claims could face legal penalties if they violate the consent agreement provided by the patient. Denny notes that All of Us is a tool, and like any tool it will be misused by someone eventually. However, I can think of no better endorsement for the All of Us policies than knowing that the loudest voices complaining about the All of Us rules are the very geneticists cited by racists and mass murderers.

Furthermore, while All of Us is primarily a genetic database, the program has made strides to incorporate environmental and social data. As one interesting example, a 2022 *Nature Medicine* publication led by Dr. Evan Brittain compared the risk of various diseases with the number of average daily steps recorded on a step counter. They saw that walking could generate anywhere from a 10-30% reduction in diseases like diabetes, heart failure, and even gout. It is important to note that the enrollees were otherwise healthy at the start of the study, so the directionality of their findings is unlikely to be one where sick people walked less.

While 10,000 daily steps have been routinely recommended as the target for weight loss, Brittain's investigation found that the activity level that protected against disease was around 8400. But it was not as if 8399 steps were useless and 8400 was a magical cure; rather, any increase in activity (such as 2000 to 4000) reduced the risk of several ailments. For a program built for genomes, the ability to generate this kind of information is extremely impressive. All of Us is also trying to fold in the same types of pollution databases that we used in our work, thus allowing any researcher inquiring about a disease or disorder to receive data on the relevant genetics, behaviors, and exposures at the same time. The program still has a way to go before it is a comprehensive tool, for example the data lacks the types of transcriptomics or metabolomics data that might validate a genetic finding. It also can't be denied that the program was birthed in the era that overvalued genetic causation. However, those currently in charge of the program seem aware that the best approach to science is to place genetics into a larger context.

~ The prior BiDil decision should be reversed and the official position of drug approvals should be that there are no *race specific* drugs given the shared human biology. Not every drug will work for every person, but there should be no governmental incentive for a company to make the spurious claim that they discovered the "Black asthma" inhaler.

~ A larger policy recommendation would be to give the drug regulatory agencies authority to assess claims made for over-the-counter drugs. Currently, only prescription drugs (or previously prescription medication that has transitioned to over-the-counter) are allowed to mention the name of the disease they intend to treat. But a far more valuable aspect of drug marketing for pharmaceutical companies is market exclusivity. If you own the patent on an FDA-approved drug on patent, then no other company in the world is allowed to sell that drug anywhere. But over-the-counter drugs don't have market exclusivity as evidenced by every pharmacy having its own brand of ibuprofen or acetaminophen. Companies that have solid evidence that their over-the-counter product treats a specific disease should be able to gain approval to make this claim. This would provide market incentive against the products discussed in the poem *Storm* as having "not been proven to work or proven not to work". Current rules effectively prevent companies from lying about how well their product works, but forcing every company to speak in euphemisms means patients won't ever know who to believe. This would require an expansion of resources being allocated to these agencies but would result in better patient care and improved profits for the companies that have products that actually work.

~ As mentioned for the journals, NIH should require that RCR, clinical impact metric, and traditional citation metrics be included in the NIH BioSketch (the official resumé format of government grant applications). This way, the underappreciated impact of clinically focused researchers (whom are more commonly women and underrepresented minorities) will be more visible.

~ Since many modern diseases can be traced to chemicals that come from combustion engine exhaust, rapid conversion to electric cars, expansion of public transportation, and a real dedication to walkable communities would be life improving along with potentially being humanity saving.

~ Dr. Winn told me, "Never mistake innovation for impact," echoing one of my favorite quotes from famed UCLA Coach John Wooden when he said, "Never mistake activity for achievement." Impact means that someone out there suffering from a disease suffers less because of your discovery. Innovation without impact might hold promise but remember that "potential rain yields no crops". To make an impact, scientists need to think about policy. Scientists are often taught to avoid policy talks, in part, because of how policy can be inherently political. But researchers like me cannot pretend that we can tell people a specific set of toxins are causing their child's eczema and then claim neutrality on what society might consider doing about it. Just like geneticists can't claim that genes account for 80% of education and pretend policy makers will ignore their (spurious) implications.

Each scientist does not have to work to implement policies around their discoveries, but every scientist should keep a thought for how their discoveries might influence policy. The most underfunded aspect of medicine is what has been termed *implementation science*. A new drug holds only potential value until it finds its way to the patients who need it. We have a lot of potential energy in science that we need to turn into kinetic energy. Dr. Winn's program at VCU has begun training people in implementation science techniques in parallel with the basic and clinical sciences. If the trainees of today are to become the leaders of tomorrow, they should be given the skills not just to come up with new ideas, but to transform those innovations into impact.

For Everyone Else

Finally, for the main audience I see for this book, the students learning about biology, or anyone looking to better understand the news reports about finding *a gene for* some disorder—for you, I made this handy guide:

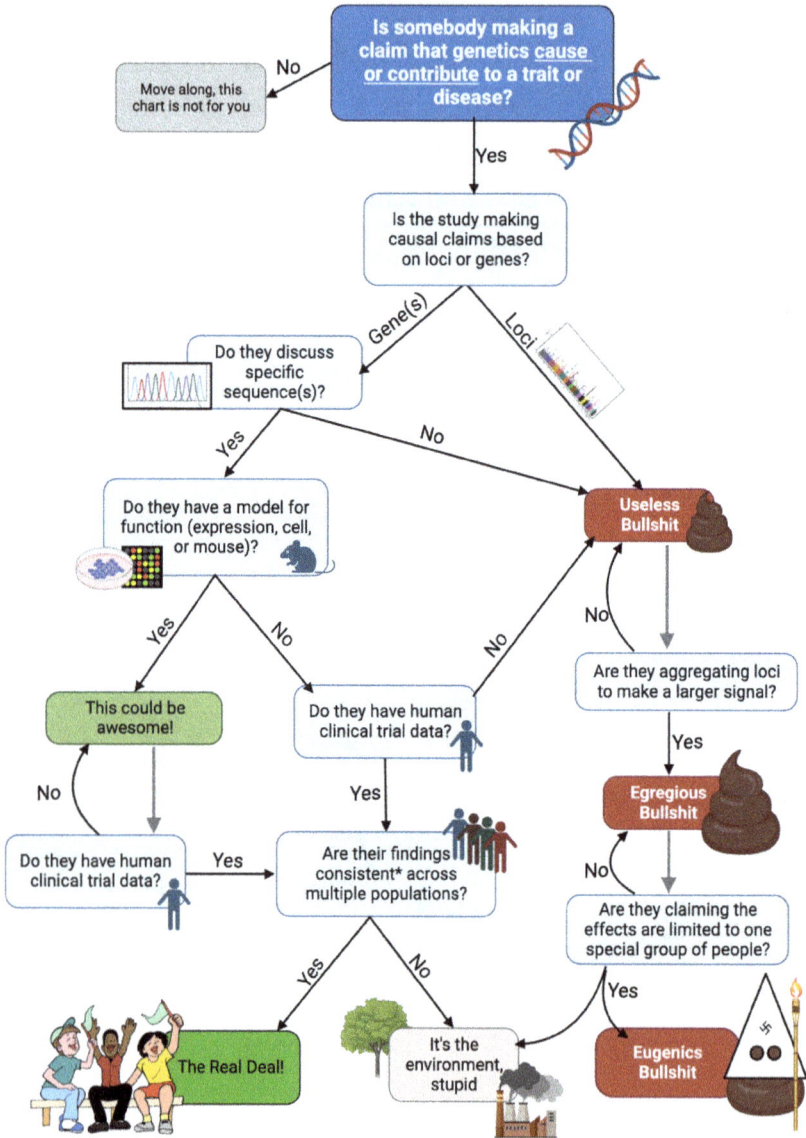

If someone is making a claim that genes *cause* something, first ask if they are talking about a specific sequence of a specific gene, or just loci. If loci, the claim is bullshit. Not bullshit in the profane definition but bullshit as defined by Harry Frankfurt in *On Bullshit* in which he outlines that liars know the truth and try to hide it, whereas bullshitters don't care what the truth actually is and just want you to believe what they say is true. If talking about specific genes, ask if they have a model or maybe even clinical data. If not, then the claim is bullshit. But if they do, the final question is whether the findings are consistent across populations. The nuance here, of course, is that genuine differences in allele frequency could indeed create a situation in which a population effectively devoid of such alleles was not impacted. But if the gene is present, even within the limits of penetrance and variance, people of all backgrounds should be impacted—otherwise, it is either bullshit or a reflection of unmeasured environmental differences.

CHAPTER 22
A Final Thought Experiment

In biomedical research, we grow accustomed to getting *informed consent* prior to enrolling a participant in our studies. The goal is to make sure people have as good an idea as possible about what is involved in, and what might result from, the study they are considering enrolling in. If you have ever had a surgery, you hopefully had someone explain to you the risks and benefits of the procedure. The surgeon might have told you something like "we are going to put you to sleep and remove your gallbladder. We expect everything to go well, but the risks include pain, bleeding, and infection". Then, they had you sign a legal document asserting you understood what they discussed with you.

Considering I made my career discussing with parents the idea of spraying their child's inflamed skin with live bacteria, I'm familiar with discussing crazy sounding research ideas. But allow me to leave you with a question of whether you would sign up for my newest study proposal.

We plan to perform a study on why certain people develop diseases and others do not. In this study, we plan to do the following:

- We plan to take the main food ingredient your body requires to regulate the immune system and remove it from everything you eat.
- We will replace it with a form of sugar that nourishes inflammatory bacteria and is mildly addictive.

- We will put preservatives in your foods that kill off the type of bacteria you need to keep your immune system regulated.
- We will fill your air with neurotoxins. And skin toxins. And lung toxins. And some heart toxins, too.
- We will fill your water with all those toxins, as well, but make sure not to skimp on the toxins that harm the brains of developing children.
- We will make your kids' clothing and bedsheets out of chemicals that are the same toxins we find in car exhaust.
- We will make the sealants on your floors or the glue on your wallpaper out of chemicals that are the same toxins we find in wildfire smoke.
- Those preservatives we put in your food? We will put those in your skincare products, as well, killing off the microbes your skin needs to function.

So, you want to sign up? You want to enroll your children? Do you think that being rich would change your risk in this study or do you suspect it wouldn't matter? If the researcher told you, "It sounds scary, but don't worry, you are white and so it won't harm you," would you believe them? What if I said your PGS for reacting to poison was low?

This is the experiment we are living in under our modern style of industrialization.

That does not mean that every exposure is one to be feared or that I plan to move to the forest and take up organic, subsistence farming. But I would put forth that it would be wise for each modern exposure to be evaluated. If a substance that *cures* a disease in mice will be studied for the potential to do the same in humans, then it seems to me that any chemical proven to *cause* a disease in mice should also be studied for the potential to do the same in humans. Any study claiming to evaluate why some humans suffer different fates than others that fails to take this overarching experiment of industrialization into account will always fail to make an impact. Sorting through all of the various environmental exposures that might cause disease will require a massive shift in focus from the dead fish to the lake and may require substantive improvements in technology. However, only fools or liars would tell you that they can impute our extremely complex environment using only the genetic similarities between parents and children.

Every early culture and human civilization believed that the Earth was the center of the solar system, if not the center of the universe. Every single day, this belief was partially validated as the sun consistently rose on one side of the sky and went down on the other. It wasn't until people started looking *beyond* what seemed obvious that they realized that the geocentric view of the world was inconsistent with the patterns of the heavens. So, too, for some aspects of genetics, there have been patterns of success that provide a daily level of validation. Yet placing the gene at the center of the biologic universe has proven to be inconsistent with the patterns of epidemiology, biochemistry, sociology, anthropology, toxicology, and basic common sense.

As cautionary tales go, GATTACA is a great movie, but the film belongs in the fantasy section, not science-fiction. Population genetics will never provide the insights the geneticists hope for, nor will it provide the validation of discrimination that the bigots hope for. Much like Godzilla did for the risks of radiation, GATTACA never depicted a realistic risk of genetic determinism. Of course, if you want to see movies about the real harms of genetic determinism, those stories would be found with the documentaries.

Population geneticists have served to foster more ethnic cleansing than therapeutic cures because they inherited the sins of their eugenics forefathers. Their failings have been passed down through poor study design and faulty assumptions, not dysgenic nucleic acid sequences. If we want to move beyond our current societal experiment, we will need to shift our paradigm and our priorities away from the genetic determinist model back to a form of medical research fully centered on improving the lives of those in need. To do so, everyone is going to need to demand more out of the medical research community, demand a cleaner environment for our children, and refuse to accept a world where innate defects are always the first suspect for the cause of disease.

ACKNOWLEDGMENTS

I'd like to thank my family for their support, especially my wife for having to listen to my rants on genetics at home and all of her priceless work copyediting the book. A special thanks is due to all of the patients that have participated in my clinical trials – without whom none of my work would be possible. I'd like to thank everyone that helped foster my career, and the research that helped drive this work. In chronological order, my parents, Dr. Robert Winn, Dr. Uday Nori, Dr. Sandip Datta, Dr. Patricia Valdez, Dr. Mihalis Lionakis, Dr. Steven Holland, CDR Ashleigh Sun, Dr. Ronald Owens and Dr. Hannah Valentine. I'd like to thank all the members of my lab over the years that have helped me along the way with various experiments. Those I did not name in the book include, Dr. Portia Gough, Momodou Jammeh, Nathan Pincus, Elim Cho, Arhum Saleem, Ali Alishahedani, Jobel Martiz, Jacquelyn Spathies, Brandon D'Souza, Timmy Tran, and Krystal Nguyen. Thank you to the support staff at NIH such as the delivery drivers, animal technicians, and everyone else that keep the science flowing. In Chapter 4, I limited the discussion to the scientists that spearheaded each project. However, each of those projects required the time and dedication of dozens of scientists deserving of recognition. Thank you to those that gave their time to be interviewed for this book, each named along the way. I would especially like to thank Angela Saini for your insights into the publishing process. Finally, a huge thank you to those that were willing to read an advanced copy of this book and provide their insights including my parents, my wife Jennifer, Dr. Amy Hsu, Dr. Cristan Farmer, Adam and Amber Stowe, Andrew Zalesak, Grace Ratley, Dr. Kevin Bird, Dr. Christopher Donohue, Dr. Monica Hooper, Lucie Tran, Dr. Portia Gough, and Ashleigh Sun. I would like to acknowledge Mehdi Hasan's book *Win Every Argument*, for giving me ideas on how to structure some of my writing. I would also like to thank the professional services provided in proofreading by Sana Abuleil, Saqib Arshad for type setting and formatting, and Herrick Destin for narrating the audiobook.

ABOUT THE AUTHOR

Dr. Myles was born and raised in Colorado. He graduated from Colorado State, obtained an M.D. from the University of Colorado, completed an internal medicine residency at The Ohio State University, completed his fellowship training in allergy and clinical immunology at The National Institutes of Health, and then obtained a Master's of Public Health through George Washington University. He worked under the mentorship of Dr. Sandip Datta investigating the mechanistic details of susceptibility to *Staph. aureus* skin infections. He is now a commissioned officer in the United States Public Health Service Commissioned Corps and has supported several USPHS missions, from the Ebola virus vaccine trial in Liberia to congressional Gold Medal Ceremonies at the U.S. Capitol. In 2019, Dr. Myles became the head of the newly formed Epithelial Therapeutics Unit to evaluate the efficacy and safety of a topical, live bacterial treatment for atopic dermatitis (eczema) supported by both Lasker Clinical Research Scholars program and the Distinguished Scholars Program. In addition to new treatments, his work has identified the chemical pollutants that cause eczema. This work is independent of his work for the government and therefore questions should be directed to GATTACAHASFALLEN@gmail.com

www.ingramcontent.com/pod-product-compliance
Lightning Source LLC
Chambersburg PA
CBHW062110020426
42335CB00013B/913